Innovation Spaces in Asia

This book is dedicated to:

The Sten A. Olsson Foundation for Research and Culture

Innovation Spaces in Asia

Entrepreneurs, Multinational Enterprises and Policy

Edited by

Maureen McKelvey

Professor of Industrial Management, School of Business, Economics and Law, University of Gothenburg, Sweden

Sharmistha Bagchi-Sen

Professor of Geography, State University of New York, Buffalo, USA and Visiting Professor at University of Gothenburg, Sweden

Edward Elgar
PUBLISHING

Cheltenham, UK • Northampton, MA, USA

Published by
Edward Elgar Publishing Limited
The Lypiatts
15 Lansdown Road
Cheltenham
Glos GL50 2JA
UK

Edward Elgar Publishing, Inc.
William Pratt House
9 Dewey Court
Northampton
Massachusetts 01060
USA

A catalogue record for this book
is available from the British Library

Library of Congress Control Number: 2014954946

This book is available electronically in the Elgaronline
Business subject collection
DOI 10.4337/9781783475681

MIX
Paper from
responsible sources
FSC® C013056

ISBN 978 1 78347 567 4 (cased)
ISBN 978 1 78347 568 1 (eBook)

Typeset by Servis Filmsetting Ltd, Stockport, Cheshire
Printed and bound in Great Britain by T.J. International Ltd, Padstow

Contents

Figures

Tables

Contributors

Sharmistha Bagchi-Sen, Professor, Department of Geography, State University of New York at Buffalo, US and Visiting Professor at Institute of Innovation and Entrepreneurship, Department of Economy and Society, School of Business, Economics and Law, University of Gothenburg, Sweden.

Evangelos Bourelos, Broman Post-Doctoral Scholar, Institute of Innovation and Entrepreneurship, Department of Economy and Society, School of Business, Economics, and Law, University of Gothenburg, Sweden.

Xiangdong Chen, Professor, The School of Economics and Management, Beihang University, China.

Guangyu Cheng, Associate Professor, Chinese Academy of Science and Technology for Development, China.

Enrico Deiaco, Director at The Swedish Agency for Growth Policy Analysis and Affiliated Researcher at Royal Institute of Technology, School of Industrial Engineering and Management, Stockholm, Sweden.

Peilei Fan, Associate Professor, School of Planning, Design, and Construction, Michigan State University, US.

Ethan Gifford, PhD student, Institute of Innovation and Entrepreneurship, School of Business, Economics, and Law, Department of Economy and Society, University of Gothenburg, Sweden.

Stine Jessen Haakonsson, Associate Professor, Department of Business and Politics, Copenhagen Business School, Denmark.

Marcus Holgersson, Broman Post-Doctoral Scholar, Institute of Innovation and Entrepreneurship, Department of Economy and Society, University of Gothenburg, and Department of Technology Management and Economics, Chalmers University of Technology, Sweden.

Carl Jeding, Head of Europe Analysis, Swedish Agency for Growth Policy Analysis, Stockholm, Sweden.

Jun Jin, Associate Professor, School of Management, Zhejiang University, China.

Peter Kedron, Assistant Professor, Department of Geography, Ryerson University, Canada.

Katarina Lagerström, Associate Professor, Department of Business Studies, Uppsala University, Sweden.

Astrid Heidemann Lassen, Associate Professor, Department of Business and Management, Center for Industrial Production, Aalborg University, Denmark.

Rasmus Lema, Assistant Professor, Department of Business and Management, Aalborg University, Denmark.

Xiaoqing Liu, PhD student, The School of Economics and Management, Beihang University, China.

Maureen McKelvey, Professor of Industrial Management, Institute of Innovation and Entrepreneurship, Department of Economy and Society, School of Business, Economics and Law, University of Gothenburg, Sweden.

Teuku Arckyansyah Meraxa, PhD student, Department of Geography, State University of New York at Buffalo, US.

Torsten Schunder, PhD student, Department of Geography, State University of New York at Buffalo, US.

Roger Schweizer, Associate Professor, Department of Business Administration, School of Business, Economics and Law, University of Gothenburg, Sweden.

Dmitrij Slepniov, Associate Professor, Department of Business and Management, Center for Industrial Production, Aalborg University, Denmark.

Yong Wang, MSc student, The School of Economics and Management, Beihang University, China.

Olof Zaring, Associate Professor, Institute of Innovation and Entrepreneurship, Department of Economy and Society, School of Business, Economics, and Law, University of Gothenburg, Sweden.

Zhengyi Zhang, PhD student, School of Management, Zhejiang University, China.

Acknowledgments

This book is dedicated to The Sten A. Olsson Foundation for Research and Culture [Sten A. Olssons Stiftelse för Forskning och Kultur]. The Foundation supports our research program "Radical Innovations for the Enhancement of the Swedish Economy," with Professor McKelvey as the principal investigator. This book is a direct outcome of this research program.

Historically, a broad spectrum of the local business and cultural community of West Sweden, including this foundation as well as entrepreneurs, foundations and multinational enterprises has supported the development of our group, the Institute of Innovation and Entrepreneurship. Their support has been vital to developing a long-term intellectual platform for advanced teaching, societal interaction and research at the School of Business, Economics and Law [Handelshögskolan] at University of Gothenburg, as well as for creating global contacts with leading international researchers and young scholars. Gothenburg is a small but cosmopolitan town, dominated by trade and the harbor and by connections between people and organizations that continue to foster creativity and local innovation for broader appeal.

Let us try to explain how the support of our business and cultural community has stimulated our work historically and today in a complex set of ways, which weave together excellent research, advanced education, and impact on society.

An important aspect is due to the historical direct support of Carl Bennet AB, SKF Group and Volvo Group. Together, they donated funds to develop the professorship awarded to Maureen McKelvey, the chair professorship in innovation and industrial management [industriell ekonomi] at the School of Business, Economics and Law [Handelshögskolan], University of Gothenburg, Sweden. Due to the initiatives of Professor Rolf Wolff, Dean of the university, these three companies – one entrepreneurial and two multinational enterprises – invested in the university and subsequently recruited Professor McKelvey in 2007 from her contemporary professorship in economics of innovation at Chalmers University of Technology, also in Gothenburg.

Another important aspect was the financing of the Visiting Professor's

Program (VPP) at the School of Business, Economics and Law. The VPP was initiated through the leadership of Professor Wolff, working together with the local Gothenburg and Swedish business and cultural community, to promote excellence. The VPP was run by the Graduate School, then headed by Professor McKelvey, as part of a strategy focusing upon scientific quality and internationalization. Most importantly for this book, Professor Sharmistha Bagchi-Sen, Professor of Geography, State University of New York, Buffalo, US, was selected and became a Visiting Professor here. She visited annually for three years and became involved in research and advanced teaching and supervision of graduate students. Professor Bagchi-Sen would also like to acknowledge the support from the State University of New York, as well as thank them for making this international exchange possible. Financiers for VPP (2009–2013) were: Carl Bennet AB; Elanders AB; SKF Group; The Sten A. Olsson Foundation for Research and Culture; Stena AB; and Volvo Group. Subsequently, Professor Bagchi-Sen was also elected to the international academic advisory committee (IAAC) here.

Many interesting individuals have contributed to stimulating discussions about what Asia – and internationalization – means for Swedish society and globally. Moreover, this local community wants professors at the university to aim higher, impact society, stimulate young people, and – as is particularly relevant to this book – understand the complex structural transformation taking place today in Asia. In turn, this book enables us to interact with society, including businesses, policy-makers and students.

This long-term support is extremely important, including an emphasis on intellectual development and a mission for excellence. The support in turn enables us to develop organizational structures to train young people and find ways to for them to contribute to the wellbeing of the broader society.

This long-term support has thus led to relevant developments in our field at the University of Gothenburg, whereby we have developed a strong team at the Institute of Innovation and Entrepreneurship (IIE) in the Department of Economy and Society. We interact regularly with local and global companies and policy-makers; we grant advanced degrees, including PhD degrees in innovation and entrepreneurship; and we run two successful Masters of Science degree programs, Innovation and Industrial Management (IIM) and Knowledge-Based Entrepreneurship (KBE). These programs are run in English, attract both Swedish and international students, and extend over two years, including the writing of the Masters of Science theses.

Finally, we would like to mention the support for specific projects and meetings by: the Volvo Group; the Institute for Management and

Innovation of Technology (IMIT); the Jan and Wallanders and Tom Hedelius Foundation; the Tore Browalds Foundation; and the Broman Foundation, in the memory of Ingemar Broman. More specific acknowledgments can be found within the chapters.

1. Introduction

Maureen McKelvey and Sharmistha Bagchi-Sen

1.1 INTRODUCTION

This book aims to provide an empirical understanding and initial conceptualization of innovation spaces in Asia, including the role of multinational enterprises, entrepreneurship and public policy. Subsequent chapters provide rich empirical illustrations about on-going processes of innovation, where the Asian context can be seen in the specific cases as having an impact upon firms and sectors with implications for economic change and socio-political transformation. This book is about Asia as the site of business, where customers, institutions and firms, both domestic and foreign, are interacting to affect global processes of innovation and production.

A few assumptions – that this book challenges – should be addressed immediately. Innovation is not necessarily high tech, such as information technology (IT) or biotechnology. The innovation process is not necessarily based on codification of knowledge with its sources in basic research. Finally, the location of innovation activities is not necessarily confined to one industry sector or one geographic region. On the contrary, innovation processes occur across different sectors and regions, and involve many different types of knowledge such as market and tacit knowledge. The innovation management literature proposes that firms in all sectors, nations and regions can engage in innovation. Moreover, business strategy is increasingly global given the increase in foreign direct investment (FDI), and there is an increasing need to improve firms' capabilities through network linkages. Therefore, this book tackles the meaning of innovation as it pertains to Asia, the processes of innovation practiced in Asia by local and/or foreign companies, and the "where" of innovation in relation to these companies' operations.

In the literature, the "where" in relation to innovation has been most profoundly explored at the regional level. A geographically centered innovation space conjures up images of regions such as Silicon Valley, Boston Route 128 and Cambridge, MA, which are all global IT and bioregions, where brilliant minds come together to push the limits of science and technology. In the most famous American example of California (Silicon Valley), path-breaking companies such as HP, Intel, Cisco, Google, Genentech and Biogen fill our imageries of the types of companies that occupy regional hotspots. They are adaptive, highly innovative and attract much attention. Sometimes, similar patterns of regional agglomerations generating positive conditions for innovation are visible in the newly emerging Asian nations of China and India, and the multiple South and Southeast Asian countries. Some regional hotspots have developed such as Bangalore, Singapore, and Shanghai, which to some extent mimic the dominant ideal type, and imageries of Silicon Valley.

However, focusing on regional aspects in the imitation of Silicon Valley provides a very limited understanding of how Asia is impacting global innovation. This book argues that not only the regional innovation systems but also national innovation systems matter, as do networks, global innovation networks or value chains and the strategies and capabilities of firms, which are involved in one or multiple geographical contexts in Asia.

This book builds upon the proposition that changes underway in Asia will help stimulate the creation of new opportunities, by a series of socio-political-economic trends such as an increase in skilled labor, shifting customer demands, upgrading of firm capabilities, and above all, the experimentation and attempts of firms in Asia and in the West to try to innovate in reaction to these trends. The rise of Asia starting with Japan, followed by South Korea and now China and India, challenges us to rethink the role of Asia in driving innovation in a global economy. And if they are not just imitating and using Western technology, then what is going on?

The analysis of innovation spaces in this book brings in analysis that, through individuals and firms, links the Asian context with the rest of the world. The empirical studies in this book include three different types of studies, helping to define the boundaries of innovation spaces. Chapters 1 to 9 address the on-going processes within specific Asian countries and Asian firms, while Chapters 10 to 14 detail examples of Asian firms moving to Western countries. Chapters 13 and 15 to 17 also describe Western firms moving into Asia to access recourses and innovate. Each chapter has a main focus as classified, and indeed, Chapter 13, on base of the pyramid, is classified twice. Moreover, all chapters emphasize the open nature of innovation and entrepreneurship, including networks and flows into and out of Asia. This is further discussed in Chapter 18.

Using one of these three approaches, each chapter in this book provides a finer grain analysis as firms explore innovation spaces into Asia – and out of Asia through firms to the rest of the world. Many insights, empirical illustrations and data are provided in this book, and they help define the choices that innovation managers have to make, and the context within which those choices are made.

Note also that the concluding chapter of this book, Chapter 18, returns to the issues of what we have learned about a conceptualization and empirical understanding of innovation spaces in Asia, and especially the roles of multinational enterprises, entrepreneurship, and public policy.

Next, Section 1.2 explains how this book may help nuance our imagery and stereotypes about innovation in Asia, while Section 1.3 provides an overview of each chapter.

1.2 NUANCING MYTHS AND STEREOTYPES ABOUT INNOVATION IN ASIA

There used to be a perception – which is starting to fade but still persists in some circles – that Asia (excluding Japan) is not innovative. Asia imitates, but does not innovate. Asia is not involved in innovation, but still focuses upon low-cost and low-quality products, that are imitative of the West.

This perception is becoming outdated – as demonstrated in many ways in this book – for many reasons. Indeed, the end of this section stresses the emergence of new opportunities, driven by the dynamics within Asia, which are causing innovation to happen within Asia, as well as global linkages to and from Asia to the rest of the world to access knowledge, resources and market. The changes propelling innovation and entrepreneurship within innovation spaces in Asia are multi-dimensional. One aspect is rapidly shifting market demand. There is rise of the middle and upper classes across Asia but particularly in countries with large populations like China and India. These consumers are demanding state-of-the-art technologies, products and services. At the other end of the spectrum, companies are trying to innovate for the base of the pyramid, where large number of consumers have limited resources and express different characteristics of demand. Another aspect is the availability of skilled labor in these countries, and these knowledge workers are linked globally either through the networks of entrepreneurs with international experience or through the organizational structure of large multinational enterprises.

Still, this statement that Asia is not innovative helps set the scene for the contributions of this book. It reflects a common perception, albeit one which is slowly fading, that this book sets out to nuance, and perhaps debunks in some ways, given the richness of current trends and depth of transformation underway.

A few myths and stereotypes about innovation in Asia are summarized below.

- *There is a lot of talk about Asia, but it is only talk. Government policy and national institutions are not supportive of technological development and innovation*: Asian governments may talk about innovation, technology and knowledge being the key attributes driving the economy, but investments through public policy are minor and Asian institutions of the political economy discourage change. The lagging public policy and institutions negatively affect both Asian home-country firms as well as companies wishing to locate there.
- *Firms from Asia tend not to be entrepreneurial*: There is a lack of entrepreneurial firms, either due to regional and country conditions, or due to lack of resources like human capital and capital. Entrepreneurship is not expected, and especially not in high-tech industries.
- *Large firms from Asia tend not to be innovative. They focus on low cost, and not on investing in resources to compete through technology and innovation*: Large Asian firms are primarily about price competition, especially low-cost and low-quality products and services, and their main advantage is low-cost labor. Large Asian firms are primarily about latecomers, who make imitative products, generics and reverse engineered existing products – which were developed in the innovative West. The Asian firms have few capabilities or resources and do not invest in innovation processes to access capabilities internationally.
- *Western firms can easily move to outsource customer development, technological development and research and development to Asia*: Most Asian customers primarily want low-cost products. Middle-class Asian consumers want the same products and services as Western consumers, so from the global company's perspective, it is just to make more of the same through mass production and minor modifications. Western firms can move to outsource, develop technology and to innovate in Asia. This process is reasonably easy and straightforward, and can follow a plan based upon lower costs. The shifting of resources into Asia does not change the "core" qualities and structure of the firm found in the West.

Table 1.1 shows how the chapters in this book provide more nuanced and interesting insights, which in some ways call into question the various myths and stereotypes about innovation in Asia.

Table 1.1 Overview of book chapters

Chapter	Contribution
Myth: There is a lot of talk about Asia, but it is only talk. Government policy and national institutions are not supportive of technological development and innovation.	
Chapter 2	Analyzes the uniqueness of the Chinese model of innovation, but questions whether this model will enable China to become a world-class innovator
	Describes specific characteristics of China's development of its innovative capacity in science and technology and relates this to dynamic effects in the economy
Chapter 3	Explains and analyzes how the intellectual property rights systems in China and India allow us to define the contours of innovators' global linkages, including specializations and flows
	Analysis of flows in and out suggests the increasing importance of India and China on the global scene as well as potentially extensive international linkages of inventors as well as the most important technological classes
Chapter 4	Focuses within the Chinese innovation space, by providing a description of the patent system as well as analysis of patent pools as a consortium mechanism to understand technological competition
	Analyzes firms and patents involved in patent pools within the audio and video technology and the telecommunication sector
Chapter 5	Explores the development of the electrical bike in China, as a co-evolutionary process involving entrepreneurs, product innovations for niche market, and policy, where this eco-innovation meets the new demands of customers
	Provides insight into the overall development of e-bike industry and diffusion in China as well as the leading company Lvyuan
Myth: Firms from Asia tend not to be entrepreneurial.	
Chapter 6	Develops a theoretical framing of the development of innovation spaces as a relationship between entrepreneurship and national innovation spaces in China
	Analyzes the development of a entrepreneurial company turning into an emerging market multinational enterprise, which is active in the wind turbine industry. The case study is based upon a conceptual framing of the relationships between the individual entrepreneur, the venture turning into a multinational enterprise and the national innovation system
Chapter 7	Applies and further develops the theoretical framing of the relationship between entrepreneurship in China and innovation spaces found in the previous chapter

Table 1.1 (continued)

Chapter	Contribution
Myth: Firms from Asia tend not to be entrepreneurial.	
Chapter 7	Analyzes a case study of knowledge intensive entrepreneurship in nanotechnology in South Korea
Chapter 8	Analyzes the interaction between knowledge intensive entrepreneurship as a process engaging individuals and their context in China and internationally to access knowledge and resources Provides case studies of four firms in nanotechnology in Suzhou, China
Chapter 9	Role of returnees to China as entrepreneurs and their ability to access global resources in high tech sectors Examines six case study firms within IT related to internet infrastructure and services, health biotechnology, and telecommunication equipment
Myth: Large firms from Asia tend not to be innovative. They focus on low cost, and not on investing in resources to compete through technology and innovation.	
Chapter 10	Addresses how Indian pharmaceutical companies are expanding their innovation spaces through a variety of strategies such as partnerships and acquisitions Frames in terms of research on pharmaceutical global value chains and foreign direct investment and provides information about Indian pharmaceutical companies making advanced investments abroad
Chapter 11	Examines innovation spaces developing in Indonesia, through cases of multinational enterprises, which leverage country-specific advantages Provides case studies of the Indonesian firms Adaro Energy, Mayora, and Astra to illustrate and apply a taxonomy of firms from emerging economies
Chapter 12	Discusses the challenge and opportunities when Chinese multinational enterprises access knowledge resources in Sweden Provides case studies of two mechanisms for accessing technology from the West, namely acquisition and hiring skilled labor. The companies are in automobiles (Geely/Volvo Car Corporation), respectively telecommunication (Huawei/Ericsson)
Chapter 13	Addresses emerging innovation spaces in India, driven by new types of customer demand and markets, specifically for the poorer segments of the economy Develops an overview of the concepts and key examples, drawn from a literature review

Table 1.1 (continued)

Chapter	Contribution
Myth: Western firms can easily move to outsource customer development, technological development and research and development to Asia.	
Chapter 14	Develops an understanding of how emerging innovation spaces as created, by focusing upon problem-solving and tasks in the Indian software industry
	Describes the global software-outsourcing industry especially the supply platform in Bangalore (India) and provide three cases in primary software industry, secondary software (automobile) and in electronics
Chapter 15	Analyzes how South Asia has recently become a relevant innovation space, offering a national innovation system with relevant skilled labor and knowledge assets.
	Provides a long-term case study of a Swedish company in software. This company has moved from a home-based entrepreneurial venture to a home based multinational enterprise with global links to a multinational enterprise with significant innovation in Sri Lanka
Chapter 16	Addresses an interactive process between headquarters, subsidiary and external conditions, when a European multinational enterprise shifts from adaptive R&D to R&D to support global markets in innovation spaces in India
	Provides a case study of the R&D unit of SKF, a Swedish-based but heavily internationalized, multinational enterprise
Theoretical and concluding chapters	
Chapter 17	Provides a conceptualization of innovation spaces
	Discusses the ramifications for understanding innovation spaces Asia, based upon an evolutionary theoretical framework for understanding how firms search within innovation spaces
	Provides insights about developing qualitative and quantitative approaches for future research
Chapter 18	Reflects upon the key concepts of this book, for example, innovation spaces, entrepreneurs, multinational enterprises and policy
	Revisits these concepts in light of what we have learnt from this book

Indeed, one of the reasons for writing this book is that many companies and public policy-makers have found out in recent years that these myths and stereotypes do not hold in reality. Firms are struggling about how to do innovation in this new context and public policy-makers are struggling to know how to react to the on-going shifts.

So let us go further into the context of business in Asia and its effects upon global innovation. Many of the on-going trends in Asia, both negative and positive aspects, have been discussed in public debate. All major newspapers and magazines regularly run features about the changes underway. Various criticisms are often raised about the institutional structure in Asian countries. In this book, a few of the chapters include reflective comments about contended aspects of socio-political institutions such as the role of state-owned enterprises (SOEs), the role of the state, the lack of democracy and corruption.

1.3 TRENDS OF EMERGENCE OF KNOWLEDGE AND OPPORTUNITIES IN INNOVATION SPACES IN ASIA

1.3.1 Asia: Countries and Sectors in this Book

Our approach differs and chapters in this book only focus on specific contemporary processes and phenomena related to innovation and entrepreneurship, with more qualitative analysis of dynamics while quantitative data is used to capture trends in industries and nations over time. This book primarily addresses the two largest economies: China and India. The geographical region is collectively addressed as Asia in this book, and Asia as used here is a broad concept, and the concept covers many countries, including East, Southeast and South Asia. Each country is unique in its own way. Many Asian countries are not represented in this book, and indeed the Asian Development Bank (ADB 2014) provides statistics and analysis on 45 economies, which they further categorize into Central Asia, East Asia, South Asia, Southeast Asia and the Pacific. Of course, a similar problem exists with using "West" or "Western," which here primarily refers to American and European firms. These limitations of using the concept of "Asia" and "West" are fully acknowledged.

Asia is also referred to in the chapters at various spatial scales – or geographical scales – from local to region to national, and with geographical references. Many chapters define processes at a specific scale, in order to describe trends and analyze processes related to innovation. For example, several chapters use the juxtaposition of local context and international networks and the role of national innovation systems in affecting the development of an entrepreneurial venture. Several chapters also span across an Asian country and Western country, and a few indicate processes and events from one country to neighboring Asian countries.

While geographical scale and space helps us understand the places and spaces, firms also operate across multiple types and dimensions of operational space. The firms studied here include a wide range of type. They may be multinationals from developed countries; they may be emerging market multinational enterprises; or they may be small entrepreneurial ventures.

In terms of this book, ten chapters include material relevant to innovation issues in China. China is by far the most dynamic economy at the moment, and of a size to have global impact, thereby somewhat justifying this emphasis upon China. Five chapters address innovation issues related to India, whereas one chapter each addresses issues related to Sri Lanka, Indonesia and Korea.

The chapters have only touched the surface of the very broad socioeconomic trends identified below, if one considers the perspective of economic and historical analysis of each specific country. Many books with excellent country studies of Asian economies have been published by scholars. This includes books at country level (Lee 2013; Fransman 1995; Okimoto 1989) as well as cross-country analysis (Mani and Nelson 2014; Zhang and Dodgson 2007; Whitley 1992) and studies of specific countries and industries (Malerba and Nelson 2012; Amann and Cantwell 2012; Nayak 2011; Sauvant et al. 2010; Ramamurti and Singh 2009; Evans and Wurster 2000; Peng 2000). The deeper structural differences such as national institutional arrangements and areas of industrial specialization within specific Asian countries are only briefly addressed in chapters, and only to the extent that they affect innovation and entrepreneurship.

There is also the question of whether the trends identified in this book are new, or simply a continuation of historical trends. Some would argue that Asia and especially China have also historically impacted technical development, even if their industrialization and economic development followed a different path from the West (Zhang 2013; Rosenthal and Wong 2011; Needham 1969).

In terms of sectors, Table 1.2 outlines the countries and sectors discussed in different chapters.

These chapters were chosen because they could provide in-depth empirical understanding of what is happening within particular innovation spaces. The main emphasis is on describing and explaining underlying processes – such as choices, strategies and resources – that are used by the entrepreneurs and multinational enterprises.

Table 1.2 Specification of chapters addressing empirical material relevant to specific sectors and countries

Sector / Country	Sri Lanka	South Korea	Indonesia	India	China
Emerging markets of e-bike, nanotechnology wind turbine		Ch 7			Ch 5 Ch 6 Ch 8
Large-scale engineering, such as automobiles, traditional energy and ball-bearings and audio/video technology				Ch 16 Ch 14	Ch 4 Ch 12
Multiple technologies and sectors			Ch 11	Ch 3 Ch 13	Ch 2 Ch 3 Ch 17
Pharmaceuticals and health biotech				Ch 10	Ch 9
Software and IT	Ch 15			Ch 14	Ch 9
Telecommunication				Ch 14	Ch 4 Ch 9 Ch 12

1.3.2 Eight Trends

Innovation is quickly arising within Asia, and in flows between countries that are driven by entrepreneurs, multinational enterprises and public policy. Hence, an initial starting point for writing this book is to get a handle on how the multidimensional changes on-going are affecting a deeper transformation, which affects innovation and entrepreneurship, and ultimately, competitiveness and economic growth.

The hype about Asia – and especially about China and India – has almost made it a truism that emerging markets will influence economic and market growth for decades to come.

The eight trends identified below are indicative of structural transformation in the global economy. The reason why this book deals with some specific points, capturing aspects of these broader trends, is that Asian developments should affect the trajectories along which future innovations will be developed, as well as affect firm strategy and public policy initiatives around the world. Most likely, these shifts are leading to new sources of new business opportunities in Asia – and with operations and networks at the company level which link Asian firms to the rest of the world.

To analyze such issues, different theories are used, as specified in each chapter in this book. Each chapter in this book draws upon its specific theoretical framework, especially literature in innovation and entrepreneurship; international business; public policy including science and innovation policy; evolutionary economics; and economic geography. As compared to a book focused upon regional or national innovation systems, this book places individuals and firms – and firms of different types – at the center of the analysis in understanding innovation spaces. Since no standardized approach exists, different theoretical approaches to understanding individuals and the firm are developed in this book, with each chapter focusing upon a specific question. Hence, the following eight trends impact innovation.

Size and growth of Asian economies

Asia is outstanding in terms of size and growth of the economy. The expected continuing growth in the economies suggests the rising fortunes of companies either headquartered there or active there.

China and India have led productivity growth of developing countries, growing 10 percent and 6 percent, respectively, between 2003 and 2012. In high-tech manufacturing, the United States has a global share of 27 percent, closely followed by China. China's high-tech manufacturing industries have grown exponentially from a global share of 4 percent in 2000 to 24 percent in 2012 (UN 2013).

Relative importance of these economies in a global economy in crisis

There has also been a shift in recent years, with an increasing relative importance of these economies. Due to the slow-down in recent years, with the global recession, there has also been relatively more emphasis on those parts of the world that do have economic growth and market growth, and less emphasis in countries where growth is slower.

Labor productivity growth in the United States and other developed countries slowed from 2 percent in the early 2000s to negative growth during the global recession, before rising to 1 percent (UN 2013).

Increasing competitiveness of business

There is also the increasing competitiveness of business located in various countries. Business in Asia has dramatically shifted and improved in the past decades. *The Economist* (2014) analyzes many of the trends in business in Asia today. A few highlights are that Asia now accounts for 27 percent of world market capitalization, and the biggest Asian conglomerates and firms like Samsung, PetroChina, Tata and NTT Docomo are similar in size to their global competitors.

Global ties of innovation spaces in Asia

Global ties of innovation spaces in Asia are also increasing. This is visible through investment statistics as well as the role of individual entrepreneurs and companies in stimulating the flows of knowledge, resources and markets. Asian countries are attractive as destinations (inward flows) and as sources (outward flows) of FDI. The figures suggest the attractiveness of China rather than India.

The Chinese FDI is one side. China's FDI inward flows grew enormously from US$47 billion in 2001 to US$124 billion in 2011. Outflows of FDI rose continuously over the last decade reaching in 2012 around US$87.8 billion representing 6.3 percent of the global outward FDI flow (UN 2013).

Indian FDI shows a different pattern and size. Inward FDI flows in India grew significantly from 2001 (US$3 billion) onwards reaching a peak in 2011 with around US$23.5 billion. Overall, US$18 billion was directed towards India in 2012 which represents 1.3 percent of the global FDI inflow in 2012. Outflows of FDI rose until 2008 reaching a peak of over US$20 billion but declined until 2012 to around US$11 billion representing 0.8 percent of the global outward FDI flow (UN 2013).

Moreover, as demonstrated in Table 1.1, several chapters in this book provide detailed understanding of how individual entrepreneurs as well as large companies are able to navigate globally.

Growing markets and new types of demand

Growing markets are an implication of the sheer size of these potentially huge new markets, with expected future market growth in middle class as well as the lower-income segment. The possible demands of these customers and consumers are for products and services with different attributes than those mass produced in the West.

Asia is estimated to have about 56 percent of the global population, currently around 4 billion people. Many comparisons have been made between India and China. While China currently has a larger population, that is expected to reverse by around 2025 due to the aging population, one-child policy, and the surge of population in India.

There is also the focus upon the lower end, and new types of demand. New concepts like "chinnovation," "base of the pyramid," "jugaad," and "frugal innovation" are being launched to express the need to rethink the foci and outcomes of innovations, in terms of these new customers. However, there are some cautionary tales. Tata Motors launched their low-priced car Nano, which failed, as compared to US$10 billion in profits on their luxury brands Range Rover and Jaguar (*The Economist* 2014).

Shifting positions of these countries in global value chains

There is a shift, and expected change in the future, in the position of these countries in global value chains. Asia, and especially China is thought to have opportunities to develop manufacturing and services, due to the globalization of components and growth of business-to-business sales in global value chains.

An estimated 60 percent of the total world trade of US$20 trillion is thought to be in trade in intermediary goods and services, used in production and sales of final goods (UN 2013: 122).

In terms of domestic value-added trade shares in global value chains, China tops the league. In the top ten countries, additional Asian countries include Korea, Hong Kong, Singapore, India, Taiwan, Malaysia and Thailand (UN 2013: 133).

Increasing share of global R&D

There is an increasing share of research and development (R&D), which is a proxy for innovative activities within manufacturing. This may also have impacts upon global products and services. There are likely effects of the emergence of new types of markets on the relative share of global markets, product design and the functioning of the global political economy.

The total estimate of global R&D was US$1.435 trillion in 2011. The United States accounts for about 30 percent of the 2011 while the East/Southeast and South Asia (including China, India, Japan, Malaysia, Singapore, South Korea and Taiwan) accounted for 34 percent in 2011. China (15 percent) and Japan (10 percent) led the Asian group. Note that Asia represents an increase of 9 percent since 2001 while the US has a decline of 7 percent (NSF 2014).

Large and growing supply of skilled labor and of scientific and technological assets

Asia is also widely identified as a large and growing supply of skilled labor and scientific and technological assets. This includes outputs measured by proxies of publications (for science) and of patents (for technology). The projected massive numbers of graduates in science and engineering will be providing an educated work force in Asia in the future.

China grew the fastest among larger developing economies, with its share rising from 3 percent to 11 percent. China has become the world's third-largest producer of scientific articles, after the European Union (EU) and the United States. In terms of the worldwide total of 5.5 million first university degrees awarded in science and engineering in 2010, the share of the EU and the United States is about 17 percent and 10 percent, respectively, while students in China earned about 24 percent of the total (NSF 2014).

1.4 CHAPTER SUMMARIES

Taken together, our primary reason for studying innovation spaces is that Asia is – and will increasingly become – a particularly important innovation space to understand.

Chapter 2 focuses upon developments of innovation space within China, as driven by policy and the development of scientific and technological capabilities. A first-order political objective for the Chinese leadership is to become a more innovative economy. The model which has served China well so far – moving labor out of a low-productivity agricultural sector into relatively unqualified manufacturing – is not deemed long-term sustainable for growth. This chapter provides insights and critical reflections about key characteristics of innovation policy in China, which will influence domestic and foreign firms. A first characteristic is that the Chinese model seems to be the opposite of frugal innovation, and indeed "opulent innovation" describes the tendency to maximize output through maximizing inputs. A second characteristic is the focus on identified challenges and elite centers. Finally, even though China is positioned in global R&D networks, a more long-term challenge is to engage in activities that will increase the "stickiness" so that future activities and competencies remain in China.

Chapter 3 examples the development of innovation spaces in relation to intellectual property, and demonstrates the specialization of activity within China and India as well as flows between these countries and the rest of the world. A first question is: who patents in China and India? The results show that while only one- fifth of all patent applications to the Chinese Patent & Trademark Office has non-resident origin, as much as four-fifths of all applications to the Indian Patent Office has non-resident origin. The largest share (35.2 percent) of non-resident Chinese applications has Japanese origin, followed by the US (25.7 percent), Germany (10 percent) and Korea (7.2 percent). The situation in India is fairly similar. The second question is: where do Chinese and Indian inventors patent? Both countries direct a majority of their foreign patent applications to the United States Patent and Trademark Office. The third question is: what do Chinese and Indian inventors patent? A proxy was made by using Chinese and Indian inventors' patenting in the US. China holds strong positions in patenting within computer technologies, electronics and information and communications technologies, while less so in pharmaceuticals, biotechnologies and chemistry. Similar to China, India holds a strong position within computer technologies, but also within pharmaceuticals, chemistry and biotechnologies. In interpreting these empirical results, the increased non-resident patenting in China and India indicates their importance for innovative products globally as well as increasing technological competition in specific sectors. Identified differences exist between China and India however, leaving questions about possibly different future pathways of development of these innovation spaces on the global scene.

Chapter 4 uses intellectual property rights and patents as an empirical measure, as the source of rich data to understand how competition among countries and firms plays out globally. Their analysis of patents in China focuses upon the mechanism of technical standards, alliances and particularly patent pools, to try to gain technological capabilities, in two sectors. Two technological classes are analyzed, namely audio and video technology, and the telecommunication sector. They were chosen because Chinese domestic firms appear rather weak in the first and more successful in the second. Comparisons between sectors are made in terms of the competitive positioning of companies inside and outside the patent pools. In the case of the audio and video technology sector, members of patent pools retain their positions, despite advantageous government policy in China. In the case of the telecommunication sector, technology changes much more rapidly, leading to new generations of products. Here, three Chinese firms – DaTang, HuaWei and ZTE – are able to obtain patents in key technologies and also strengthen their competitive positions, within China and globally.

Chapter 5 addresses the emergence of a new innovation space based on a new market within China: the development of the electrical bike (e-bike) industry and a leading firm in China. This is analyzed as a co-evolutionary process involving technologies, entrepreneurs, policy, niche market and an eco-innovation meeting new demands of customers. The chapter provides an analysis of the evolution of e-bike use and industry in China, then moving on to a case study of the strategies of the lead producer, Lvyuan. The development of e-bikes in China is linked to issues of transportation and the presence of infrastructures, but also to values and practices about biking and the regulatory framework. Producers have had to meet the demand of various categories of users, thereby promoting niche markets rather than a bandwagon effect of general diffusion.

Chapter 6 first provides a theoretical framing to explicitly relate the entrepreneur (individual) with the venture (company) to the national innovation system in China. The proposed framework defines an innovation space as a multifaceted organism not restricted geographically; but rather seen as consisting of multiple levels, including institutional, organizational and geographical aspects. This chapter also provides empirical insights of this journey from entrepreneurial venture to multinational enterprise, by focusing upon the process of expansion and internationalization of research and development of a Chinese wind turbine manufacturer. In doing so, the chapter provides a thick case description, with many details about the development of this Chinese company in the renewable energy industry. Right from its establishment in 2006, the company has been emphasizing the importance of innovation for its activities.

Chapter 7 follows the broader analytical framework developed in the previous chapter, now applied to the national context of an entrepreneur in South Korea. The framework is also further developed, in terms of process, strategy, organization and technology. A South Korean company in the nanotechnology industry is studied. The entrepreneurial venture, Park Systems, works in the manufacturing of atomic force microscopy. The chapter primarily focuses upon how this firm explores and can maneuver through innovation spaces, in an interactive process between the venture, the Korean national innovation system and global linkages.

Chapter 8 addressed an aspect of innovation spaces in China, namely the development of entrepreneurship in advanced sectors. The main focus is upon how four ventures within nanotechnology can develop, through analyzing interactions between key variables within the model of knowledge-intensive entrepreneurship. Nano represents a series of emerging technologies, and ones that are often seen to be interdisciplinary and involve a distributed form of knowledge among many different actors such as universities, existing large firms and entrepreneurial firms. The rapid development of science-based industries of nanotechnology is one based on ample technological and market opportunities, and these opportunities are being created between the entrepreneur (person), the entrepreneurial venture (organization) and the national innovation system. Analytical focus is put on the role of regional and institutional support such as the government loans and the science park as well as the specific characteristics of the founders, including international networks. The chapter focuses upon four in-depth case studies of Chinese firms, based in Suzhou Industrial Park. The firms are Beaver Nano-Technologies Co., Ltd, Poynova Materials Suzhou Ltd, Hanano and Nano-Micro Technology Company.

Chapter 9 addresses the development of global innovation spaces in Asia, by focusing upon the role of returnee entrepreneurs in developing knowledge-intensive entrepreneurship within China. This chapter demonstrates that individuals as returnee entrepreneurs – with specific skills and networks from the West – can bridge geographical boundaries when they return to Asia, and thereby stimulate entrepreneurship as well as access global resources, especially financing and public policy. This chapter provides and analyzes six case studies of knowledge-intensive entrepreneurial ventures in China, in the high-tech sectors of internet infrastructure and services, health biotechnology, and telecommunication equipment. The ventures studied are all started by returnee entrepreneurs: Asia Info, Sohu, Baidu, Sibiono, Sandia Medi Tech and UTStarcom.

Chapter 10 examines how Indian pharmaceutical companies are expanding their innovation spaces through a variety of strategies such as partnerships and acquisitions. In doing so, the on-going evolution of emerging-market multinationals from India in the pharmaceutical industry is analyzed, with an analysis of the impact of these multinationals on patterns of outward foreign direct investment into advanced economies. These firms target high technology and advanced economies early on, and also when no conventional firm-specific advantages (FSAs) are identified. An important context is the global value chain in the pharmaceutical industry, and the positions of Indian firms. This chapter provides data and insights about the evolution of the FSAs of leading Indian pharmaceutical companies in relation to the historic context of Indian industrial development, as well as the development over time as the leading Indian pharmaceutical companies overcome barriers to entry.

Chapter 11 analyzes Indonesia as an emerging innovation space, with a particular focus upon the internationalization strategy of multinational enterprises from this emerging market. Indonesia presents a particular type of political economic market, where state-owned enterprises control most of the energy and welfare industries while multinational enterprises and domestic business control most of the trading houses and services, which are less regulated. This chapter provides three cases to illustrate the evolution of Indonesian emerging country multinational enterprises: Adaro Energy, Mayora and Astra.

Chapter 12 focuses on how firms from emerging markets can access and tap into relevant external knowledge for innovation in Western countries, and thereby stimulate flows within new innovation spaces across countries. This chapter focuses on China and Sweden, and more specifically on companies in the automobile and telecommunication industries. Chinese companies are increasingly moving abroad to access both advanced technologies and new markets. This chapter presents two case studies of Chinese firms moving into Sweden: Geely's purchase of Volvo Car Corporation in the automotive industry and Huawei's hiring of Ericsson's previous employees in the telecommunication industry. The cases represent two mechanisms for accessing scientific and technological resources, namely foreign acquisition and employment of key individuals from competitors through the establishment of research and development labs.

Chapter 13 addresses emerging innovation spaces, driven by new types of customer demand and markets, and specifically new forms of innovations for the poorer segments of the economy. The chapter focuses particularly upon India. The idea of innovations for the poorer segments of the economy has gathered much attention, because a specific characteristic of emerging-market economies is that a considerable share of consumers belongs to the bottom (or base) of the pyramid (BOP). Though the individual purchasing power is low, the size of the BOP market segment in sum is enormous taken in aggregate, which makes it an attractive market segment. While these are incentives to access this market segment, companies struggle to deal with the specific market conditions with their traditional business and R&D models. This chapter aims to give an overview on foreign direct investment in R&D in emerging markets with a special focus on India and empirical illustrations of innovation strategies and modes of interaction.

Chapter 14 addresses global flows in the creation of new innovation spaces in India, but this time in software and usually through outsourcing and investments. India has been the foci of regional development in software around Bangalore, and this chapter tackles the issue of the nature and global distribution of innovation activity in software, which is reliant upon provision of services through skilled employees. Analysis of the software industry indicates there is considerable scope for innovation in emerging-market countries. It is possible as an incremental extension of routine outsourcing low-cost countries, using a global value chain argumentation. Three case studies are presented: primary software industry firm Digital Media Networks and the outsourcing of product development (OPD) to Aditi Technologies; secondary software industry firm Auto IT and the outsourcing of custom application development (CAD) to MindTree Consulting; and electronics firm Telecom Corp and the outsourcing of engineering services (ESO) to Sasken Communication Technologies. The case studies help to explain why innovative software activities, including problem-framing activities, are likely to follow the relocation of routine software production activities to emergent innovation spaces.

Chapter 15 explores the strategies and experiences of a small European software firm. The case study covers three phases in the development of the firm and its innovative activities in software in relation to financing and markets. During the period covered, the firm shifts from being a Western-based entrepreneurial venture started by software/engineering students, to a home-based MNE with global links to an MNE with significant innovation in Sri Lanka. The perspective here is how, over time, the firm interacts with the national innovation system in the host nation, in such a way as to change the location of innovative activities for reasons having to do with financing, markets and crisis. Based upon the analysis of the interwoven narrative of the development of technology, markets, financing and sourcing software development, the chapter proposes three distinct periods. They are:

1. The founding phase: building the domestic R&D base in the early 1980s to the early 1990s.
2. The expansion phase: R&D internationalization through direct investments in the early 1990s to late 2000.
3. The phase of consolidation, adaptation and internationalization of innovative activities, 2000 to 2010.

The case study illustrates how the internationalization of innovation processes can be used to maintain the competitive edge of a company as well as the necessity of having a host-nation innovation system that can provide relevant skilled labor and technological assets.

Chapter 16 addresses how a European MNE can access, but is also changed by, moving into R&D to support production and new products, within an innovation space in India. The SKF Group is the leading global supplier of products, solutions and services within its five technology platforms: roller bearings, seals, mechatronics, services and lubrication systems. The case study is of the development of R&D in India for SKF. India went from being the site of a relatively small unit, with the main task to assist local production by locally adapting existing products to becoming one of two global R&D centers that perform advanced R&D, for the global market. The case provides insights, by analyzing three mechanisms identified in the literature review: headquarters assignment; subsidiary choice; and local environmental adaptation. Moreover, the chapter indicates the importance of MNEs in creating hubs of technical competencies in emerging markets. This is seen as a dynamic and evolutionary process, based on a combination of market, technological and managerial knowledge that is crucial for the subsidiaries to achieve an extension of the mandate over time.

Chapter 17 draws upon a line of research to propose a conceptualization of innovation spaces, and focuses upon the implications for Asia. The main perspective taken in this chapter – and most of the book – is the firm perspective, requiring microdata and a deep understanding set in a context, and this can be analyzed both through case studies and through quantitative data of macro trends and later, econometric techniques to test hypotheses.

Chapter 18 reflects upon the conceptualization and empirical understanding of innovation spaces in Asia, and especially the roles of entrepreneurs, multinational enterprises and policy. This includes a theoretical conceptualization as well as relating the findings in each chapter to the overall conceptualization proposed.

NOTE

1. We wish to thank the financiers, as found in the detailed acknowledgments for the book as a whole.

REFERENCES

Asian Development Bank (ADB) (2014). *Asian Development Outlook 2014: Fiscal Policy for Inclusive Growth*. Mandaluyong City, Philippines: ADB.

Amann E. and Cantwell, J. (2012). *Innovative Firms in Emerging Market Countries*. Oxford: Oxford University Press.

The Economist (2014). "Business in Asia." 31 May–6 June, 411 (8889).

Evans, P. and Wurster, S. (2000). *Blown to Bits: How the New Economy of Information Transforms Strategy – Indonesia*. Boston, MA: Harvard Business School Press.

Fransman, M. (1995). *Japan's Computer and Communication Industry: The Evolution of Industrial Giants and Global Competitiveness*. Oxford: Oxford University Press.

Lee, K. (2013). *Schumpeterian Analysis of Economic Catch-Up: Knowledge, Path-Creation and the Middle Income Trap*. Cambridge: Cambridge University Press.

Malerba, F. and Nelson, R.R. (2012). *Economic Development as a Learning Process across Sectoral Systems*. Northampton, MA, USA and Cheltenham, UK: Edward Elgar.

Mani, S. and Nelson, R. (2014). *Trips Compliance, National Patent Regimes and Innovation: Evidence and Experience from Developing Countries*. Northampton, MA, USA and Cheltenham, UK: Edward Elgar.

Needham, J. (1969). *The Grand Titration: Science and Society in East and West*. London: George Allen and Unwin.

Peng, M. (2000). *Business Strategy in Transition Economies*. Thousand Oaks, CA: Sage.

National Science Foundation (NSF) (2014). "International comparisons of R&D performance." Available at: http://www.nsf.gov/statistics/seind14/index.cfm/chapter-6#s1 (accessed 14 October 2014).

Nayak, A. (2011). *Indian Multinationals: The Dynamic Growth in a Developing Country Context*. New York: Palgrave Macmillan.

Okimoto, D. (1989). *Between MITI and the Market: Japanese Industrial Policy of High Technology*. Palo Alto, CA: Stanford University Press.

Ramamurti, R. and Singh, J.V. (2009). *Emerging Multinationals in Emerging Markets*. Cambridge: Cambridge University Press, pp. 200–225.

Rosenthal, J.-L. and Wong, R.B. (2011). *Before and Beyond Divergence: The Politics of Economic Change in China and Europe*. Cambridge, MA: Harvard University Press.
Sauvant, K., Pradhan, J., Chatterjee, A. and Harley, B. (2010). *The Rise of Indian Multinationals*. New York: Palgrave Macmillan.
United Nations (UN) (2013). *World Investment Report 2013: Global Value Chains: Investment and Trade for Development*. New York and Geneva: United Nations.
Whitley, R. (1992). *Business Systems in East Asia: Firms, Markets and Societies*. London: Sage.
Zhang, M. (2013). "Innovation management in China." in Dodgson, M., Gann, D. and Phelps, N. (2013). *The Oxford Handbook of Innovation Management*. Oxford: Oxford University Press, Chapter 18.
Zhang, M. and Dodgson, M. (2007). *High-Tech Entrepreneurship in Asia: Innovation, Industry and Institutional Dynamics in Mobile Payments*. Northampton, MA, USA and Cheltenham, UK: Edward Elgar.

2. To boldly go . . . Characteristics of China's innovation policy

Enrico Deiaco and Carl Jeding

2.1 INTRODUCTION

This chapter addresses the development of the innovation space within China, by focusing upon innovation policy, the development of scientific and technological capabilities in research and in firms, and the implications for global competitiveness. The rise of the Chinese economy has been a focal point for much of the discussion about the global economy during at least the last decade. To a large extent this growth has come from moving labour out of an extremely low-productivity agricultural sector into relatively unqualified manufacturing. The dual forces of removing political/regulatory barriers to trade and investment, and the increased outsourcing and offshoring of goods production have made China a leading producer of manufactured goods in the world.

This growth model has worked well, and it is probably still viable for a decade or so of relatively high growth (Das and N'Diaye 2013). There is still a vast pool of labour of several hundred millions of people employed in the agricultural sector, which remains characterized by low productivity. For a number of reasons, however, this growth model is becoming challenged. For almost a decade now, the pressure and drive of Chinese politics has been to change China's development model, move up the value chain, and become a more innovative economy. One of the most recent expressions of this impetus was the declaration of the Central Committee of the Communist Party of China (CCPCC 2013) that: "We must accelerate the transformation of the growth model, and make China an innovative country."

In this chapter we will outline some of the specific characteristics of China's development of its innovative capacity. Some indication of where China stands today relative to other countries is given in the 2013 INSEAD Global Innovation Index. To take a few figures, there China ranks 35th on the global innovation index, 75th on the information and communication technologies (ICT)-infrastructure, 55th on the ease of getting credits, 98th of knowledge-intensive employment and 69th on creative goods and services.

The rapid development of technology and innovation that is currently underway in China is quite unprecedented in scale and scope for a middle-income country, and there are no ready models to follow. In this chapter we argue that China has chosen a model that has a few distinctive features that we aim to highlight and to discuss and assess the pro and cons in closing the innovation gap to advanced economies.

First, the Chinese model seems to be the opposite of frugal innovation. Perhaps "opulent innovation" could describe the tendency to maximize output through maximizing inputs. In terms of manpower, intellectual resources, natural resources and money, China relies mainly on its riches rather than their efficient use as a way to develop its innovative capabilities. Such a strategy is only an alternative for a country so large and so rich in resources, but even for China its style of opulent innovation is clearly costly. In this chapter we will point to some of these costs when climbing closer to the science and innovation frontier.

Second, Chinese research and innovation policy is highly focused. The resources are directed towards specific identified challenges, towards the later stages of the innovation chain rather than basic research, and towards elite centres rather than broad institutional capacity building. This may be a good strategy being far from the technology frontier, but picking winners is inevitably more risky the closer to the frontier you get (Aghion et al. 2005). Given the extremely rapid increase of resources devoted to research, the focused approach also raises problems of absorptive capacity. Efficiency of use of resources is an issue, related of course to the Chinese style of opulent innovation.

Third, China's rise to become an economic superpower has been closely interlinked with the international development of value creation in global value chains. The fragmentation of production has allowed China to become the most important player in some segments of these value chains. As will be shown, China is increasingly also positioning themselves in global research and development (R&D) networks. However, to attract and, more importantly, to retain the more profitable innovative segments of global value chains in R&D, China will have to display "stickiness" which includes implementing various microeconomic incentives that shape the sophistication of company operations and the quality of an innovative business environment (Delgado et al. 2012). Which factors create this stickiness for innovative activities is less than perfectly understood, but it is probably not the same factors that create stickiness for manufacturing.

The main thesis is therefore that China does not seem to have found a superior model of its innovation system for catching up with players at the innovation frontier, and the specific features of the model we believe may hamper the desired ambitions of being a world-class innovator. Section 2.2 presents some overall figures of the rapid development of China's investments into science and technology. Sections 2.3 to 2.5 describe and analyse the three specific characteristics of the Chinese innovation system model in more detail. In the last section, 2.6, we discuss the pros and cons of the model and assess some implications for catch-up and prospects for technological leadership.

2.2 SOME INDICATORS OF SCIENTIFIC AND TECHNOLOGICAL DEVELOPMENT IN CHINA

The pace and scale of China's economic transformation have few historical precedents. In 1978, and before the large-scale reforms took place, China was one of the poorest countries in the world. Its gross domestic profit (GDP) was one-fortieth of the US level compared to one fifth in 2010. Clearly, much of the growth has been driven by relocating from agriculture to manufacturing, as well as injecting huge amounts of capital investments (Zhu 2012). However, since the reforms in 1978 there has also been a tremendous increase in science and technology spending and the following sections highlight some overall macro indicators of capacity building in China, focusing mainly on the input side. In these sections, the policy model behind the overall picture will be spelled out in more detail and discussed more thoroughly.

2.2.1 Investments in Skills, Science and R&D

China is now a world player in mobilizing resources for human capital and science and technology (S&T). There has been a rapid growth of all components of the HRTST pipeline, from university enrolment to graduate studies in various programmes. In 1978, China supplied almost no PhDs in science and engineering, but by 2010, they were producing 25 per cent more than the United States (Freeman 2009). The development is large even when normalizing according to population; China produces 2.2 new graduates per 1000 in the population aged 25–34 compared to 2.7 in the European Union (EU) (IUS 2014).

The evolution has been most rapid in the last decade. Various studies show that the percentage of young cohorts enrolling in university curricula between 2000 and 2008 rose from 11 per cent to 35 per cent and the number of graduates increased from 1.7 million to more than 7 million (Zhang et al. 2013). Since 1998 the share of GDP devoted to the expansion of education has tripled and the number of universities and colleges has almost doubled. Thus, the share of active population with university qualifications is almost comparable to the average of the eurozone (INSEAD 2013).

In contrast to other East Asian developing countries, China invested in university research quite early in the economic development process (Zhang et al. 2013). In terms of number of researchers, China is now number two in the world, after the United States. Although large in absolute numbers, the Chinese figures still lag behind wealthier countries when put in relation to population figures. In 2009 the number of researchers in R&D per million citizens was 863 in China, compared with 3850 in Germany or 5046 in Sweden (World Bank 2014).

China's rapid growth in terms of scientific publication is now recognized even in comparison with that of other large developed countries. Various projections indicate that publication may even outstrip the US as the biggest source of scientific efforts by 2020. Scientific publications have been growing particularly fast in engineering, chemistry and physics, which are longstanding and strong footholds in many of the Western countries (Veugelers 2013). Thus, scientific capacity building has increased the Chinese world share of scientific resources and is equal to the US and just behind Europe. During recent years, some spectacular investments in various strategic areas of sciences have been reported and have received a lot of international attention. For instance, to name a couple of the prominent examples cited, it is claimed that China has installed more DNA sequencing capacity than the US, and has recently installed the world's most powerful supercomputer (Jonquières 2013).

Still, the US is by far the biggest spender on R&D and accounts for about 32 per cent of the global total, although the share is in decline (Veugelers 2013). Of all the Asian countries China has made the biggest leap forward and is now the second biggest spender in the world, accounting for about 12 per cent of global R&D expenditure (Figure 2.1). This means that absolute R&D expenditure is almost on the same level as Germany, France and Italy combined and where Japan has been pushed to third place on the world ranking (IUS 2014). The acquired position is impressive since it is a consequence of rapid growth of only ten years and covering the years 1990 to 2012, as shown in Figure 2.1. During the period China was growing by 22 per cent annually. As a comparison, the growth of South Korea was huge in an international comparison but averaging only 12 per cent over the same period. This achievement is all the more striking, given that it has taken place at the same time that Chinese GDP has grown by more than 10 per cent annually. Thus, comparing GDP, China is overperforming compared to other countries.

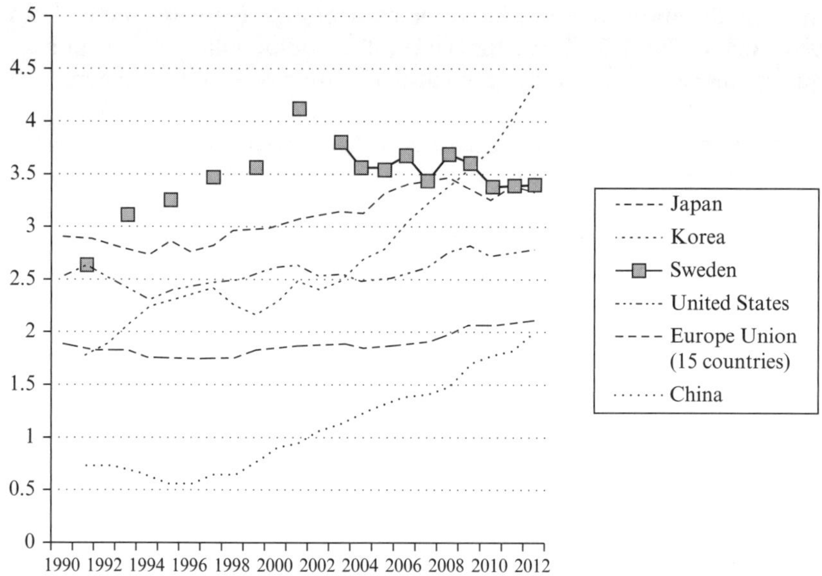

Source: Wu (2013)

Figure 2.1 R&D expenditures as share of GDP, selected countries, 1990–2012

Another distinguishing feature of the Chinese R&D development is that the business sector is the main spender on R&D, thus resembling some of the countries on the world technology frontier. Almost 72 per cent of national R&D comes from business compared to 33 per cent in UK, 28 per cent in Germany and 31 per cent in the US (INSEAD 2013). However, compared to these countries the overall share of business R&D in China accounted for by state-owned enterprises (SOEs) is considerable. In addition, the Chinese state's influence also over non-SOEs is much stronger than in Western economies (Atkinson 2012). China's ambitions are far-reaching, and the Chinese state takes a central part in this.

In summary, human capacity building, and science and technology development, has risen rapidly in China over a short period of time. The sheer input and scale is impressive. In addition, Chinese emergence in science has been uniquely fast for a country that ranks in 197th place in terms of income per head (World Bank 2014), suggesting that China's development contradicts conventional wisdom of economic development in successive stages and where after decades of development and increase in income countries begin to create new-to-the world technology (Puga and Trefler 2010).[1] But can one assess China's innovation potential on the basis of its past and current S&T performance? And what are the specific characteristics as well as the pros and cons of the chosen innovation model? These are questions that the following sections will explore more deeply.

2.3 OPULENT INNOVATION

Given the large expenditure on R&D, an interesting question is what kind of output is generated by this input of resources. One way at looking at this is through how many scientific papers and research articles are published, and how influential they are in terms of citations.

In number of publications, China displays a rate of growth at least as impressive as that of total R&D expenditure. In 1980, China's share of the world's total scientific publications was 0.16 per cent. In 2011 the share had increased to 12.17 per cent, second only to the United States (Fu et al. 2013). However, China's share of world citations does not match its share of publications, indicating problems of quality in its scientific publishing.

During the period 1995 to 2006, the gap between Chinese researchers' publications and their citations widened considerably, which has brought scholars to suggest "publication inflation". Institutional factors paving the way for such inflation have been strong financial and other incentives for researchers, where a quite common bonus for the individual researcher equals up to two months' salary for publishing a paper. Also, on an organizational level, the number of publications is a major indicator for appraisal, ranking and funding of research bodies in intense competition with each other. These strong incentives are coupled with a lack of "peer review in its true sense, due to lack of immediate as well as excellent peers, and the strong cultural resistance to assessment, particularly at the individual level." (Fu et al. 2013). Plagiarism and outright scientific fraud is a well-known problem in Chinese research, urging *The Lancet* in an editorial to call on the Chinese government to reform its academic sector to improve research ethics (*The Lancet* 2010). While these problems are in no way universal, they are sufficiently widespread to warrant calls for systemic reform.

Another output measure of research and innovation activity is patent filings. Here, too, Chinese growth has been spectacular. Having adopted its first patent law in the 1980s, China is today the world leader in numbers of patents (see Table 2.1) and the State Intellectual Property Office of the People's Republic of China (SIPO) is the largest patent office in the world (WIPO 2013).

Table 2.1 Patent applications by country, 2010

	Number of national patent applications	Number of international (PCT) applications	Percentage international (PCT) patents
China	391177	16403	4.2
Japan	344598	38873	11.3
United States	490266	48896	10.0

Source: WIPO (2012), WIPO IP Facts and Figures

As with scientific publications, there is a lively discussion about the quality of Chinese patents. The rapid growth of the number of patents has certainly been stimulated by institutional factors and incentives relating to the number of patents held by organizations and individuals. On national, provincial and even municipal levels, Chinese firms are incentivized to file and obtain as many patents as possible. Financial remuneration, tax breaks and other incentives are coupled to the filing of patents, which may of course lead to a focus on patent quantity rather than quality. In support of this argument, one could point to the widespread use of utility model (UM) patents in China. Such applications are subject to far less rigorous controls than proper patents, and are cheaper and faster to obtain.

Measuring the quality of patents is difficult, but one indicator is international patenting. If an inventor comes up with a brilliant idea, they may want to protect it in many markets, for instance via an international patent application through the Patent Cooperation Treaty (PCT) system. If the objective is to collect a Chinese state subsidy, such a process is probably not worth the effort and expense. A comparison between some of the world's largest patent offices indicates that there is a substantially lower degree of international patents emanating from China.

The problem of patent quality is recognized by the Chinese government, and in its Promotion Plan for the Implementation of the National Intellectual Property Strategy in 2013, one of the goals is to raise "the quality of intellectual property rights (IPRs) and innovation efficiency [. . .] encouraging the IP creators to transfer the focus from the quantity to the quality" (SIPO 2013).

Even if China is in the early stages of development, new research has shown that advanced nations are granting rapidly growing numbers of patents to inventors in countries like China and India. A specific feature is that US patents are granted to local Chinese inventor teams working for foreign multinationals. An important fraction of these patents contain intellectual inputs from researchers outside China, a trend identified as international co-invention. What it suggests is that, just like manufacturing has become a fragmented and globally dispersed activity, so innovation is increasingly being done in global networks. China, with its enormous scientific and engineering talent pool, can tap into and become an important link in these global innovation chains (Branstetter et al. 2013).

There are numerous and well-described problems with the enforcement of intellectual property rights (IPR) in China (Suttmeier and Yao 2011). Many authors regard the weaknesses of Chinese IPR regime as teething problems, and point to similar claims made about Japan and South Korea when those countries were in similar stages of economic development. As those countries moved from low-cost manufacturing to R&D and design, their IPR protection strengthened. A similar development can be expected in China as Chinese companies start voicing their demand for improved IPR protection, according to the argument (Simon 2012). While this argument could be true, it does not take into consideration the strong role of the state in all parts of Chinese industry and the political influence over the Chinese judicial system. The absence of democracy and rule of law make China different from Japan and South Korea. In short, what those factors imply is that the state has other ways to protect those Chinese enterprises that it favours than through strong and transparent enforcement of IPR legislation. Essentially, justice isn't blind in China; judicial process is in many ways subordinate to political objectives. Much-published rhetoric from the new leadership indicates that the strengthening of the rule of law is a prioritized area for development and reform (CCCPC 2013). The process to introduce the rule of law in a society where it is lacking is a difficult one, and the leadership's announcements signal the severity of the problem as much as a willingness to solve it.

China's drive towards becoming a world leader in innovation has given a number of high-profile successes. The world's fastest supercomputer, Tianhe-2, or the Chinese space programme including the satellite navigation system Beidou, illustrate that China is by no means stuck in a role of only cheaply reproducing technology developed elsewhere, but rather is capable of producing world-leading innovations. A number of Chinese universities are ranked among the top 100 in the world. In some areas of research, Chinese centres and researchers are among the world leaders. There is, however, reason to believe that the rapid expansion of Chinese R&D has put quantity before quality (Bound et al. 2013). This has given rise to an innovation system where the average quality of output, in terms of, for instance, research publications or patents, is low despite the existence of instances of excellence. The result is a system that could be described as the opposite of frugal innovation. Enormous resources are put into the system, which generates some high-quality output but where systemic efficiency in terms of high-quality innovation per unit of input resources is low.

Two drivers of this systemic inefficiency seem plausible. The first is a common feature of planned economies with low transparency. Such systems tend to rely on simple quantitative measures with little or no bearing on quality or efficiency. Examples of this kind are rife. For instance, the local government in Hangzhou has set as one of its goals for the upgrade of its economic structure that the total floor area of its science parks shall exceed one million square metres (Hangzhou.gov.cn n.d.). Incentives of the kind reported in relation to academic publishing and patent applications exemplify the same kind of focus on quantitative factors alone.

A second driver has to do with absorptive capacity. Capacity building and competence development of highly complex activities such as academic research is a slow process. Beginning from the relatively low starting level of the Chinese R&D sector it is far from surprising that not all of the huge influx of new resources can find efficient uses. Given that healthy incentive structures are put in place and a systematic and broad capacity building occurs, the absorptive capacity of the sector should be expected to increase over time.

2.4 A POLITICALLY PLANNED, HIGHLY FOCUSED APPROACH

After the founding of the People's Republic of China in 1949, its science and technology sector was to a large extent modelled on the Soviet system of centralized S&T planning. Its central players were the Chinese Academy of Sciences (CAS), concerned with broad research issues and basic research; ministerial research institutes focusing on mission-oriented objectives set by the respective ministries; and research institutes under the provincial governments acting primarily to support the local economy and industry. The system has been described as characterized by "a disconnection between research and commercialization on the one hand, and between research and education on the other hand, and by priority-setting clearly geared towards national defence-related projects in strategically targeted areas, like the hydrogen bomb, and the man-made satellite" (Benner et al. 2012).

After the near-total halt of research and higher education during the Cultural Revolution (1966–76), the Chinese S&T system has been subject to several waves of reform, each introducing some degree of connection between research and commercialization, and of introducing competition between researchers and/or research institutions. The reformation of the S&T sector has, as noted previously, gone hand in hand with massive increases in funding and resources allocated. At the heart of this is a strong belief within the Chinese government that scientific development is the key to China's economic progress, prosperity and societal development. In particular, natural sciences and engineering have high status in this regard, with many of the top leadership sporting academic backgrounds in those fields.

One consequence of the high level of attention from the political sphere to matters of science and technology is that there is also a strong involvement of China's top leaders in S&T decision-making. The setting of priorities and strategies for a Chinese research agenda is done within a framework of politically decided documents. From the very top of the political hierarchy with the Central Committee of the Communist Party and the Leading Group on Science, Technology and Education of the State Council (chaired by Prime Minister Li Keqiang) comes directions on objectives, focus and instruments of the Chinese S&T sector. These are further specified at ministerial level in long-term plans and five-year plans, as well as medium- and long-term plans for science and technology. A further layer of concretization and detail is provided at provincial government level and in national S&T programmes and research bodies such as CAS. Even within individual research organizations and universities, research agendas are set from a political (Communist Party) perspective and circumscribed by the planning documents handed down from higher levels of the hierarchy. Despite all the reforms carried out, the Chinese scientific sector must be described as highly politicized and subject to extensive planning.

One striking feature of the Chinese public S&T sector, which is probably a result of the high degree of political steering, is its very strong orientation towards missions. Resources are directed towards achieving certain identified objectives, dealing with specific challenges or developing relatively narrowly defined scientific areas. These missions are identified in an abundance of plans and programmes such as the 863 Program, Key Technologies R&D Programme, mega engineering projects or strategic key industries (Table 2.2). To a large extent, the missions and sectors identified in these programmes are overlapping, which means that research within those areas receives parallel streams of large-scale funding.

Another striking feature of the Chinese system is its focus on building up centres of excellence. As noted above, the average quality within the Chinese S&T system is relatively low. The tendency of Chinese S&T policy is not, however, to make broad investment in capacity building and general improvement of the system. Rather, the inclination of

Table 2.2 Focus areas for the 863 Program and key strategic industries

Technologies in the 863 Program	Key strategic industries
Information technology	Next-generation IT
Biotechnology and medical technology	Biotechnology
New materials	New materials
Advanced manufacturing technology	High-end manufacturing equipment
Advanced energy technology	Energy-saving and environmental protection
	New energy
Environment technology	
Marine technology	
Advanced agriculture technology	
Advanced transportation technology	New energy cars
Earth observation and navigation technology	

funding bodies is to concentrate resources on building up elite centres of excellence. The top five Chinese universities' share of both national research funding and of scientific publication is high and rising over time (Fu et al. 2013). The centrally decided programmes 211 and 985 both aim at providing resources to a select number of elite universities and research institutes, in order to raise them to "an international advanced level". One of the instruments in building centres of excellence are the special funds dedicated to State Key Laboratories within identified research fields. Of the 220 State Key Laboratories, 96 per cent are located at the universities identified in the 211 programme, and more than half of them are located at universities in Beijing or Shanghai (Benner et al. 2012).

One problem associated with the high concentration of resources, both to institutions and to certain research fields, is a lack of absorptive capacity in the receiving entities. Successful professors receive massive amounts of funding, but do not have the capacity to carry out the funded projects themselves. Thus, projects are carried out by junior researchers and students in the professors' names.

In general, the basic salaries of scientists are very low. In order to earn a salary to live on researchers take part in as many externally funded projects as possible. This results in researchers overstretching themselves in terms of projects, which is detrimental to high-quality research. Most projects, however, do not fund salaries but are dedicated to financing equipment. This results in huge overinvestment in equipment and an underfunding of staff. Private consulting assignments in projects that never reach the research institutes is an important source of income for researchers, further reducing the capacity to carry out high-quality research in the institutes' projects.

An overall picture of the Chinese S&T system emerges as one where resources are highly concentrated to elite research institutions and research fields selected in a highly politicized environment. In some fields, the result has been the development of excellent R&D in remarkably short time. The capacity to deliver research output in parity with the input resources is, however, questionable. The politicized nature of the prioritization and agenda-setting process leads to a very strong focus on research that can be expected to directly deliver results in fields that are identified as important and prioritized. Picking winners is a risky business, which can lead to overinvestment in certain fields that prove unsuccessful, simultaneously making it difficult to finance research and development in fields not picked by the political system (Breznitz and Murphree 2011). The focus on the direct "utilitarian" approach to research funding is clear from the fact that an extremely small share of over-all Chinese R&D expenditure is on basic research: 4.7 per cent in 2011, whereas 83.5 per cent went to development (MOST 2012). With this relative weakness in basic science follows a risk that more applied research and development, which builds on results from basic research, will also suffer.

2.5 CREATING STICKINESS IN THE CHINESE INNOVATION SYSTEM

History shows that periods of extremely rapid growth will eventually slow down. Although, there are still a lot of misallocation and market distortions left in China, which if reduced will increase total factor productivity to a large extent in the forthcoming decade (Hsieh and Klenow 2009). More of Chinese growth in the future will depend on the ability of firms to pursue innovation.

It seems, though, that the literature is more pessimistic about the Chinese innovative capability compared to the ability of moving labour from low productivity in agriculture to low-skilled jobs in manufacturing or maximizing R&D inputs. Although policy plans are bold – for example, with an aim that S&T and innovation will contribute to 60 per cent of GDP growth in 2020 (it was 40 per cent in 2011) – it is often claimed that Chinese policy-makers underestimate the challenge of being a fast follower compared to being a country at the innovation frontier.

Thus, if China is to capture and retain more profitable and innovative segments in global value chains it has to display a high degree of stickiness. This feature of an efficient innovation system is defined here as the various microeconomic factors that shape the ability to create, attract and anchor innovative activities of both domestic and foreign firms and include, among others, the quality of the national business environment, the sophistication of company operations and the ability of creating high-performing innovative ecosystems. So how has stickiness developed so far, and what policy strategies and organization models have been planned and taken to enhance it?

2.5.1 Some Aspects of Stickiness in the Innovation System

Although various positive developments and policy actions have taken place or are being planned in China, the transformation from an imitator to an innovator is still relatively thin and fragile, lacking the sort of stickiness found in developed countries at the innovation frontier.

On an aggregate level, the Chinese national innovation system is still lagging behind in creating innovations compared to maximizing technology input, although it is ever more inventive. Rankings compared to other Asian countries show that China ranks 6th in the innovation inputs but 8th in the outputs and where Singapore, Hong Kong, South Korea and Japan have higher rankings (INSEAD 2013).

The national innovation system literature argues that providing an enabling environment as well as connecting the various functions of the innovations system is essential in terms of performance (Edquist 1997). Some single studies, based on canonical regression techniques, seem to indicate that China ranks first compared to their Asian competitors as being efficient in turning innovation inputs into outputs, but more studies with better data are needed in order to confirm the efficiency scores on the aggregate level (Sheng and Zhao 2012). More qualitative and firm-level studies, however, paint a much bleaker picture on the innovative efficiency of Chinese firms, particularly among the SOEs and the majority of small to medium-sized enterprises (SMEs) (see Görg and Liu 2012; Nolan 2014; Xingyuan et al. 2014).

The SOEs still dominate many Chinese markets under a regime of monopoly or oligopoly. They generate almost 40 per cent of industrial output in markets that the Communist Party has claimed as strategic, such as banking, steel, aerospace, air transport, telecommunication and energy. It is recognized that the SOEs are heavily subsidized and also inefficient. Some figures show that the average returns in fact have been negative over the period 2001 to 2009 (Sheng and Zhao 2012). And many of the sectors where SOEs are present suffer from fragmented production and chronic overcapacity. For instance, more than a hundred Chinese car companies struggle for market share and survival, and the 2010 capacity utilization fell from 90 per cent to below 70 per cent in 2013, suggesting that a reckoning is not far off (Jonquières 2013; *Financial Times* 2014). Thus, few of the SOEs can be regarded as innovative except a few examples such as Huawei, ZTE and Haier.[2]

The absorptive capacity of firms is an important factor behind creating stickiness. There are some promising examples of this capacity such as the observed catching-up in sectors such as high-speed trains, aerospace and defence, automobiles, electronics and clean energy where China (and India) are coming nearer to superpowers like the US and Japan on a range of industry-specific performance indicators such as talent availability and some aspects (but not all) on the cost of doing business. But they still have some way to go when it comes to innovation (Orzag et al. 2014). However, various studies have shown that even if raw talent is immense these talent pools still contain, in absolute terms, relatively few researchers and entrepreneurs who have become capable of directing world-class R&D efforts in key areas of technology (Branstetter et al. 2013). It also seems, according to survey results, that the outlook for private enterprises remains less than brilliant and has continually deteriorated in recent years due to, among others, weaker legal securities for privately owned companies, barriers of liquidity and market access, and inefficient internal management of many SMEs (Bloom and Van Reenen 2010; Xingyuan et al. 2014).

Surely, there are a few examples of where sustainable conditions for stickiness have been created. The case of clean energy is an interesting example of the specific innovation model taken and where Chinese manufacturers not only have managed to increase innovativeness from mere imitation but have also developed their own model of innovative behaviour. A recent study by Lahm (2014) argues that even if many of the technologies in the wind and solar sector originate from the US and Europe, the Chinese have mastered and been innovative in the engineering of these technologies. In 2011, according to Lahm, five of the ten largest solar manufacturers and four of world's top wind producers were from China. Thus, in less than a decade Chinese firms were competing in global markets providing novel technologies with high-cost performance ratios.

There are many factors behind this development and Lahm points to an innovative manufacturing ability as a key factor behind the observed stickiness. In particular, Chinese manufacturers have improved product design, fast-paced product cycles and built scale economies, which have allowed the renewable energy sectors to evolve into a global industry and where Chinese firms have been enablers of re-engineering design, blueprints and prototypes for mass manufacturing. An additional novel feature is that the Chinese firms have encouraged new collaborative modes of product development and where, and in particular, Chinese manufactures and German suppliers have worked together to bring new products to markets.

Anchoring is an important factor in creating stickiness and foreign direct investment (FDI), as is well known, has certainly been an important driver in the Chinese development of technology. Parts of China have been especially successful and the province of Chengdu has developed into hosting several of the world's largest multinationals. Again, one can observe the embryonic development of a growing and specific Chinese innovation model. As indicated in Section 2.3 the close cooperation (or co-invention measured as patents granted) between inventors located in China and inventors located in foreign multinationals has created an embryonic international division of labour within global R&D networks (Branstetter et al. 2013). In addition, recent studies show that China (and to a lesser extent India) is the most important destination of R&D-related projects worldwide for applied research in the form of cross-border greenfield investment projects. But it is still on a low scale compared to domestic R&D investment by foreign multinationals (Chaminade et al. 2014).

Although the dependence of FDI will continue to be important for technology development, the policy ambition up to 2012 is to reduce dependence of foreign technology to less than 30 per cent in 2020 compared to 41 per cent in 2011 (Ljungwall et al. 2014). Thus, quite a few analysts have been pessimistic as regards this ambition or techno-nationalism, as some commentators have framed it, together with the reduction of market access that has developed for some time (Fabre and Grumbach 2012). Others are more convinced of large possibilities of innovation in China. For instance, the American chemical company Dow has 500 scientists working in their R&D facilities, generating more patents per scientists than in other locations in the company (Fabre and Grumbach 2012).

Even if there are some promising examples of innovative sectors and firms it is not enough to sustain China's historically rapid growth. One of the most important features of stickiness is the ability to grow large and internationally renowned domestic companies that will be able to catch up with the innovative global giants.

Again, the pessimists regarding China's potential seem to be in majority. Not many Chinese companies have yet made it to a high international standing. Chinese firms still have to overcome several important capability constraints in order to compete on world markets (Nolan 2014). The lack of internationally recognized consumer brands, knowledge spillovers via customer supplier linkages and inefficient management functions that shape innovativeness and export success are still not present at the firm level. A large part of the Chinese export industry is still made up of subcontractors that add and capture little value to China as shown by the famous iPhone example (Kraemer et al. 2011) and by Growth Analysis (2013).

The challenge for the few global top Chinese companies is equally fierce. The case of Lenovo is an example of the challenges ahead both for the company and for the Chinese business climate. Lenovo has recently bested HP to become the top peddler of desktop computers worldwide (*The Economist* 2014). It is also rising fast in in the global smartphone market. But in order to stay competitive in a market where the demand for PCs is shrinking, and with formidable competitors such as Samsung and Apple in smartphones, it has to innovate. But then the question remains whether it has the imaginative and transformative innovations that organizations like Apple or Samsung have, or the quality of a rich innovative eco-system of talent and entrepreneurs in order to catch up with these giants. The development, so far, seems to suggest a rather long way to go.

Summarizing, one can see signs that point to China becoming an innovator rather than a successful imitator. However, the signs are still thin with only a few truly global players in a limited set of business sectors. The transformation from an imitator to an innovator is another game compared to competing with low manufacturing costs. In addition, China still has unfavourable conditions and policies for being a truly innovative and sticky business environment. The following section will assess some of the policies planned and implemented to enhance innovation.

2.5.2 The Role of Innovation Policy in Creating Stickiness

The many economic reforms implemented over the last decade have on the margin enhanced general and favourable innovative conditions and thus increased stickiness. What other reforms would be needed to make the indigenous innovation agenda more trustworthy?

The 12th five-year plan (2001–15) emphasized innovation by stating the goals of moving up the value chain of the entire economy, supporting an increase in R&D spending and increasing incentives for innovation. Innovation was also prioritized in the latest plenum of the Political Bureau of the Communist Party of China in establishing a framework conducive to innovation. Some examples that have a special bearing on creating stickiness are:

- Legal and regulatory environments: IPR legislation; promotion of academic ethic.
- Knowledge infrastructure: university science parks; more special economic zones; technology transfer centres.

- Innovation financing: government private and foreign venture capital fund; simplify the creation of high tech start-ups.
- FDI and mobility: better conditions for foreign companies to establish, invest and grow than trends; incentives for Chinese overseas students and researchers to return; Thousand Talents Program.

Each of these points represents a real challenge, and where the challenge is as always in the details of future implementation. One example is on legal and regulatory issues. A common view is that China partly fails to innovate because it lacks strong and stable protection for intellectual property. In 2001, China joined the World Trade Organization (WTO) and therefore is supposed to follow these rules, although the compliance is far from perfect as many examples show (Raustiala and Sprigman 2013). In addition, the planning document in 2006 and the definition of indigenous innovation include the following wording: "enhancing original innovation through co-innovation and re-innovations based on the assimilation of imported technologies." This has been interpreted as an official green light for piracy. The future will tell whether compliance to international rules will increase. But as the examples from the wind and solar industry show, imitative development together with collaboration and sharing of technology can be a critical part of creativity and innovation, and can benefit both Chinese and Western companies.

Central to innovation stickiness is the access and sophistication of financing. Clearly, there are many examples of failures in the Chinese innovation system in this respect (Cao et al. 2013). The existence of open capital markets is an essential function of an efficient innovation system and in that respect the system has been dysfunctional for years. A free capital market does not yet exist and it is difficult to establish a large-scale company without state finance.[3] Although changes to the system are discussed (and China has recently opened the door to new private banks), the state controls all large banks and credit is often distributed according to political criteria in favour of large SOEs. The result has been a financial repression where state-owned banks have provided very cheap money, which to a large extent has been used for inefficient investment in unneeded production capacity, property developments and science parks.

There has also been much discussion among analysts about whether the Chinese state lacks the capability to practise a proactive innovation policy, which may throw China's future into doubt (Wu 2013). As argued in Section 2.4, the latest policy ambitions to create innovation capability seem to resemble ways of the previous backing of high-technology industrialization. Government-supported research institutions are spread throughout the country and building key policy research organizations and forging ties between industrial firms and research organizations have priority. But again these policies do have a flavour of old policy models, where national and sectorial plans and visionary strategies do not take into account the entrepreneurial spirit needed to be truly innovative as innovation theory suggests. This techno-nationalism trajectory also seems to be deeply ingrained in the Chinese priority setting policy model of enhancing R&D.

It has been questioned whether activist Chinese industrial policy models in the past have failed to the extent that is claimed (Chu 2011). Experiences from the automobile industry suggest that policies were ineffective at first, but improved by taking into account results of many of the local experiments. This adaptive learning ability combined with a strong catch-up consensus providing performance standards to establish a national automobile industry could, as is sometimes argued, be effective when it comes to catching up with other areas of technology development and innovation.

Summarizing, the verdict about these different policy ambitions and the ability to create stickiness will be determined in the future. At best, and if correct, clever and adaptive learning policies might improve innovation although whether the state-directed policy will work for the much more fuzzy innovation and entrepreneurship strategy is more of a challenge. At worst, the suggested policies, if implemented, are of questionable economic and commercial value with high opportunity costs and inefficiencies, as examples from promoting electrical vehicles seem to indicate (Tagscherer 2012). A final consideration is about the co-evolution of economic and political reform. During the last three decades China has carried out economic reforms without political reform. They have implemented institutional changes and reforms piecemeal, giving benefits to key interest groups within the state sector. As many analysts of national innovation systems would claim, this is not a strategy that will work for being at the innovation frontier.

2.6 CONCLUSIONS

"Will China make it?" That is the question asked in China itself, of course, but at least as eagerly in its competitor countries in the West, hoping that massive Chinese research spending will help push global technological to progress further (Fernald and Jones 2014). Googling the question gives an indication of just how much interest it generates. Whether China will make it in her quest to move up the value chain and become a more innovative economy is still an open question. Given historical records and recent developments both optimists and pessimists can certainly find arguments and statistics supporting their respective views. But as argued here, this binary split is too crude to capture and assess a process as complex as the dynamics of innovation. Instead, the approach taken is that the speed and scale of China's transformation makes it a real-time experiment where there is no role model to follow. To boldly go where no man has gone before, China will have to cross the river while feeling for stones, and as has been argued there are still lots of stones before China will close the innovation gap to developed countries.

The observations in this chapter point to the fact that China's economic development over the last decades has gone hand in hand with rapid and considerable improvement in scientific and innovative capability, albeit from a low starting level. But China has not yet followed the institutional paths of innovation of the West. The purpose of this chapter has therefore been to highlight three specific features of the Chinese approach, which may inform a continued debate on their implications for future catch-up and the aspired technology leadership.

First, there are obvious weaknesses in the institutional set-up governing the Chinese innovation system. Flawed incentives seem to put obstacles in the way of developing a world-class R&D system. Insufficient absorptive capacity stemming from a weak ecosystem of innovation leads to severe inefficiencies. Despite this, the sheer size of China's investments and human capital supply in the R&D sector undoubtedly contribute to a catching-up in technology and innovation. Progress and development happens, although at a high cost. If a sufficient amount of the invested resources are devoted to improving the quality of institutions and general capacity building, one can also expect the absorptive capacity and efficiency in the system to increase over time.

Second, part of the problem of whether indigenous breakthroughs and innovation might be achieved is due to the deep-rooted political inclination to focus on mission and excellence-driven research over institution building and diffusion-related objectives. The benefits for this focused strategy have served China well when being far from the scientific and innovation frontier. Moving closer to the frontier the question is whether these focused investments will be absorbed by a thriving ecosystem of skilled entrepreneur and global companies that are able to transform knowledge that can be sold on competitive world markets. For this to happen much more creative destruction and focus on basic research will be needed.

Third, Chinese innovation systems have produced a new form of stickiness that may be a basis for a more efficient innovation system. It rests on close cooperation between local inventors in China and multinational innovative companies. The rapid development in wind and solar sectors from small, niche producers to global industries illustrates an interesting collaborative mode of product development between manufacturers and foreign suppliers. The effect of this mode of collaboration in emerging countries in general and in China in particular has been a small but growing international division of R&D, where multinational enterprises can now provide innovative technologies to global markets at a lower cost and introduce products more suitable both for emerging as well as for advanced markets. In a world where R&D is also fragmented into global innovation chains, China may hold a number of competitive advantages in the stages of innovation that rely on rapid and incremental adaptation of products. However, that will require some bold policy reforms to include stronger entrepreneurial and market mechanisms in order to catch up.

The increase in R&D and innovation activity in emerging economies and the following division of R&D labour may very well lead to a win-win outcome with a more integrated global innovation system. Given the size and importance of China in the world economy, China's successful transformation into an innovative economy is an outcome to be hoped for outside of China as well as within. This positive outcome is coupled to China's willingness and ability to continue improving its innovation system and integrate fully into an open and liberal global world order.

NOTES

1. Puga and Trefler (2010) coined this phrase about China's impressive development: "Wake up and smell the ginseng."
2. Zhihong et al. (2011) suggest that only a handful of companies in the ICT equipment sector account for the overwhelming share of Chinese patents during 2000 and 2006. Of all foreign filings, 29 per cent were in ICT.
3. Sheng and Geng (2014) report that China's State Council recently announced a "blueprint" for capital reform with the objective to "support open, fair, and integral market processes, and to protect investors, particularly the legal rights of small investors".

REFERENCES

Aghion, P., N. Bloom, R. Blundell, R. Griffith and P. Howitt (2005), "Competition and innovation: an inverted-U relationship", *The Quarterly Journal of Economics*, 120(2), 701–28.

Atkinson, R.D. (2012), "Enough is enough: confronting Chinese innovation mercantilism." The Information Technology and Innovation Foundation (ITIF). Available at: http://www2.itif.org/2012-enough-enough-chinese-mercantilism.pdf (accessed 19 October 2014).

Benner, M., L. Liu, and S. Schwaag Serger (2012), 'Head in the clouds and feet on the: research priority setting in China', *Science and Public Policy*. Available at: http://spp.oxfordjournals.org/content/early/2012/02/17/scipol.scs006 (accessed 19 October 2014).

Bound K., T. Saunders, J. Wilsdon and J. Adams (2013), "China's absorptive state: research, innovation and the prospects for China–UK collaboration." Available at: http://www.nesta.org.uk/sites/default/files/chinas_absorptive_state_0.pdf (accessed 19 October 2014).

Branstetter, L., G. Li. and F. Veloso (2013), "The rise of international co-invention." National Bureau of Economic Research. Available at: http://www.nber.org/chapters/c13028.pdf (accessed 19 October 2014).

Breznitz, D. and M. Murphree (2011), *Run of the Red Queen: Government, Innovation and Economic Growth in China.* New Haven, CT: Yale University Press.

Bloom, N. and J.V. Reenen (2010), "Why do management practices differ across countries?" *Journal of Economic Perspectives*, 24(1), 203–24.

Cao, C., L. Ning, L. Xia and L. Liu (2013), "Reforming China's S&T system", *Science*, 341(6145), 460–2.

Central Committee of the Communist Party of China (CCPCC) 2013), Decision of the Central Committee of the Communist Party of China on Some Major Issues Concerning Comprehensively Deepening the Reform, adopted at the Third Plenary Session of the 18th Central Committee of the Communist Party of China on November 12, 2013, p. 17. Available at: http://www.china.org.cn/china/third_plenary_session/2014-01/16/content_31212602.htm (accessed 20 October 2014).

Chaminade, C., D. Castellani and M. Plechero (2014), *The Emergence of China and India as New Innovation Power Houses – Threat or Opportunity.* Entreprenuership Forum. Available at: http://entreprenorskapsforum.se/wp-content/uploads/2014/02/GloFo_The_Emergence_of_webb.pdf (accessed 20 October 2014).

Chu, W.W. (2011), "How the Chinese government promoted a global autmobile industry", *Industrial and Corporate Change*, 20(5), 1235–76.

Das, M. and P. N'Diaye (2013), "Chronicle of a Decline Foretold: Has China Reached the Lewis Turning Point?" International Monetary Fund Working Paper WP/13/26.

Delgado, M., C. Ketels, M. Porter and S. Stern (2012), "The Determinants of National Competitiveness." National Bureau of Economic Research Working Paper 18249. Available at http://www.nber.org/papers/w18249 (accessed 20 October 2014).

The Economist (2014), "The rise of the frugal innovation", 24 May, p. 55.

Edquist, C. (ed.) (1997), *Systems of Innovation: Technologies, Institutions and Organizations*. London: Pinter Publishers/Cassell Academic.

Fabre, G. and S. Grumbach (2012), "The World Upside Down – China's R&D and Innovation Strategy." Fondations maison des sciences de l'homme, Working Paper Series, No. 7, April.

Feng, X., C. Ljungwall and G. He. (2014), *The Ecology of China's Private Enterprises*. Series on Chinese Economics Research: Vol. 14. Singapore: World Scientific.

Fernald, J. and C. Jones (2014), "The future of US economic growth", *American Economic Review*, 104(5), 44–9.

Financial Times (2014), "China's Automotive Sector Needs a Shakeout to Thrive", 28 May.

Freeman, R (2009), "One Ring to Rule Them All? Globalization of Knowledge and Knowledge Creation." National Bureau of Economic Research Working Paper 19301. Available at http://www.nber.org/papers/w19301 (accessed 20 October 2014).

Fu, J., U.R. Frietsch and U. Tagscherer (2013), "Publication Activity in the Science Citation Index Expanded (SCIE) Database in the Context of Chinese Science and Technology Policy from 1977 to 2012", Fraunhofer ISI Discussion Papers Innovation Systems and Policy Analysis, No. 35.

Growth Analysis (2013), "Outsourcing och offshoring av FoU i globala värdekedjor." WP/PM 2013:10. Available at: http://www.tillvaxtanalys.se/sv/publikationer/pm/working-paper-pm/2013-08-30-outsourcing-och-offshoring-av-fou-i-globala-vardekedjor.html w19301 (accessed 20 October 2014).

von Görg, H. and Liu W-H. (2012), "China's role in the global economy and technological catch-up." Kiel Institute Focus 17. Available at: https://www.ifw-kiel.de/media/kiel-institute-focus/2012/kiel-institute-focus-17/view (accessed 20 October 2014).

Hangzhou.gov.cn (n.d.), "2.The two-wheel driven strategy accelerating the upgrade of economic structures."Available at: http://www.hangzhou.gov.cn/main/zpd/English/goal/gov/T307091.shtml (accessed 20 October 2014).

Hsieh, C.T. and P.J. Klenow (2009), "Misallocation and manufacturing TFP in China and India", *The Quarterly Journal of Economics*, 124(4), 1403–48.

INSEAD (2013), "The Global Innovation Index 2013." Available at: http://www.globalinnovationindex.org/content.aspx?page=gii-full-report-2013#pdfopener (accessed 20 October 2014).

Innovation Union Scoreboard (IUS) (2014), "EU more innovative but the differences between member states are still high and diminish only slowly." Available at: http://ec.europa.eu/enterprise/policies/innovation/policy/innovation-scoreboard/index_en.htm (accessed 20 October 2014).

Jonquières, de G. (2013), "Who's Afraid of China's High-Tech Challenge?" ECIPE Policy Brief 07/2013. Available at: www.ecipe.org/media/publication_pdfs/PB7.pdf (accessed 20 October 2014).

Kraemer, K.L., G. Linden and J. Dedrick (2011), "Capturing value in global networks: Apple's iPad and iPhone." Available at: http://pcic.merage.uci.edu/papers/2011/Value_iPad_iPhone.pdf (accessed 20 October 2014).

Lahm, L. (2014), "Sino-German Collaboration and the Development of Global wind and Solar Industries, 1990–2012." MIT Working Paper No. 2013-38.

The Lancet (2010), "Scientific fraud: action needed in China," 375(9709), 94.

Ljungwall, Christer and Patrik Gustavsson Tingvall (2014), "Is China different? A meta-analysis of the growth-enhancing effect from R&D spending in China", RATIO Working Paper No. 233, http://ratio.se/media/248167/cl_pt_r_d_china_233.pdf.

Ministry of Science and Technology of the People's Republic of China (MOST) (2012), *China Science and Technology Statistics Data Book*.

Nolan, P. (2014), *Chinese Firms, Global Firms: Industrial Policy in the Age of Globalization*. London: Routledge.

Orzag, M., U. Shriram and D.J. Snower (2014), "Economic performance index: an industry-centric measurement approach." VOX. Available at http://www.

voxeu.org/article/industry-centric-economic-measurement?quicktabs_tabbed_recent_articles_block=0 (accessed 20 October 2014).
Puga, D and D. Trefler (2010), "Wake up and smell the ginseng: international trade and the rise of incremental innovation in low-wage countries", *Journal of Development Economics*, 91(January), 64–70.
Raustiala, K. and C. Sprigman (2013), "Fake it till you make it: the good news about China's knockoff economy", *Foreign Affairs*, July/August.
Sheng, H. and N. Zhao (2012), *China's State-Owned Enterprises: Nature, Performance and Reform (Series on Chinese Economics Research)*, Singapore: World Scientific. Available at http://www.amazon.com/Chinas-State-Owned-Enterprises-Performance-Economics/dp/9814383848.
Sheng, A. and X. Geng (2014), "Harnessing China's competitive streak." Project Syndicate, 29 May. Available at: https://www.project-syndicate.org/commentary/andrew-sheng-and-geng-xiao-emphasize-chinese-leaders--need-to-balance-growth-enhancing-competition-with-stabilizing-controls (accessed 20 October 2014).
Sheng, H. and N. Zhao (2012), *China's State-Owned Enterprises: Nature, Performance and Reform (Series on Chinese Economics Research)*. Singapore: World Scientific.
Simon, D.F. (2012), "China's new S&T reforms and their implications for innovative performance, testimony before the US-China Economic and Security Review Commission." Available at: http://www.uscc.gov/sites/default/files/5.10.12Simon.pdf (accessed 20 October 2014).
State Intellectual Property Office of the People's Republic of China (SIPO) (2013), "Promotion plan for the implementation of the national intellectual Property Strategy in 2013." Available at: http://english.sipo.gov.cn/news/official/201303/t20130326_789188.html (accessed 20 October 2014).
Suttmeier, R.P. and X. Yao (2011), "China's IP Transition: Rethinking Intellectual Property Rights in a Rising China." National Bureau of Asian Research, NBR Special Report #29.
Tagscherer, U. (2012), "Electric Mobility in China: A Policy Review." Discussion Paper, Fraunhofer Innovation and Policy Analysis No. 30.
Veugelers, R. (2013), "The world innovation landscape: Asia rising?" Bruegel Policy Contribution, February. Available at: http://www.bruegel.org/publications/publication-detail/publication/766-the-world-innovation-landscape-asia-rising/ (accessed 20 October 2014).
World Bank (2014), "Researchers in R&D (per million people)." Available at: http://data.worldbank.org/indicator/SP.POP.SCIE.RD.P6 (accessed 20 October 2014).
World Intellectual Property Organization (WIPO) (2013), *World Intellectual Property Indicators – 2013 Edition*. Available at: http://www.wipo.int/ipstats/en/wipi/ (accessed 20 October 2014).
Wu, H. (2013), "China's Growth and Productivity Performance Debate Revisited." The Conference Board Economics Program Working Paper Series. Available at: https://www.conference-board.org/pdf_free/workingpapers/EPWP1401.pdf (accessed 20 October 2014).
Zhang, H., D. Patton and M. Kenney (2013), "Building global-class universities: assessing the impact of the 985 Project", *Research Policy*, 42(3), 765–75.
Zhu, Xi. (2012), "Understanding China's Growth: Past, Present and Future", *Journal of Economic Perspectives*, 26(4), 103–24.

Zhihong, Y, M. Eberhardt and C. Helmers (2011), "Is the Dragon Learning to Fly? An Analysis of the Chinese Patent Explosion." VOX. Available at: http://www.voxeu.org/article/china-imitator-innovator (accessed 20 October 2014).

3. The role of intellectual property rights in innovation spaces: the cases of China and India

Marcus Holgersson

3.1 INTRODUCTION

This chapter describes the development of China and India as innovation spaces linked to the global economy through specializations and flows of technical development.[1] The focus is upon developments of the Chinese and Indian patent systems and on the flows of patents into and out from these countries. Patent systems and patent-like rights have been used in various nations and to various degrees since the fourteenth and fifteenth centuries (Granstrand, 1999; Guellec and Potterie, 2007). However, China's first patent law was adopted as late as 1984, which can be compared to the adoption of formal patent laws in 1623 in England, in 1790 in the United States and in 1856 in India (based on the British patent law) (Holgersson, 2012).

Many Asian countries, not the least China, have historically had limited protection of inventions against imitation. Even when patent systems have been available, the actual protection given by those systems has been limited since the enforcement of patents has been difficult or impossible for inventors. Therefore, inventors (individuals, firms and other types of agents) have often resorted to other types of mechanisms to deter imitation and to appropriate sufficient returns from innovation investments, such as by technological specialization, by secrecy or by establishing personal relationships [*guanxi*] (Keupp et al., 2010). The limited enforcement and/or use of intellectual property rights (IPR) systems by developing and newly industrialized countries is not surprising, however, since allowing domestic imitation of foreign innovators enables catching-up in relation to more innovative and industrialized countries (Granstrand and Holgersson, 2014).

During the 2000s the situation started to change with both China and India focusing increasingly on building up institutions and competences related to IPRs. These changes have partly come as a result of the creation and enforcement of the Agreement on Trade Related Aspects of Intellectual Property Rights (TRIPS) by the World Trade Organization (WTO). China joined the WTO on 11 December 2001, while India was one of the original members (from 1 January 1995).[2] Subsequently, the situation has progressed further, with innovation and intellectual property (IP) policies being adopted at the national level.

The 12th Chinese Five-Year Plan (2011–2015),[3] for example, states that: 'China should upgrade its capabilities in indigenous research and innovation in science, technology and administration, train more innovative talents and improve education for workers. In a word, we will strive to speed up the construction of an innovation country.'

In terms of IP, China issued the Outline of the National Intellectual Property Strategy[4] in June 2008, here exemplified by a couple of strategic foci that were included:

> (13) Revise laws and regulations to punish infringements on IPRs and strengthen judicial punishment. Help right holders to improve consciousness and capacity to protect their own interests. Lower the cost of right enforcement. Increase the cost of infringements. Curb infringements effectively. [. . .]
> (15) Strengthen the knowledge propagation on intellectual property right and increase the awareness of intellectual property right in the whole society. Carry out the ordinary intellectual property right education extensively.

In November 2010, the State Intellectual Property Office (SIPO) in China published the National Patent Development Strategy (2011–2020),[5] describing how the previously issued outline was to be implemented, including a number of quantitative goals:

> The capacity and level to create patents will be improved by a large margin. The annual quantity of applying for patents for inventions, utility models and designs will reach 2 million. China will rank among the top two in the world in terms of the annual number of patents for inventions granted to the domestic applicants, and the quality of patents filed will further improve. The number of owning patents every one million people and the number of overseas patent applications filed by Chinese applicants will double. The proportion of patent applications in industrial enterprises above designated size will reach 8% and the quantity of owning patent rights will significantly rise.

These quantitative goals are being supported by incentives such as cash bonuses, tax cuts, resident permits and academic tenures. These incentive systems have then been criticized for their focus on the quantity of Chinese patenting rather than the quality of it (Lohr, 2011; Tung, 2011).

Policies in India have made a similar transition to a more

innovation-related focus. This can be illustrated by a statement in the early 2000s by the president of India at the time, Dr A.P.J. Abdul Kalam, in which he highlights the planned shift from imitation to innovation:[6]

> Today India has become one of the strongest in the world in terms of scientific manpower in capability and maturity. Hence, we are in a position not only to understand the technologies that we may have to borrow, but also to create our own technologies with extensive scientific inputs of indigenous origin.

As compared to the policies of China, Indian policy documents have raised more concerns about the intense privatization of knowledge through the use of IPRs internationally, and placed less emphasis on increasing domestic patenting and strengthening IPRs, as exemplified by the Science and Technology Policy 2003:[7]

> The process of globalisation is leading to situations where the collective knowledge of societies normally used for common good is converted to proprietary knowledge for commercial profit of a few. Action will be taken to protect our indigenous knowledge systems, primarily through national policies, supplemented by supportive international action. For this purpose, IPR systems which specially protect scientific discoveries and technological innovations arising out of such traditional knowledge will be designed and effectively implemented.

Nevertheless, the Indian government has declared 2010–20 the decade of innovation and one of the key elements of the Science, Technology and Innovation Policy 2013 is the positioning of 'India among the top five global scientific powers by 2020' (p. 4). This goes in line with China's goal to become a research and development (R&D) and innovation-based nation by 2020, which was stated in *The National Medium- and Long-Term Plan for the Development of Science and Technology (2006–2020)*, published by the Chinese government in early 2006.

Given these nations' increased focus on innovation, and the role of IPRs in this process, this chapter aims to shed some light on China and India as innovation spaces. The chapter especially focuses on the specific institutional aspect concerning the patent system and the use of it.

Skepticism has been raised about the IPR systems in these countries, for example regarding the large quantities of low-quality patents being granted in China (*The Economist*, 2010). However, a substantial amount of the patents filed in these countries stems from foreign inventors, not the least from Western countries. Further, Asian inventors do not limit themselves to patenting in their domestic markets. In fact, from 1996 to 2009 China climbed from the 27th to the 8th position on the ranking of top foreign countries patenting in the US, while India climbed from the 30th to the 16th position (Granstrand and Holgersson, 2014).

This chapter aims to bring some clarity to the current developments of China and India as innovation spaces, especially in terms of their IPR developments. More specifically the chapter aims to answer three questions related to this development: (1) Who patents in China and India? (2) Where do Chinese and Indian inventors patent? (3) What do Chinese and Indian inventors patent?

3.2 CONCEPTS AND DEFINITIONS USED IN THE DATA COLLECTION

A patent grants its holder the right to exclude others from commercially exploiting the patented invention without authorization during a limited time period. The purpose of patent systems is to incentivize private innovation investments by enabling innovators to appropriate sufficient returns on investments. In return the patent applicants pay patent fees and let the patent documents be published, thus stimulating not only the creation of technical knowledge (technology), but also the diffusion of it.

Focus is here on 'invention patent applications' as opposed to utility model patents and design patents, which are used as complements to invention patents in some countries. Invention patent applications are in the following mentioned simply as 'patent applications.' Patent applications are submitted to patent and trademark offices (PTOs), which are typically national entities (exceptions include the World Intellectual Property Organization – WIPO – and the European Patent Office – EPO). These PTOs evaluate the patent applications upon the underlying inventions' characteristics of novelty, non-obviousness and usefulness/technical character. If the requirements for patentability are fulfilled and fees are paid, a national patent is granted, which then also needs to be enforced in the national market and institutional system in cases of infringement. This national patent may typically be complemented with patents for the same invention in other nations, and this set of related national patents for the same invention is typically called a patent family.

The national origin of patents and patent applications is ambiguous, as the underlying inventions could be created by inventors across national borders, and/or by inventors working in multinational firms or by multiple inventors across multiple firms with various nationalities. However, the residence of the first named applicant (or inventor) is typically used as an indicator of the national origin of a patent. This approach is used

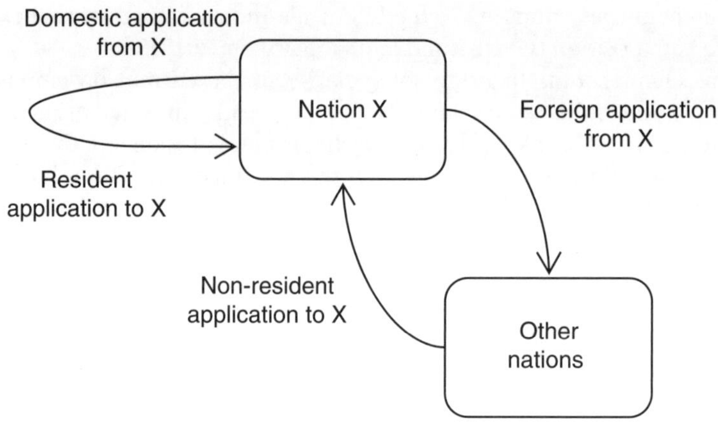

Figure 3.1 Classification of incoming and outgoing patent application streams to and from a country

here as well, especially since it is the approach used in the WIPO statistics database.

Based on the origin, the incoming and outgoing patent application streams related to a specific nation can be classified, see Figure 3.1.

First, *outgoing applications* from a specific nation are either domestic, if they are filed at the domestic PTO, or foreign, if they are filed at any foreign PTO. Second, *incoming applications* to a specific nation's PTO are either resident, if they origin from the same nation, or non-resident, if they origin from any other nation. Thus, any domestic application from a nation is also a resident application to that nation's PTO, while any foreign application from a nation is also a non-resident application to some other nation's PTO.

3.3 PATENTING IN AND FROM CHINA AND INDIA

An investigation into the details is needed because, as described above, the policies of China and India have shifted with an increasing focus on innovation and IP (at least to some extent). This section will study the patenting in and from China and India, based on statistics from WIPO.[8] As also described above applicants to any given national PTO can be divided into resident applicants, that is, applicants with the same nationality as the PTO, and non-resident applicants, that is, applicants with another nationality than the PTO.

With respect to the share of resident vs. non-resident applicants, China and India exhibit quite large differences. Actually, while roughly one-fifth (21 percent) of all patent applications to the Chinese PTO in 2011 had *non-resident* origin, roughly one-fifth (again, 21 percent) of all patent applications to the Indian PTO had *resident* origin (the latter is a share that has been fairly constant throughout the 2000s). The low share of non-resident applicants in China comes as a result of the fact that resident patenting in China has been heavily promoted during the early 2000s. The absolute number of resident Chinese patent applications has grown from 25000 (2000) to 416000 (2011), that is by a multiple of 17. Although the number of non-resident Chinese patent applications has also been growing with a high pace, the lower growth rate has meant that the share of non-resident applications has dropped from roughly one in two to roughly one in five between 2000 and 2011.

Let us now turn to focus, on the one hand, on non-resident patent applications to China and India, respectively, and on the other hand, on foreign patent applications from China and India, respectively. The reason for this is that this enables the illumination of the importance of China and India in international patenting, without being biased due to domestic/resident patenting streams (which has major influence on statistics, as indicated above by the Chinese developments).

Relative growth of non-resident applications to the Chinese and Indian offices have been larger than the relative growth of the total worldwide patenting, as illustrated by Figure 3.2, in which the total worldwide patenting has been normalized by dividing it with a constant (19.4) to end up at the same 2011 value as China. Thus, referring to massive amounts of resident applications behind the increase in patenting at the Chinese and Indian PTOs is only partially correct for China, and hardly correct for India. In Section 3.3.1 the non-resident application streams to China and India are investigated, more specifically to probe the question of which nationalities patent there (RQ1).

Foreign patenting from China and India has also increased steeply, although at lower absolute levels. This indicates the increasing role of these nations in creating inventions that are not only good enough to be covered by patents at the domestic PTOs, but also at foreign PTOs. In Section 3.3.2, the destinations of the foreign patenting streams from these countries are identified (RQ2). Not surprisingly it turns out that the PTO in the US (USPTO) is the most important receiving office. The US is therefore used as a reference market to indicate the technological areas in which China and India focus their patenting, which also indicates the focus of their R&D efforts. Section 3.3.3 presents results on this, both on the most important technologies in terms of absolute size and on the most important technologies in terms of technological specialization (RQ3).

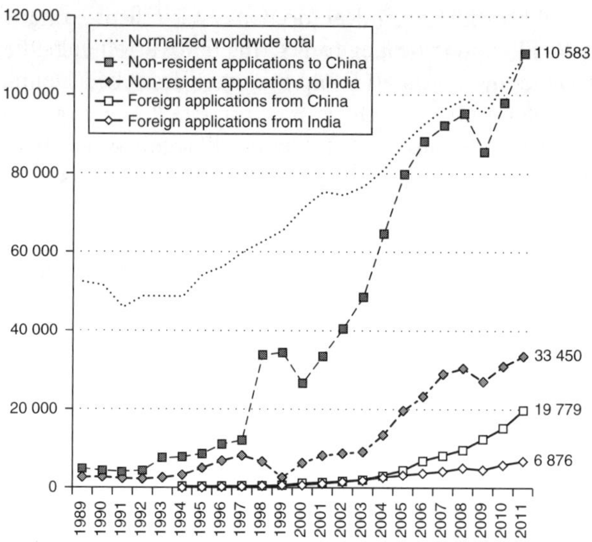

Figure 3.2 Yearly patent applications to and from China and India and normalized worldwide total (legend in order of 2011 magnitude)

3.3.1 Who Patents in China and India?

In terms of non-resident applications, which in 2011 amounted to roughly 111 000 in China and 33 000 in India, an interesting question to probe is what the origins are of these applications to the Chinese and Indian PTOs, respectively. This is investigated here by looking at the share of non-resident applications that originate from various nations, and by looking at the revealed market advantage (RMA) that various national origins have on the two markets. The RMA measure was introduced by Granstrand and Holgersson (2014), building upon the measures of revealed comparative advantage (RCA) (Balassa, 1965) and revealed technological advantage (RTA) (for example, Soete, 1981, 1987).

The RMA indicates whether the share of foreign patent applications to a specific PTO from a certain national origin is larger than the share of total foreign patent applications to that PTO from all national origins. Thus, it indicates whether a national origin is more ($1<RMA$) or less ($0\leq RMA<1$) focused on a specific PTO for its foreign patent applications than the worldwide focus on that PTO. The formal definition of RMA of country *a* in market *i* is:

$$RMA_{ai} = \frac{p_{ai} / \sum_{i=1}^{N} p_{ai}}{\sum_{j=1}^{N} p_{ji} / \sum_{j=1}^{N} \sum_{i=1}^{N} p_{ji}}$$

Here p_{ai} is the number of foreign applications from country a to receiving office i, and N is the total number of countries.[9] For statistical purposes, it is advisable to use the symmetric version of this index (here used for correlation calculations), namely the revealed symmetric market advantage (RSMA), as defined below (for a motivation, see Granstrand and Holgersson, 2014):

$$RSMA_{ai} = \frac{RMA_{ai} - 1}{RMA_{ai} + 1}$$

Looking at the main origins of non-resident patent applications to the Chinese PTO (over a three-year period to decrease impact from temporary variations) it is clear that Japan and the US contributed to more than 60 percent of these applications during 2009–11 (see Table 3.1).

This might not be surprising, as these are by far the main foreign patenting nations worldwide (US applicants filed 549 000 foreign applications during 2009–11 while Japanese applicants filed 532 000 foreign applications during the same period). Japan and the US are focusing more than other nations on the Chinese output market for patents not only in absolute terms, however, but also in relative terms, indicated by RMA indices above 1 for both of them. Actually, there are only ten nations that have RMAs above 1 in China in the studied period, and the only major ones of these are Japan, the US and Korea. Further, there is a significant positive correlation between the total number of foreign applications from nations active in foreign patenting during 2009–11 and the RSMA of these nations in China.[10] A plausible explanation is that the largest IP 'producing' nations have identified the importance of holding a competitive IPR position also in China. However, this is subject to large variations, as indicated by the variation in RMAs below and above 1 among the main origins (see Table 3.1), which apart from Japan and US mainly include European countries (Korea being the exception).

Now, looking at patent applications to the Indian PTO, non-resident patent applications are much more frequent than resident applications, as described above. The main origins of non-resident patent applications to the Indian PTO are again dominated by the US and Japan (see Table 3.1), and patent applications to the Indian PTO from US applicants (30 000 during 2009–11) actually outnumber the domestic applications from Indian applicants (25 000 during 2009–11). A number of mainly European

Table 3.1 Share by main origins, of non-resident patent applications in China and India, 2009–2011

Origin of non-resident patent applications in *China*	Share	RMA	Origin of non-resident patent applications in *India*	Share	RMA
Japan	35.2%	1.45	United States of America	32.9%	1.32
United States of America	25.7%	1.03	Japan	13.5%	0.56
			Germany	11.8%	0.90
Germany	10.0%	0.77	Switzerland	5.3%	1.31
Republic of Korea	7.2%	1.14	France	5.1%	0.93
France	3.6%	0.65	Netherlands	4.5%	1.33
Netherlands	3.1%	0.91	United Kingdom	3.4%	0.86
Switzerland	2.6%	0.65	Sweden	2.8%	1.27
United Kingdom	1.8%	0.45	Republic of Korea	2.4%	0.38
Sweden	1.8%	0.81	China	2.4%	1.10
Italy	1.2%	0.59	Italy	2.1%	1.05
Finland	1.0%	0.84	Canada	1.7%	0.67
Other origins (below 1% share)	6.9%	–	Finland	1.7%	1.42
			Unknown	1.4%	–
			Denmark	1.2%	1.18
			Israel	1.1%	0.85
			Belgium	1.1%	0.90
			Australia	1.0%	0.87
			Other origins (below 1% share)	4.6%	–

countries (with a few exceptions) are then also important origins for Indian non-resident patenting. Although the US as the largest applicant origin also has an RMA above 1, the positive correlation between the total number of foreign applications from various nations active in foreign patenting during 2009–11 and the RSMA of these nations in India is smaller than in the case of China.[11]

3.3.2 Where Do Chinese and Indian Inventors Patent?

As illustrated in Figure 3.2, Chinese foreign patenting has increased steeply during the last decade. Table 3.2 presents the PTOs that during 2009–11 received the largest shares of the foreign patent applications from China (and India), together with China's RMAs in those nations.

Table 3.2 Share of main receiving offices, of foreign patent applications from China and India, 2009–2011

Receiving offices of foreign patent applications from *China*	Share	RMA	Receiving offices of foreign patent applications from *India*	Share	RMA
United States of America	54.8%	1.64	United States of America	67.6%	2.02
European Patent Office	13.3%	1.37	European Patent Office	7.2%	0.74
Japan	7.2%	0.97	China	2.9%	0.22
India	4.7%	1.12	Japan	2.8%	0.37
Republic of Korea	3.6%	0.69	Australia	2.4%	0.80
China, Hong Kong SAR	2.9%	1.74	Canada	2.2%	0.53
			Republic of Korea	1.8%	0.35
Canada	2.0%	0.48	South Africa	1.4%	1.67
Russian Federation	1.8%	0.94	Brazil	1.4%	0.79
Australia	1.8%	0.58	Mexico	1.3%	0.69
Other offices (below 1% share)	7.8%	–	New Zealand	1.0%	1.55
			Other offices (below 1% share)	8.0%	–

It is quite clear that China focuses its foreign patenting on the US, both in absolute terms and in relative terms (compared to other nations' focus on the US, as indicated by the RMA above 1). Similar to the incoming patents to China being mainly from the US, Japan and European nations, the outgoing patents are mainly directed to these nations. However, large receivers of Chinese patent applications also include India, Korea, Hong Kong, and so on, as seen in Table 3.2.

Indian foreign patenting has also increased steeply during the last decade, as illustrated in Figure 3.2, and the focus on the US market is even more evident than in the case of China. Table 3.2 shows the main receiving offices of Indian foreign patent applications, and the USPTO receives 68 percent of these, meaning that India focuses twice as much as the general worldwide focus on the US. The EPO is the second-largest receiver of Indian patent applications, while Japan receives a relatively low share of foreign applications from India. Due to index constructions, the massive focus on the US downplays the RMAs of India in other countries, and worth noting in light of this is the relatively large shares of applications (and thereby high RMAs) from India that are still received by South Africa and New Zealand. These countries do, similar to India (see Section 3.3.3), invest and patent extensively in pharmaceuticals, biotechnologies, medical technologies and chemistry, which can be compared to Japan, where there is limited focus in these areas. The technological distance between two countries is a likely determinant of how frequent inventors in one country patent in the second country, and vice versa. This needs further research, however.

3.3.3 What Do Chinese and Indian Inventors Patent?

A related question to that of where Chinese and Indian inventors patent is the question of *what* they patent. This is probed in this chapter by using US patent data as a proxy for patenting activities from these countries, since the US has traditionally been the most important market for foreign patenting from other countries (and, as proven above, it is the most frequent one for foreign patenting from China and India). Although this approach has been common in previous research (for example, Patel and Pavitt, 1994; Granstrand et al., 1997; Granstrand and Holgersson, 2014), it is related to some limitations, as country choices for foreign patenting are, for example, dependent upon the activities of the inventor and its competitors, as well as upon the (potential) customer base and the institutional environment in the specific country in question.

Nevertheless, data on patenting in the US gives a useful indication of the technological focus of various actors/nations. In fact, there is a fairly high correlation between the technology shares (defined as the share of a country's patents in a specific technology) as measured by EPO patenting and the technology shares as measured by USPTO patenting, and this holds for both China and India.[12] This is then an indication of the validity of using US patent data as a proxy for foreign patenting.

Apart from using technology shares – that is, the shares of Chinese and Indian patenting at the USPTO in different technological areas – the revealed technological advantage (RTA) is here used as an indicator of these nations' relative focus in various technological areas as compared to all non-resident applicants' focus in various technological areas when patenting at the USPTO. The formal definition of RTA of country a in technology i is:

$$RTA_{ai} = \frac{p_{ai} / \sum_{i=1}^{M} p_{ai}}{\sum_{j=1}^{N} p_{ji} / \sum_{j=1}^{N} \sum_{i=1}^{M} p_{ji}}$$

Here, p_{ai} is the number of US patent publications from country *a* in technological area *i*, *M* is the total number of technological areas and *N* is the total number of countries. To avoid biasing the results, patent publications from US patentees are excluded from the dataset, meaning that only non-resident patent publications are considered in the denominator when calculating the RTAs. Similar to the use of RMA above, a symmetric version of the RTA – that is, the revealed symmetric technological advantage (RSTA) – is calculated for statistical purposes (here used for variance calculations):

$$RSTA_{ai} = \frac{RTA_{ai} - 1}{RTA_{ai} + 1}$$

Table 3.3 presents Chinese and Indian patenting in various technological areas.

China clearly focuses on computer technologies, electronics and information and communication technologies (ICTs), both in absolute terms and in relation to other countries' focus (as measured by RTA). China focuses relatively less on, for example, pharmaceuticals, biotechnologies and chemistry.

India also focuses on computer technologies, but in contrast to China, India also focuses heavily on pharmaceuticals, chemistry and biotechnologies, see Table 3.3. India also has a very strong relative position (although fairly small in absolute sense) in information technology (IT) methods for management.

A nation's (or, more generally, any actor's) overall focus on specialization vs. diversification across technologies can be indicated by the variance of the RSTA measure across technologies, since a nation that specializes in some technologies at the expense of others will have a higher variance in RSTA than a nation that has a focus in various technologies similar to the worldwide average. As expected from the results described above, the variance of China's RSTAs is lower (0.054) than that of India's RSTAs (0.16). This means that while China has larger foreign patenting than India, it is also more similarly distributed to general foreign patenting across countries of origin, as compared to India. Thus, India is more specialized in some technological areas, such as pharmaceuticals, biotechnologies, medical technologies and chemistry, as described above.

Table 3.3 Technology shares and revealed technological advantages, 2009–2011

Technology	China		India	
	TS	RTA	TS	RTA
1 – Electrical machinery, apparatus, energy	11.6%	1.54	2.0%	0.27
2 – Audio-visual technology	9.3%	1.25	3.2%	0.43
3 – Telecommunications	6.4%	1.59	2.8%	0.69
4 – Digital communication	10.2%	2.47	5.0%	1.22
5 – Basic communication processes	1.5%	0.89	1.7%	1.02
6 – Computer technology	15.2%	1.20	23.5%	1.87
7 – IT methods for management	0.6%	0.78	3.7%	4.62
8 – Semiconductors	3.5%	0.53	0.6%	0.09
9 – Optics	2.6%	0.52	0.2%	0.05
10 – Measurement	3.0%	0.74	1.6%	0.38
11 – Analysis of biological materials	0.3%	0.52	0.7%	1.24
12 – Control	1.6%	1.06	1.4%	0.89
13 – Medical technology	2.0%	0.54	2.3%	0.61
14 – Organic fine chemistry	2.2%	0.70	16.3%	5.23
15 – Biotechnology	1.4%	0.66	4.3%	2.00
16 – Pharmaceuticals	2.4%	0.71	16.3%	4.86
17 – Macromolecular chemistry, polymers	0.6%	0.39	1.1%	0.67
18 – Food chemistry	0.3%	0.37	1.0%	1.10
19 – Basic materials chemistry	0.9%	0.63	1.9%	1.27
20 – Materials, metallurgy	0.9%	0.85	0.9%	0.83
21 – Surface technology, coating	2.4%	0.78	1.5%	0.48
22 – Micro-structural and nano-technology	0.0%	0.43	0.0%	0.95
23 – Chemical engineering	1.8%	0.97	1.0%	0.55
24 – Environmental technology	0.5%	0.54	0.3%	0.33
25 – Handling	1.5%	0.74	0.8%	0.39
26 – Machine tools	1.9%	0.85	0.5%	0.23
27 – Engines, pumps, turbines	1.2%	0.55	1.2%	0.55
28 – Textile and paper machines	0.6%	0.33	0.6%	0.32
29 – Other special machines	1.4%	0.70	0.9%	0.44
30 – Thermal processes and apparatus	2.2%	2.23	0.5%	0.50
31 – Mechanical elements	2.5%	1.00	0.5%	0.21
32 – Transport	1.2%	0.44	0.4%	0.15
33 – Furniture, games	2.8%	1.67	0.3%	0.17
34 – Other consumer goods	1.5%	1.15	0.7%	0.53
35 – Civil engineering	2.0%	1.32	0.5%	0.30

3.4 ANALYSIS: WHAT HAS BEEN LEARNED

The research underlying this chapter has been guided by three interrelated questions. These questions have been probed by studying and analyzing worldwide patent statistics as provided by WIPO. This section will recall the questions, and briefly summarize the results that have been presented in more detail above.

The first question of who patents in China and India was studied by analyzing the origin of patent applications to China and India, respectively. First, the results show that while only one-fifth of all patent applications to the Chinese PTO has non-resident origin, as much as four-fifths of all applications to the Indian PTO has non-resident origin. This probably has more to do with the high absolute numbers of resident applications in China than a low interest among non-resident inventors to apply for patents there. The largest share (35.2 percent) of non-resident Chinese applications has Japanese origin, followed by the US (25.7 percent), Germany (10 percent) and Korea (7.2 percent). The situation in India is fairly similar, where the largest share (32.9 percent) of non-resident Indian applications has US origin, followed by Japan (13.5 percent), Germany (11.8 percent) and Switzerland (5.3 percent). Clearly, Japan and Korea focus more on China than on India in their foreign patenting, which probably partly has to do with both geographical and technological proximities between these nations and China.

The second question was regarding where Chinese and Indian inventors patent. The results show that both China and India direct a majority of their foreign patent applications to the USPTO. While a little more than half (54.8 percent) of China's foreign applications are submitted to the USPTO, as much as two-thirds (67.6 percent) of all foreign applications from India goes to the USPTO. Further, the EPO is a large (although much smaller than the USPTO) receiver of foreign applications from both China (13.3 percent) and India (7.2 percent). While the third and fourth largest receivers of foreign applications from China are Japan (7.2 percent) and India (4.7 percent), the third and fourth largest receivers of foreign applications from India are China (2.9 percent) and Japan (2.8 percent).

The third question asked what technological areas Chinese and Indian inventors patent in, and the analysis was probed by using Chinese and Indian inventors' patenting in the US as a proxy for the general patenting of China and India. The results show that China holds strong positions in patenting within computer technologies, electronics and ICTs, while less so in pharmaceuticals, biotechnologies and chemistry. Similar to China, India holds a strong position within computer technologies, but also within pharmaceuticals, chemistry and biotechnologies.

3.5 DISCUSSION AND FUTURE RESEARCH

As illustrated in this chapter, innovators and patentees do not necessarily only cover their national product/service markets with patents. They could (and should) also focus on covering other important markets, such as nations with large and important output markets and nations in which they localize manufacturing activities, with the aim to inhibit rivalry production. China is already the largest manufacturing nation in the world, and the importance of covering this manufacturing market will increase even further while the massive market potential is realized and while the IP laws are strengthened in line with the TRIPS (Trade-Related Aspects of Intellectual Property Rights) Agreement. Further, increasing indigenous innovation activities in China will provide yet another reason for non-resident innovators to patent there, although R&D nations are typically less important for innovators to protect by patents than production nations and output market nations (Holgersson, 2012). Increased focus on innovation in China will also lead to even further strengthened IP positions throughout the world, for example in terms of large portfolios of patents held by Chinese actors in various jurisdictions.

The IP future for India is not as clear, although the sheer market size will probably attract foreign patentees with increasing pace in parallel with a strengthened IPR system. While China's focus on electronics and ICTs have probably spurred patenting and patent numbers a lot, since these industries are of a *complex* character in which patent propensities and patent frequencies are high (for example, Holgersson, 2013), Indian innovators have mainly focused on *discrete* types of technologies. Products and services built upon discrete technologies are based on few patentable inventions (Cohen et al., 2000), thus focusing on such technologies limits the patentable output and patent numbers. Nevertheless, at present India seems to have lagged behind China in terms of IP developments. For example, China but not India has gained a position in the group of the five largest PTOs (also including EPO, USPTO and the Japanese and Korean PTOs), a group called IP5 that jointly discusses how to improve international collaboration and efficiency in the patent examination process.

While this chapter has focused on the development and use of IPRs in China and India, it relates to the development of these nations as innovation spaces more generally. The increased non-resident patenting in China and India indicates the increased importance of these national (input and output) markets for innovative products from around the world. Analogously, the increased foreign patenting from China and India indicates the increased importance of these nations as innovators, especially within certain technological areas. These parallel developments create new opportunities and challenges for innovators, not only for Chinese and Indian ones but maybe even more so for foreign ones, and especially for incumbent technological leaders.

This chapter started with stating three research questions, which have been investigated throughout the chapter. However, the investigation of these questions has, as most commonly is the case in research, led to new questions that need additional research. The chapter now ends by stating questions, which could provide some guidance for further discussions and research.

Clearly, more analysis is needed to understand what these observed differences will mean in the future. For example, what does it mean for the future competitiveness of China and India, respectively, that Chinese inventors hold such a strong domestic position in IPRs, as opposed to the Indian situation? Further, this chapter hypothesizes that the patenting from one country to another country is to some extent determined by the technological distance between the two countries. Why might that (not) be true? Moreover, this chapter focuses on the increasing importance of two large quickly developing countries as origins and destinations of innovations and patents. What might be the parallel development of small developed countries, such as Finland, Sweden and Switzerland, and especially what might be the future role of their domestic PTOs for resident as well as non-resident patent applicants? How might that impact the domestic innovation systems in such small nations? Similar questions could be probed also for small countries in Asia.

NOTES

1. Marcus Holgersson has carried out this research at the Institute of Innovation and Entrepreneurship, University of Gothenburg, during his postdoctoral scholar period. We would like to acknowledge and thank the Broman Foundation for supporting the research through the postdoc stipendiums.
2. As a developing country, India was, however, allowed to adopt a transitional stage of the application of TRIPS until 1 January 2000, after which legislation had to be adjusted.
3. China's 12th Five-Year Plan has been translated by the Delegation of the European Union in China, and the translation is available here: http://www.britishchamber.cn/content/chinas-twelfth-five-year-plan-2011-2015-full-english-version.
4. An English translation of the Outline of the National Intellectual Property Strategy is available here: www.wipo.int/edocs/lexdocs/laws/en/cn/cn021en.pdf.
5. An English translation of the National Patent Development Strategy (2011–2020) is available here: http://graphics8.nytimes.com/packages/pdf/business/SIPONatPatentDevStrategy.pdf.
6. As cited on the website of the Indian Department of Science and Technology: http://dst.gov.in/stsysindia/stp2003.htm.
7. Available on the website of the Indian Department of Science and Technology: http://dst.gov.in/stsysindia/stp2003.htm.

8. Note that there are some limitations in this dataset, especially for recent years in which data may not yet have been properly reported from various PTOs. For example, the total number of non-resident patent applications to all PTOs around the world should equal the total number of filings from all countries to PTOs abroad. However, while the former number is roughly 755000 for 2011, the latter is roughly 816000. Nevertheless, the data is useful, as comparisons within the same datasets are more interesting than comparisons across datasets for the purpose of this chapter.
9. Note that since p_{ai} is the number of foreign applications from country a to receiving office i, the following holds since a domestic patent application is not a foreign application: $p_{ai} = 0$ if $a = i$.
10. The Pearson correlation is 0.32. The Spearman correlation is 0.70, which might be of more interest in this case (since the linear relationship included in the Pearson correlation is not necessarily of interest). Both are significant at the $p = 0.01$ level.
11. The Pearson correlation is 0.11 and the Spearman correlation is 0.40. Only the Spearman correlation is significant at the $p = 0.01$ level.
12. For China, the Pearson correlation is 0.59 and the Spearman correlation is 0.75. For India, the Pearson correlation is 0.67 and the Spearman correlation is 0.61. All four correlations are significant at the $p = 0.01$ level.

REFERENCES

Balassa, B. (1965). Trade liberalization and 'revealed' comparative advantage. *The Manchester School of Economic and Social Studies*, 33(June).

Cohen, W.M., Nelson, R.R. and Walsh, J.P. (2000). 'Protecting their intellectual assets: appropriability conditions and why US manufacturing firms patent (or not)'. National Burea of Economix Research Working Paper 7552.

The Economist (2010). 'Patents, yes; ideas, maybe'. October 14. Available at: http://www.economist.com/node/17257940 (accessed 20 October 2014).

Granstrand, O. (1999). *The Economics and Management of Intellectual Property: Towards Intellectual Capitalism*. Northampton, MA, USA and Cheltenham, UK: Edward Elgar.

Granstrand, O. and Holgersson, M. (2014). 'Multinational technology and intellectual property management – is there global convergence and/or specialisation?' *International Journal of Technology Management*, 64(2/3/4): 117–47.

Guellec, D. and Potterie, B.v.P.d.l. (2007). *The Economics of the European Patent System*. Oxford: Oxford University Press.

Holgersson, M. (2012). *Innovation and Intellectual Property: Strategic IP Management and Economics of Technology*. Gothenburg, Sweden: Chalmers University of Technology.

Holgersson, M. (2013). 'Patent management in entrepreneurial SMEs: a literature review and an empirical study of innovation appropriation, patent propensity, and motives'. *R&D Management*, 43(1): 21–36.

Keupp, M.M., Beckenbauer, A. and Gassmann, O. (2010). 'Enforcing intellectual property rights in weak appropriability regimes'. *Management International Review*, 50(1): 109–30.

Lohr, S. (2011). 'When innovation, too, is made in China'. *The New York Times*, January 2, 2011. Available at: http://www.nytimes.com/2011/01/02/business/02unboxed.html?_r=3&# (accessed 20 October 2014).

Patel, P. and Pavitt, K. (1994). 'Uneven (and divergent) technological accumulation among advanced countries: evidence and a framework of explanation'. *Industrial and Corporate Change*, 3(3): 759–87.

Soete, L. (1981). 'Technical change, catching up and the productivity slowdown'. In O. Granstrand and J. Sigurdson (eds), *Technological and Industrial Policy in China and Europe*. Lund, Sweden: Research Policy Institute.

Soete, L. (1987). 'The impact of technological innovation on international trade patterns: the evidence reconsidered'. *Research Policy*, 16(2–4): 101–30.

Tung, A. (2011). 'Experts critique China's patent filing system'. *China Daily*, January 25. Available online: http://usa.chinadaily.com.cn/epaper/2011-01/25/content_11912788.htm (accessed 20 October 2014).

4. Technological competition in the Chinese market: an investigation based upon patents

Xiangdong Chen, Xiaoqing Liu and Yong Wang

4.1 INTRODUCTION

This chapter focuses upon innovation spaces in China, based upon an investigation of patents.[1] The act of patenting facilitates the competitiveness of industries and firms, based upon the ownership of technology in regional and international markets. However, until recently, the leading actors in patenting have been based in developed countries, rather than in developing countries. Moreover, there are interesting questions to explore about the emergence of national innovation systems, and particularly of institutions related to intellectual property rights (IPRs). They are just emerging in countries like China and India, where the patented technologies are also thought to be important to future competitiveness, in being able to capture the value of future innovations sold in markets.

This chapter focuses on how technological competition may play out in patent pools, which may also help identify the strategic position and relative competitiveness of specific technologies within China and for firms and industries within the Chinese market. This chapter first discusses patents and patent pools. The chapter then goes on to study two types of patent pools in China, namely audio and video technology, and telecommunication technology. On the one hand, patent pools may strengthen the members' competitive position in international markets. On the other hand, they seem to also stimulate *local* competition in the Chinese markets. An increased number of patents in the technologies are observed, taken also by companies outside the patent pool, and taken both by Chinese and foreign firms in China. This chapter also discusses aspects of a faster pace of technological innovation and rapid expansion of vast market volume as a route for latecomers to catch up and compete.

4.2 PATENTS AND PATENT POOLS

In this chapter, patent alliances or pools, are related to technical standards and specifications in two sectors. Patent pools entail a type of collective assets of patents, which are each in turn held by specific owners, be they individuals, entrepreneurs or multinational enterprises (MNEs). A patent pool, first of all, may be seen as a positive solution for accessing patented technology. The research by Heller and Eisenberg (1998) on patents over biotechnologies started off this debate. They suggested that when there were too many obstacles, numerous patents actually led to harmful effects on social welfare or public interest, as it would hinder society to adopt a new technology. In their view, there is a solution – the anticommons theory, initiated by Michelman (1982) and developed by Heller and colleagues. The core concept of anticommons is that if ownership of a certain body of technological knowledge is divided into too many shares, it may not be used effectively and thoroughly.

Indeed, along with practical problems of patent accumulation in typical sectors such as semi-conductors, biotechnology and computer software, this phenomena of a large volume of existing patents being owned by too many owners may prevent individual companies with particular patents from adopting the new technology. This is due to additional necessary patents, owned by other holders, that the individual company must fight their way through – often described as a "patent thicket."

Cross-licensing and patent pooling are two effective methods to overcome patent thickets, although there might be different types of pooling, such as complementary or substitutive – but only the former may help in the long run that patent pooling can be effective (Shapiro, 2000). Shapiro conducted research on patent thickets and found that complementary patent pools are better than substitutive ones in terms of social welfare concerns. Lerner and Tirole (2004) continued this line of research, stating that patent relationships among member licensors or stakeholders can change in the two extremes between complete substitution and complete complementariness. It is concluded that the strong nature of complementariness among patents is more beneficial to social welfare.

Thus, major research work in developed countries reveals that patent alliances, or patent pools, are an effective mechanism to create commonly accepted platforms and standards. This should decrease transaction costs, prevent companies from licensing blocks, and avoid expensive litigation, hence increasing cost-effectiveness in economic terms, especially for those patents complementing each other in the pool (Aggarwal and Walden, 2005). However, in the case of developing countries, patent pools might be ineffective to those companies outside of the pool, in terms of technology diffusion. So far, research on these issues in a developing-country context is rare.

An important aspect of patents is standard-oriented patent clusters. While industrial standards are issued traditionally by government organizations, or by industrial associations, recent development proves that large companies in related fields can create large-scale industrial-based standards. These alliance-created standards can be found in the case of the patent pool. In fact, the patent pool can be seen as a strategically oriented emerging organization, which can successfully develop early-stage products into matured and well-established ones (Rice and Galvin, 2006).

For operational issues inside the patent pool, Lerner et al. (2004) conduct their theoretical and empirical research on 63 patent pools between 1995 and 2001. They conclude that: (1) an individual license may be less possible when there are substitutive patents inside the pool; (2) when the number of members in the pool increases, an individual license might be easier to realize; and (3) when larger-sized pools have more control over infringement, they therefore usually attract smaller pools into their own. Of course, negative impacts also exist during early phases of industrial development, but they won't be delved into here.

Two major functions of patent pools can be identified. One is to share patents owned by members of the pool, and to license these patents as a standardized package to a third party *outside* of the pool. In this case, a patent pool can be an effective way to tackle the practical problem of patent thickets in the adoption of newly developed technologies. Secondly, patent pools provide an efficient way of sharing relevant technology resources and of promoting social welfare.

However, these two functions of patent pools can also bring high costs or even implement barriers for others trying to adopt relevant cutting-edge technologies.

4.2.1 Empirical Studies and China

International studies suggest that the current IPR system may provide too much protection and special monopolistic power to patent owners, and that corresponding spillover and technology diffusion in developing countries can be blocked to certain sectors. Chinese companies (as do other developing-country firms) usually operate outside patent pools and are thus under pressure from the pools and their owners, especially in sectors like information and communication (ICT), which are dependent upon technology and patents.

Overseas patenting in Europe and in the US can be seen as a powerful barrier to Chinese overseas exports. Claims of patent ownership declared on Chinese firms by overseas companies have been increasing, and they bring high licensing fees, which in turn influence the Chinese firms' exports to overseas markets. The higher licensing fee charged by those companies who are inside an alliance or patent pool can be seen as negative or even destructive upon Chinese exporters (Zhao and Wu, 2010).

In recent years, local companies in China, especially state-owned companies, have begun to take a strategic approach to patents. Several patent-based alliances have been organized among Chinese companies. These are spread over sectors such as audio and video digital coding, electrical pressure cookers, color TVs, construction materials, light-emitting diodes, natural gas equipment, porcelain, and others (Guo and Gao, 2006; Qin, 2013). For example, the Audio Video Coding Standard Workgroup of China (AVS Workgroup) was established in June 2002. This is the Chinese industrial standard for high-definition digital broadcasting, high-density digital storing media, wide-band multimedia telecommunication and wide-band internet media. As another instance, in October 2002, the industrial alliance TD-SCDMA was organized with eight member companies in the patent pool. And another – in February 2009, China-based Huawei joined the WiMax patent pool, becoming a collaborative licensors with Alcatel-Lucent, Cisco, Clearwire, Intel, Samsung Electronics and Alvarion.

Milestones can be identified for China's continual implementation of opening up towards patenting. After 11 December 2001 (when China formally entered into the World Trade Organization [WTO] as one of 154 member countries and regions), investment of overseas companies, primarily dominated by larger MNEs, into the Chinese market increased dramatically. This developed significantly after the mid-1990s, with a sharp increase in patenting volume by overseas MNEs, which dominated telecommunication (overseas patented technology accounting for 91 percent), computing (93 percent in information storage technology), medical and pharmaceutical, electronic, and home appliances (76 percent in basic electronic circuits).

However, it should be noted that between 1985 and 2001, major investment by larger MNEs was concentrated in manufacturing industries, and hardly related to technological development and patenting. This changed after the late 1990s and especially during the year 2000, when competition in the Chinese market was increasingly keen. This shift is highly relevant to understanding the dynamics of industrial sectors.

Based on these research findings, this chapter will investigate three questions related to patent pools in China: (1) Do patent pools bring positive or negative effects to social welfare? (2) Is there any competitive place left for latecomer companies, outside of the pool? (3) Does such a pooling mechanism favour companies in developed countries, or is there still space left for companies in developing countries, such as China?

4.3 TECHNOLOGICAL COMPETITION IN THE CHINESE MARKET, SEEN THROUGH PATENTS

4.3.1 Audio and Video Sector

A first aspect of the audio and video sector to discuss is technological competition backed by patents in the DVD product market. In December 1994, Sony and Philips initiated the high-density Multimedia CD format and technology standard (MMCD) as the first DVD technology formulation. On 24 January 1995, Toshiba initiated the Super Density Disc System (SD-DVD). The intense competition between the above two standards, MMCD and SD-DVD, resulted in a unified standard in September 1995, which led to the establishment of the DVD Consortium for so-called DVD 1.0, beginning in April 1997. This consortium, a major international organizational body, designs, maintains and develops the industrial standard known as the DVD Forum. The members of the consortium include ten key companies: Toshiba, Matsushita, Sony, Mitsubishi, Hitachi, Pioneer, JVC, Philips, Thompson and Time Warner. Other members (about 230, including IBM, Intel, Samsung, LG, Sharp, ITRI or Taiwan Industrial Research Institute, NEC, Microsoft, Sanyo and Disney) are active participants but do not have the decision-making power that is held by the original ten members.

Since there were diversified but key patented technologies shared by different controlling firms, two separate patent pools were formalized based on the DVD Forum. One was initiated in 1998 by Sony, Philips and Pioneer Japan, and is known as DVD 3C. The other was initiated in 1999 by Hitachi, Matsushita Electric Industrial, Mitsubishi Electric, Time Warner, Toshiba and Victor Company in Japan, and is known as DVD 6C. These pools were mainly formulated around the Redray industrial standard system, to be contrasted with the Blu-ray system.

However, for the next generation of the DVD industrial standard system, agreement over technical standards was difficult, and there were arguments among the member firms. In May 2002, nine companies – Sony, Philips and Matsushita, collaborating with Hitachi, Pioneer, Samsung, LG, Sharp and Thompson – led the formation of another industrial standard for the Blu-ray Disc (BD), known as the Blu-ray Disc Founders (BDF). Almost at the same time, Toshiba produced its own standard, passed in November 2003 by the DVD Forum with eight to six votes, to confirm HD (high-definition)-DVD as one of the next-generation DVD standards, simultaneous with the development of HD TV.

On 18 May 2004, BDF changed its name to the Blu-ray Disc Association, in order to accept more members. Due to the larger disc volume and faster reading speed of Blu-ray, more renowned international companies were increasingly willing to participate in the association, these including Apple, Dell, TDK and HP. In this way, the industrial-standard-backed competition was developing between the two groups of multinational enterprises, specifically the Blu-ray system dominated by Sony, and the HD-DVD group led by Toshiba (also joined by Time Warner and others, including Sanyo, Microsoft and Intel).

Eventually, the Blu-ray system took the lead when Toshiba announced in 2008 that the company would withdraw from the HD-DVD sector. This development shows clearly that in the field of audio and video technology – DVD products in particular – technology resource, although diversified between a number of companies, is dominated by a few leading firms.

The implications here are that this dominance within patent pools by multinational enterprises left very limited space for latecomers to enter the sector and compete. Therefore, it is expected that Chinese companies coming into this sector late would face barriers in developing their own technologies, even if they have a larger number of patents than ever before.

A second, interesting aspect of the audio and video sector is the competition between patented technologies on the local Chinese market. Here we use patent records at SIPO (State Intellectual Property Office of the People's Republic of China) to investigate competition among companies in China.

In DVD technology, patents are developed around the optical disc structure, information storing methods, and corresponding materials, defined collectively here as 'optical storage technology' (OST), which can be further divided into core technology and applied technology in six important technical areas (T1–T6), as classified by the International Patent Classification (IPC) code (Guo and Qiao, 2005) (see Table 4.1). On the other hand, the EPO (European Patent Office) records on patents in the field of DVD patent pools can be used as an important reference for related technical fields, assuming that the Chinese companies would patent their technologies in the same areas as registered in the EPO.

The six technical fields used in this study (as per the IPC) are the same technical areas as those patents inside the pool, as formally announced and registered in the EPO. These are used as the proxy, in order to compare the technical nature and developing trend of patent-based competition in China's market between companies located inside the pool and those outside of the pool. The sample data applied to this study represents

Table 4.1 IPC areas on optical storing technology (OST) and total patenting shares, 1994–2006

Classes	Tech fields	IPC subclass	Description for IPC sub-classes	Patenting volume (%)
Core Technology	T1	G11B7/24	Optical recording media: structure and materials	2,170 (17.62%)
	T2	G11B7/007	Optical recording media: signal / information ordering (micro-structure)	1,057 (8.58%)
Applied Technology	T3	G11B7/004	Recording, remerging, removal method; erasing circuit	1,491 (12.11%)
	T4	G11B7/08	Sensor or power installation for recording position	655 (5.32%)
	T5	G11B20/10	Digital recording or remerging format for discs	4,468 (36.28%)
	T6	G11B33	Accessories / disc storage device	2,473 (20.08%)

Sources: Fields reference: Guo and Qiao (2005); Data selected through website of SIPO

patents in the six fields, covering the time period between 1994 and 2006. In total there were 3530 patents in the six fields, as seen in Table 4.1.

The recording structure and related materials (T1) and information routing/ordering (T2) are fundamental core technologies, which usually determine a series of industrial standards along with the development of corresponding markets. Other types of technologies, such as writing circuits (T3), storing format (T5) and optical source installation (T6) are practical technologies for engineering and production, reflecting active market demand and technical competition. These technologies are based on new standards, newly developed disc materials and updated micro-structure. Also based on the technical-field information, we can indicate that the technologies in this kind of product, the DVD player, are concentrating on information writing and storing, which is very different to other products, such as telecommunication technology (discussed later).

Three important groups of overseas multinational enterprises active in China in the audio and video sector are examined in this study, namely companies in DVD 3C, DVD 6C and Thompson. Another group is defined as 'non-pooled overseas companies in the Chinese market,' and

Table 4.2 Companies in different typical groups for investigation in this study

Groups	Companies in the sample group
DVD3C	Sony, Philips Electronics N.V., Pioneer, LG
DVD6C	Hitachi, Matsushita Electric Industrial, Mitsubishi Electric, Time Warner, Warner Brothers, Toshiba, Victor Company of Japan, Intl. Business Machines, Sanyo Electric, Sharp, Samsung Electronics, Thomson
Mainland CHINA	Firms from Chinese Mainland, including Hong Kong and Marco.
TAIWAN	Firms from Chinese Taiwan
FOREIGN	Firms from foreign countries, not including firms in DVD pools, such as DVD3C or DVD6C groups.

there is a reference group of local Chinese Mainland firms. Taiwanese companies are also examined as a reference group. All sample groups of companies are listed in Table 4.2.

Table 4.1 shows patent distribution for the sample groups (T1–T6) over the related technical fields. Prior to 1995 when industrial standards for DVD players were first established, patented technology was rather scarce. For example, patented technologies for optical source installation and distribution (T4) and data-storing format (T5) were only three or four in a given year, there was no patent in reading (T3) and only one in 1992 on disc-storing installation technology (T6). However, right after the establishment of the DVD industrial standard in 1995, there was a rapid growth in T4 and T5, with almost ten per year. There was also fluctuation during 1997 and 1998, when the DVD-R and DVD-RAM formats were implemented, which further speeded up the pace of technological competition in practical disc technology. In the meantime, the volume of invention patenting was still limited in technology areas T3 and T6.

Until 1999, when the three independent licensing organizations (DVD3C, DVD6C and Thomson) were operating more independently, there was significant growth in DVD-related technologies. Among them, optical storing structure and material technology (T1) and writing circuits (T3) were major parts. While T1 is closely connected with the development of functional disc standards, T3 was heavily involved in large industrial production under relevant standards. Patenting in other technology fields, such as information ordering (T2), optical source distribution (T4) and optical storing installation (T6) also steadily increased during those years.

The competitive strength of the firms inside the pools in comparison to the non-pooled patentees can be examined using patent volume and related growth rate in particular technical fields. Since the Blu-ray DVD technology-based alliance was formed in 2002, invention patenting before that year (between 1985 and 2002) can be considered as primarily Redray DVD technology, while patents after 2003 can be considered as a combination of Redray and Blu-ray DVD technologies.

The changes can be summarized as three features. First, shares of patents in the six fields of key technologies decreased in the second time period (after 2002) for those inside-pool member companies, except in T1; nevertheless, those areas continued to be dominated by the pooled groups, especially in fields T2, T3 and T5. Second, there are typical areas where technologies are only partially controlled by inside-pool companies. These technological patents may be heavily influenced by non-pooled companies, especially non-pooled overseas companies in fields T1, T4 and T6. In fact, non-pooled groups control more than 50 percent of patenting volume in both T1 and T6 from the first period (1985–2002), and T1, T6 and T4 in the second session (after 2002). Third, companies from mainland China appear competitive in the second time period, especially in fields T1 and T4–T6.

Although the two DVD technologies, Blu-ray and Redray, control a larger share of in the six technical areas in terms of patenting volume, there is evidence of non-pooled groups competing with those in the pool. However, there is no doubt that in-pooled patents do possess higher control and are able to patent *within* China.

A third aspect of the audio and video sector to look at is the economic impact of patent pools upon Chinese exports of DVD players as well as the initiation of a local DVD patent pool in China. In 2001, the total exports of DVD players reached 10.5 million sets, with a 275 percent increase on the previous year, and China became the world's largest country to export MPEG II-based DVD players. However, since 2002 when patent alliances DVD3C and DVD6C started to charge Chinese producers a patent licensing fee for DVD players, a turning point in DVD production in China can be noted.

Since almost all necessary technologies for DVD production are owned by overseas companies, the total royalty claims meant that large numbers of local companies in mainland China switched their production resource to other sectors, which in turn decreased the DVD supplies. Table 4.3 provides detailed royalties required by corresponding pools.

In this case, Chinese exports of DVD players to Europe, where major patent pools of DVD-related technologies are registered in the EPO, are greatly affected. According to statistical data from Chinese customs, between January and May 2005, exports of DVD players from Shanghai harbor dropped 78.6 percent, while foreign-capital-backed companies

Table 4.3 Licensing fee by DVD patent pools for Chinese DVD producers

Starting time	Name of the pooled alliances	Royalty fees
April, 2002	DVD6C	4.0 USD per set.
Oct. 2002	DVD3C	5.0 USD (3.5 USD) per set
2004	DVD1C, Dobby system	4.1 USD per set
2002	DTS	2.0 – 3.0 USD per set
2006	MPEG-II	4.0 USD (2.5 USD) per set
Notice: DVD 3C changed its royalty charge to 3.5 USD per set and MPEG Group also changed the royalty fee to 2.5 USD after July 2005		Total: 20.1 USD per set before 2005 16.1 USD per set after 2005

Sources: based on Lu and Hao (2010). Also refer to: http://www.dvd6cla.com/royalty rate.html

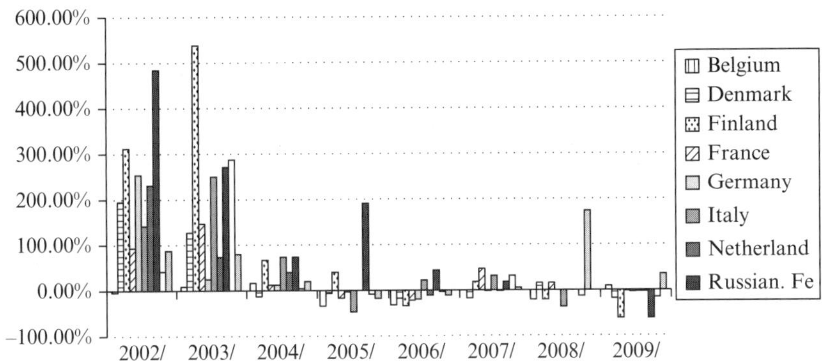

Figure 4.1 Changes in exports of DVD players from China to Europe

in China increased their exports by 10.7 percent. Figure 4.1 indicates that since 2004, the export of DVD players from mainland China to five European countries has dropped dramatically due to the larger licensing payment. In order to keep running in the market, most of the local companies have to choose affiliate partners from larger global companies (being legal users or licensees for those patents) for processing trade (exporting the complete product by operating under an overseas brand and importing the necessary components), so that they can still produce DVD players, but with a thinner export margin. In 2005, the export volume to the top ten European markets reached 31.22 million sets, accounting for 88.6 percent of overall exports to Europe, and 17.8 percent of total exports to the world. It is estimated that among the 2,787,000 DVD players exported from Shanghai, 2,563,000 were produced by companies operating under foreign capital or overseas-owned companies in China – only 193,000 (6.9 percent) were produced by locally owned firms.

Due to these competitive pressures, tremendous efforts were adopted in China to develop locally owned technology in key audio and video technology fields. Enhanced Versatile Disc (EVD) technology was developed by 15 Chinese companies, even before 2002. The EVD technology was actually an upgraded version of DVD product, and based on Redray technology. Enhanced Versatile Disc players can accept a larger volume of recording content and provide five times higher definition than the original DVD players. Plus, EVD players can be matched with HD TV systems. Patents in EVD technologies are dominated by local Chinese companies, who hold patents for more than ten key technologies. By adopting such technologies and by producing EVD players, the Chinese producers can benefit from paying a lower royalty fee (only about US$4 per set).

However, there were a number of practical obstacles for Chinese and international market development regarding EVD technology. First, these products are heavily dependent upon large international producers of entertainment programs and content suppliers to make their programs in a standardized way. While the EVD technologies were designed to compete with Blu-ray technologies, some core technologies and related major components in DVD players were still necessary for the EVD products. Therefore, the total costs were slightly higher than for the DVD players. In this case, EVD players in the Chinese market have not yet been developed successfully enough to compete in high-definition audio and video product fields.

Finally, there is the issue of the Audio-Video Service (AVS) patent pool of Chinese companies. The formation of the AVS patent alliance is an important organizational action in China. The patent pool was initiated by primarily Chinese companies. In 2002, when the Blu-ray alliance was going to be launched by a larger international company group, the Ministry of Information and Industries of China approved the Chinese working team for digital audio and video coding technology standard (the AVS working team). The team initiated and organized a national standard (the AVS standard). In May 2005, the AVS Industrial Alliance was formed, including 12 pioneer member companies from China (including TCL, Skyworth, Hisense, Waves, Shanghai Broadcasting & Electronic, ZTE Corporation and Huawei).

This was another step up for those companies operating in the Chinese market to firmly support the AVS standard-based industrialization of related products. In September 2007, China's first HD DVD product was produced, accompanied by the China High-Definition Disc Industrial Development Alliance (CHDD), which includes hardware and software manufacturers and content suppliers. The formal name of this group of products is CH DVD (China High-Definition DVD), based on the HD DVD technology standard system, with key modifications of the orientation system. Major advantages of the HD DVD standard are that such a system can accept a wider range of discs, including the Redray system,[2] together with lower costs, which may be a key factor influencing Chinese local companies. On 22 April 2009, CHDD launched a Blu-ray system product (CBHD) in the local market, using local Chinese technologies. The CBHD is market based and only a standard for member companies, rather than a national standard. Although technically the pool successfully controls the necessary patented technologies (typical patents in this group are owned by members in the alliance[3]), in terms of market share, the competitive power of the group is still very limited.

In order to contrast dynamic change of the competitive position of those patents by different owners inside and outside of the pool, effective measures, namely RTA (revealed technology advantage) and TS (technology share) are applied in this study. RTA (refer to formula 4.1) indicates competitive strength (or weakness) by certain companies in specific areas:

$$RTA = 100 \times \tanh\left(\ln\left(\frac{T_{ij} \Big/ \sum_{i} T_{ij}}{\sum_{j} T_{ij} \Big/ \sum_{ij} T_{ij}}\right)\right) \tag{4.1}$$

where i denotes corresponding sample companies; j denotes investigated technology areas; T_{ij} represents relevant company i's patent volume in the specific technology area j.

Data was then selected for granted patents for companies in the pool and patents in the same fields owned by companies outside of the pool, between 1985 and the end of 2009, in order to cover the time period before and after the pool was formed. The major technical fields the AVS Group covers are: security arrangements for protecting computers (G06F21); television systems (H04N7); components of television systems (H04N5); and arrangement in tele-control, or telemetry systems (H04Q9). According to the results of data processing and distribution of the practical values of RTA and TS, the competitive position of each member company in the market was not greatly changed during the formation of the pool. Those domestic companies in a dominant position (where the values of both RTA and TS are higher) or an important position (where the value of

TS is higher, while RTA's value may be lower) can control their position and competitive power in current technical fields defined by the pooled patents. On the other hand, our results also indicate that local member companies inside the pool may improve their position, in terms of patenting volume-based RTA measures, over some overseas firms in the pool as well as other overseas companies outside of the pool. However, due to a smaller market size of relevant products in the adoption of the related technologies, the impact upon overseas competitors by Chinese companies through formation of AVS patent pool is still limited.

4.3.2 Telecommunications Sector

The first aspect of interest in the telecommunications sector is technological development. Key technologies in telecommunication sectors are usually clustered into wireless access and wireless transmission. Development and significant changes in these technologies can usually refer to generations of products. For example, the second-generation mobile telecommunication products emerged from the first-generation product market, and now mainly deal with signal system, which has transferred from analog to digital, while later technologies of multiple access were transferred from single-frequency division multiple access (FDMA) to a combination of frequency division multiple access, time division multiple access (TDMA), and even coding division multiple access (CDMA).

While the third generation of products (represented by wider bandwidth) developed from the second generation, the mobile telecommunication product incorporated not only a wider frequency spectrum, but also multiple accesses in coding division type as mainstream (Wen et al, 2010). Regarding the upcoming fourth-generation technologies, the most important improvement is described as Long-Term Evolution (LTE), considered as a complete renewal of air interface. Two groups of technologies are key, namely single-carrier frequency-division multiple access (SC-FDMA) and virtual multiple-input multiple-output, known as Virtual MIMO (Du, 2011). Based on the rapid development of information and technical research in telecommunication fields, there are typically ten key technology areas in current telecommunication fields. They are: source coding and data compression; multiple access and spread spectrum communication; channel coding and interleaving; digital modulation and demodulation; power control technology; diversity technique and rake receiver; multi-user detection technology; multi-carrier transmission technology; smart antenna technology; and software radio technologies (Chang, 2011).

According to Derwent Innovation Index (DII), patents on 3G-related areas are primarily distributed among countries in Europe (24 percent), the US (22 percent), China (10 percent), Korea (10 percent) and Japan (9 percent) (as reported by Luan and Yi, 2009). In China, according to the SIPO database, the biggest owner of patents in the telecommunication sector is Japan (about 30 percent). Local Chinese owners are in second position, while the US and Korean companies are in third place. Most patents in 3G areas in China are still held by overseas multinational enterprises.

Technologies involved in the telecommunications sector are perhaps the most dynamic of any sector. The fast pace of product renewal as well as technological innovation can be decisive for market competition. Although Chinese companies own limited resources of key patented technologies, their corresponding records and shares of the technologies have been quickly upgraded. For example, Chinese companies are now developing patents in LTE, which is considered as the newly formed pool related to the fourth generation of mobile telecommunication products. Meanwhile, while the fifth generation of technologies is already under way, Chinese companies actively compete and collaborate with other international companies, not only in technologies, but also in industrial standards. The standard for mobile communication, TD-SCDMA, established by Chinese companies, has been accepted as one of the international telecommunication standards by the ITU (International Telecommunication Union).

In the fourth-generation period, Chinese telecom companies have increasingly controlled larger shares of technologies in the industry, especially in TDD-LTE areas. Huawei, ZTE and DaTang are typical successful leaders in China. In the area of LTE, patents owned by Chinese patentees have been booming in recent years. According to SIPO data, up until May 2010, there were 431 pieces of patenting applications in LTE-related fields, among which 81 percent was from local Chinese companies. Only six months later in December 2010, LTE-related patenting volume had jumped to 680 with 82 percent of total applications from local sources.

Moreover, the large market in China may put local Chinese companies in a better position to compete with overseas companies and develop their own products in more appropriate ways for local consumers, especially the younger generation. For example, according to a statistical report in China,[4] up until the end of 2013, the total volume of mobile phone customers in China reached 390 million, among them 110 million were added during 2013. Specifically, users of Time Division (TD) among 3G customers increased by approximately 46.8 percent by the end of 2013. On a global scale, the mobile telecommunications market in China accounts for 14 percent of the world total, positioning China in second place, behind the market size of the US. The size of the local market is an obvious factor promoting the local success of Chinese companies.

Another interesting aspect of the telecommunications sector is competition. In the first and second-generation periods, Chinese telecommunication firms were at a learning stage and did not have any control over patenting. During the third-generation stage, Chinese local firms in telecom industries did control some core technologies, and were successful in producing one of the key worldwide industrial standards (TD-SCDMA).

The research in this chapter is conducted based on data from the China National Intellectual Property Rights (CNIPR), established in 1999 by the National Intellectual Property Right Publishing House. All data sourcing is based on two keywords. For example, the combination of 'mobile communication' plus '(keywords in technology areas)' is used to select patent data. Nine detailed key technology areas are selected based on previous analysis of typical technology fields in telecommunication industries: multiple access technology; channel encoding; digital modulation; power control; diversity technique; multi-user detection; multi-carrier; smart antenna; and software radio technologies.

Based on their patent volume, three Chinese companies, three Japanese and Korean companies, and six European and US companies in the telecommunications industry are compared in terms of their competitive advantage on technologies in the above nine technical areas.

Figure 4.2 visualizes the results, using the values of RTA introduced for audio and video technology, over the nine areas. It shows the distributions of important owners of patents by countries and technology.

Figure 4.2 shows that although six Western companies generally possess a superior position in all nine technology areas, other sample companies are also competitive in different fields. Among them, Chinese companies appear especially significant in technology areas in diversity technique, multi-user detection, multi-carrier and smart antenna, while the sample Japanese and Korean companies are more competitive in multiple access technology, channel encoding, digital modulation and power control.

Another area of interest in the telecommunication sector is the WCDMA patent pool. In 2000, the Wideband Code Division Multiple Access (WCDMA) standard, developed by European and Japanese companies, was approved as one of the international standards of the third-generation telecommunication technology by the ITU. In November 2002, the major owners of WCDMA technology under patent protection, namely NTT DoCoMo, Ericsson, Nokia and Siemens, initiated patent licensing plans, providing the necessary licensing conditions to pool outsiders. Afterwards, Fujitsu, Matsushita, Mitsubishi, NEC and Sony agreed to participate in the plan, which marked the formation of a patent pool in the WCDMA

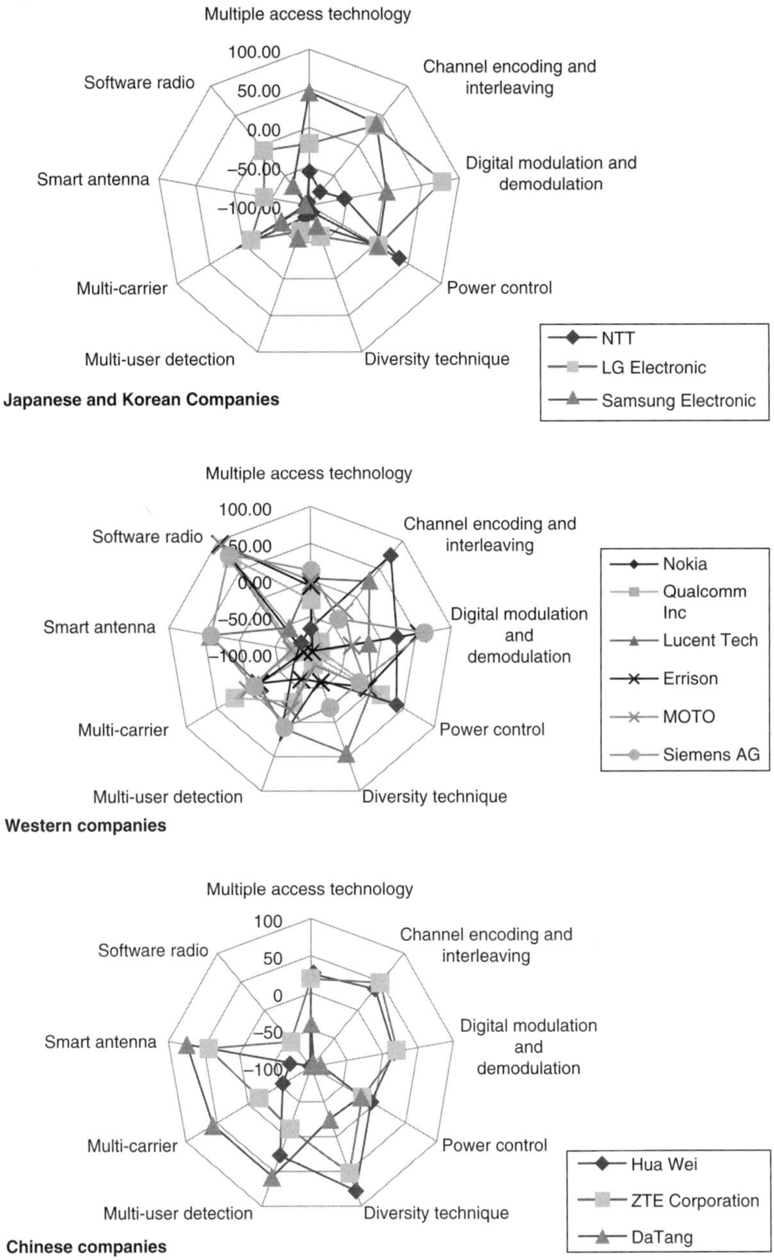

Figure 4.2 Comparison of overseas telecom companies' competitive strength with local companies in China

Table 4.4 Major groups of companies in this study

Patent Pools / Non-Pools	Member Firms	
WCDMA Pool	French Telecom (FT) Nippon Telegraph & Telephone (NTT) Fujitsū Kabushiki-gaisha NTT DoCoMo (NTTD) Royal KPN (KPN) SHARP	MITSUBISHI ELECTRIC Siemens AG NEC Corporation Panasonic (Matsushita) SK telecom (SK) TOSHIBA
Non-pooled Overseas Companies	LG, MOTO, Ericsson, Philips, Qualcomm Inc, Lucent, Nokia, Samsung, Sony	
Non-pooled typical leading companies from Mainland China	DaTang, Hua Wei, ZTE Corporation	
Non-pooled institutions from China	Other Taiwan companies (TAIWAN)	Other Mainland China companies (MAINLAND)

area. The formal pool is composed of nine companies: ETRI, Fujitsu, KPN, Mitsubishi, NEC, NTT DoCoMo, NTT, Sharp and Siemens. The pool licensed members' patents to outsiders once it was formally established in January 2004.

Now, five groups are chosen for detailed investigation, as seen in Table 4.4. According to Guo and Qiao (2005), core patents in the WCDMA can be represented by patents under H04J13 and H04B, in terms of IPC code. Under the technical area of H04B, the sub-field with the most applications to different patenting institutions is H04B7. Therefore, related technologies supporting WCDMA are selected in field H04J14 (radio transmission system) and H04B7 (code multiplex system). Within those technical areas, radio transmission system technology (H04B7) is a fundamental source of WCDMA-grouped technology – patenting movement in this area has been on the increase since 1985. However, after 2002, when the licensing plan was initiated, patenting quantity increased dramatically up to about 5000 per year. Code multiplex system technology in H04J13 is comparatively smaller than H04B7.

Our analysis indicates that companies inside patent pools, and those outside the pools, are different in terms of patenting in technical fields (see Table 4.5). Three characteristics, based upon the empirical analysis found in Table 4.5, can be identified. First, in the later time period, the share of

Table 4.5 Comparison of patenting nature between multinational enterprises inside and outside patent pools, 1996–2003 and 2004–2013

Member competing companies in groups	Patenting 1996–2003		Patenting 2004–2013	
	Ratio to all patents	Ratio to own patents	Ratio to all patents	Ratio to own patents
Companies inside patent pools				
French Telecom (FT)	0.04%	1.86%	0.22%	6.00%
Nippon Telegraph & Telephone (NTT)	7.07%	37.66%	3.98%	16.64%
Fujitsū Kabushiki-gaisha	1.57%	3.24%	2.41%	4.22%
MITSUBISHI ELECTRIC	3.58%	4.99%	1.15%	1.87%
Siemens AG	4.06%	4.45%	1.08%	1.06%
NEC Corporation	8.54%	12.43%	3.70%	8.66%
Panasonic (Matsushita)	7.82%	3.10%	4.07%	1.90%
NTT DoCoMo (NTTD)	5.23%	38.92%	3.98%	17.17%
Royal KPN (KPN)	0.01%	3.57%	0.00%	0.00%
SHARP	0.53%	1.22%	1.32%	1.61%
SK telecom (SK)	0.20%	16.67%	0.43%	16.75%
TOSHIBA	1.55%	1.99%	0.53%	0.52%
Sub-Total and average	40.18%	16.67%	22.87%	9.57%
Major competitors outside patent pools				
LG	3.85%	4.19%	6.46%	4.57%
MOTO	4.70%	11.25%	1.66%	9.29%
Ericsson	0.07%	2.19%	1.46%	5.46%
Philips	3.18%	2.04%	0.92%	0.76%
Qualcomm Inc	7.97%	25.44%	5.07%	8.30%
Lucent	0.00%	0.00%	0.61%	4.79%
Nokia	5.48%	10.22%	1.84%	4.50%
Samsung	9.60%	4.94%	7.92%	2.71%
Sony	2.40%	1.75%	2.07%	1.21%
DaTang (China)	1.16%	26.98%	6.25%	18.09%
Hua Wei (China)	14.64%	18.00%	19.33%	5.49%
ZTE Corporation (China)	6.77%	18.51%	23.54%	7.90%
Sub total and average	59.82%	17.65%	77.13%	11.00%
Total / Average	100.00%	10.65%	100.00%	6.23%

patenting volume by companies inside the patent pools is much less than firms outside the pools. Among them, the leading Chinese company group was the major owner of those non-pooled patents in the same technical fields. Second, both inside and non-pooled company groups raised their own ratio of the patenting fields to telecommunication technologies.

Third, Chinese company groups appear more significant in the patenting boom in related technical fields, indicating their strong potential to compete in the future telecommunications market.

Our analysis suggests that every specific company can be assigned to a category, based upon our empirical analysis of the RTA over time, and which we call leading position, competitive position, volume position and conservative position. Without being able to present the results in detail here, the results suggest that companies within patent pools tend to maintain their positions. Many companies outside patent pools tend to have decreasing positions. However, there are three major exceptions, namely the Chinese companies DaTang, ZTE and Huawei, which have all strengthened their patents and positions in these periods.

4.4 CONCLUSIONS AND IMPLICATIONS

This chapter provides a detailed empirical investigation based on patents in two sectors in China over a period of time. Different dynamics can be observed as to the role of standards and patent pools upon companies, particularly upon Chinese companies. Table 4.6 provides a summary.

In the video and audio sector, we can say that the technology resource is comparatively less versatile, along with a rather fixed technology path. This means that the technology can be easily controlled by members of patent pools. Because of this, companies outside the pool can have difficulty competing and catching up with the companies inside.

In this case, the dominant position over technology is valuable over a longer-term product life cycle, such as dominating owners inside the DVD patent pool group, with narrowly defined technology areas and longer-term market development. The formation of such patent alliances in this case can have a much stronger impact upon companies outside the pools. In this way, latecomers outside of the pools can find it difficult to catch up, even though there might be stronger local policy support, such as in the case of the AVS Group in China. Although the product and related technologies are now gradually moving out of the market, the practical influence of the DVD patent pool still holds meaningful lessons for other patent-based competition.

On the contrary, in the telecommunications sector, there has been a more diversified set of technology resources needed for products, and a faster pace of technology progress, with a number of generation changes of products in a rather short time. With regards to the Chinese market, the demand is huge, which may serve as a potential factor for local com-

Table 4.6 Summary of technology, company positioning and impact on competitiveness in the two sectors

Nature of patent pool	Impact	Position change before and after the pool formation	Competitiveness of the technologies and the patent pools
Audio and video technologies: DVD patent pools. Multiple pools Limited technology path/technology fields Overseas company dominated.	Positive impact especially on domestic companies outside of the pool in their patenting volume	Major companies inside the DVD pools are continually dominating and firms outside the pool find it difficult to compete.	Dominance of key patents in larger market scale, with stable technology life cycle / highly effective control by key members inside the pool
Audio and video technologies: AVS patent pool Single pool with larger numbers of domestic and foreign participants.	Minor, slightly negative impact upon overseas companies	Not significant change	There is no clear effect, due to limited market scale
Telecommunication technologies: WCDMA patent pool Single pool with nine typical technical fields, and larger numbers of in- and outside- pool companies in the fields, mixed with foreign and domestic firms	Diversified impact upon companies outside of the pool, diversified effect upon foreign and domestic companies.	Members inside patent pools remain stable in their positions, but some lost in next generation technologies. There are spaces for domestic late comers (companies outside of the pool) to upgrade, by both substitutive and complimentary patents.	Diversified technologies, faster pace of product and technology innovation, short term effective control by patent pool members.

panies to be more successful in terms of catching up in technologies and market-share growth.

In this case, technology development involves many components and changes rapidly, leading to new generations of products. Here, the patent pools, with multiple player groups or multiple technology paths and increasingly diversified technological areas, may be more difficult to control and may be monopolistically developed by companies inside the pool. Hence, any monopolistic position as granted by patent ownership can be rather temporary. This implies that companies outside the specific pool may have a chance when the next generation of technologies is emerging and the next products are under development. The telecommunications sector and related patent pool transformation from 2G through to 4G is an important example of this situation.

Implications for emerging innovation spaces in China can also be drawn. China is obviously important for both domestic and overseas companies, which are primarily multinational enterprises in these two sectors. This research implies that although companies with their pooled patents may have stronger competitive power against the outsiders, there is still space for those companies outside of the pools to develop and to upgrade their competitive position in certain technical fields.

Chinese companies appear to be making headway. In terms of both decreased technology share and technology concentration (measured by technology in designated fields to the company's own overall patented fields), the dynamic change shown in this study implies that there are both substitutive and complementary patents owned by outsiders to the telecommunications technology pool. In that industry, a technical standard or patent pool may therefore only have a limited function to control related technical products, and be even less competitive, especially in newly emerged technology fields in next-generation products, typically 4G or equivalent in the mobile phone sector, if compared with products in the 2G period. Especially when technology is developing very quickly, the results clearly suggest that even latecomer companies can catch up in further generations of product, in terms of either substitutive or complementary patent resources.

This also implies that the nature of the market development needs to be addressed as an important and decisive factor to practical effect of patent alliances. An aspect to take into further consideration is the impact of highly diversified demand for differentiated technologies and even 'local' market-based technologies. The market in China is vast, especially for the telecommunications sector, and the size effect in the sector is even more significant than the case of the audio and video sector. This may provide another interesting explanatory factor for the success of Chinese latecomers in other technologies and sectors.

NOTES

1. The authors wish to thank Professor McKelvey at Gothenburg University and her team for actively organizing this important book, which effectively connects modern theoretical research on technology innovation with practical technology competition and collaboration in emerging Chinese economies. The authors also wish to thank IDRC for its research project support on 'Patent Pools in China – Patenting Behavior of Foreign Invested Firms and Its Implication on Local Innovation Capabilities and IP Policy Challenges', project number: 104529-011, and NSFC (Nature Science Foundation of China) for its supported research project (project number: 71173009) on patent value study from which some research findings here are derived. A number of graduate students at Beihang University participated in this project, including: Nili Ha, Xiaoqing Liu, Ke Xu, Xi Yang, Bei Li and Yong Wang. Xi He and Boxiong Wang participated in data renewal and collection/processing for this chapter.
2. http://www.hc360.com, 29 June 2005.
3. http://www.eeo.com.cn/, 26 April 2009.
4. http://www.miit.gov.cn/.

REFERENCES

Aggarwal, N. and Walden, E. (2005). "Standards setting consortia: a transaction cost perspective". *HICSS*, 2005, 2014 47th Hawaii International Conference on System Sciences, 2014 47th Hawaii International Conference on System Sciences 2005, pp. 204c.

Chang, F. (2011). "Analysis on LTE-related patent technologies by ETSI". *Journal of Modern Telecommunication Science and Technologies*, 1(8), pp. 6–9 (in Chinese).

Du, Y. (2011). "Key technologies and their development on TD_LTE". *Digital Communication*, 1(9), 36–8 (in Chinese).

Guo, L. and Gao, Z. (2006). "The mechanism of the formation of patent pools and its enlightenment to China". *Science & Technology Industry of China*, 4, 41–5 (in Chinese).

Guo, W. and Qiao, D.F. (2005). "Patent situation and development tendency of DVD". In *Patent Situation and Development Tendency of Typical Industries 2004–2005*, China Intellectual Property Society, Beijing: Science Press (in Chinese).

Heller, M.A. and Eisenberg, R.S. (1998). "Can patents deter innovation? The anticommons in biomedical research". *Science*, 280, 698–701.

Lerner, J and Tirole, J. (2004). "Efficient patent pools". *The American Economic Review*, 94 (3), 691–711.

Lerner, J. Strojwas, M. and Tirole, J. (2004). "The design of patent pools: the determinants of licensing rules". *The RAND Journal of Economics*, 38 (3), 610–25.

Luan, C. and Yi, S. (2009). "Patent measurement on global 3G technologies and its implications for opportunities and challenges to Chinese companies". *Technology and Innovation Management*, 30 (6), 729–31 (in Chinese).

Lu, J. and Hao, Z. (2010). "Patent royalty charges by MNEs in China". *Journal of University of International Business and Economics*, 3, 76–82 (in Chinese).

Michelman, Frank I. (1982). "Ethics, Economics and the Law of Property." In

Nomos XXIV: Ethics, Economics and the Law. Edited by J. Roland Pennock and John W. Chapman, New York: NYU Press, pp. 3–40.

Qin, T. (2013). "Research on the current status of Chinese oatent pools". *Legal system and Society*, 6, 278–9 (In Chinese).

Rice, J., and P. G. Galvin (2006). "Alliance patterns during industry life cycle emergence: the case of Ericsson and Nokia." *Technovation*, 26, 384–95.

Shapiro, C. (2000). "Navigating the patent thicket: cross licenses, patent pools, and standard setting". *Innovation Policy and the Economy*, 1, 119–50.

Wen, W., Zhang, Y. and Li, Y. (2010). "Key technologies in current mobile communication sectors". *Networking Technologies and Application*, 1 (1), 46–8 (in Chinese).

Zhao, T. and Wu, X, (2010). "Chinese exploration and practice in the establishment of patent pools". *Electronics Intellectual Property*, 5, 37–9 (in Chinese).

5. The role of customers in the development of the e-bike industry and a leading company in China

Jun Jin and Guangyu Cheng

5.1 INTRODUCTION

This chapter explores the innovation space of China, through the interactions of demand, regulation and entrepreneurs in developing a new eco-innovation industry, namely e-bikes. Although China has undertaken continual high-speed economic growth since 1978, the environmental degradation in Chinese cities has also become more and more serious.[1] One aspect is environmental pollution from traffic congestion. As a result, eco-innovation has become an important option in China, but so far, research is not so common on eco-innovation occurring in Chinese industries other than environmental ones. This chapter therefore focuses on one type of eco-innovation that is considered disruptive, namely the electric bicycle (henceforth referred to as e-bikes). E-bikes are expected to take the place of automobiles in cities, especially in congested areas. In addition to a reduction in congestion, e-bikes are expected to reduce emissions. This chapter focuses upon China as an innovation space for the development of a new industry and also a case study of an entrepreneurial, leading firm.

As e-bike use has diffused through China, however, some conflicts have developed regarding speed limits, use restrictions in certain areas and safety. First, there is a gap between the speed requirement of users and the regulated speed limit. Second, a conflict continues to exist between citizens' need for e-bikes and metropolitan bans on their use. Finally, e-bike accidents are on the rise and strict regulations are being enforced. Facing strict regulation and bans in some cities as well as facing gaps between user demands and legal limits on e-bikes, Chinese e-bike producers are at a crossroads in determining their industry development pathways.

This chapter provides a review of the development process of e-bikes in China, which shows that a niche market has developed based on user demand for a cheap and fast mode of transportation. Users are widely recognized as one of the critical sources of innovation (von Hippel 1986, 2005). Nevertheless, eco-innovation research is rarely analysed using theories of user-driven innovation, although interesting questions include: What are the roles of users in the development of eco-innovation products? Is there a negative impact of user demand on the innovation?

This chapter will address these questions using an analysis of the evolution of e-bike use in China and the strategies of one of the lead producers Lvyuan. China has its specific consumer behaviours, visions of leisure, environmental values and different stories about transportation. Hence, while the development of e-bikes in China is linked to the issue of transportation and the presence of infrastructures, it is also linked to values and practices about biking.

5.2 LITERATURE REVIEW

5.2.1 Eco-innovation

With environmental issues becoming a global concern, the concept of eco-innovation has been increasingly accepted by academia, government and industry. Following Kemp and Pearson, an eco-innovation can be defined as "a product, production process, service or management or business method that is novel to the organization (development or adopting it) and which results, throughout its life cycle in a reduction of environmental risk, pollution and other negative impacts of resource use (including energy use) compared to relevant alternatives" (Kemp and Pearson 2007:7). This definition is similar to Rennings' definition of eco-innovation in which innovation "can be developed by firms or non-profit organizations, they can be traded on markets or not, their nature can be technological, organizational, social or institutional" (Rennings 2000:322). From this perspective, eco-innovation should not be viewed as an isolated occurrence but as the by-product of social and institutional innovation because "effective environmental policy requires understanding not only technological but also lifestyle dynamics" (Duchin 1999, quoted in Rennings 2000:323).

From the perspective of sustainable development, analysis of eco-innovation should not be reduced to a solely Schumpeterian dynamic, where technological factors are prominent, but should also integrate demand factors such as lifestyle, values, and preferences as well as the way they are mediated through social institutions. In this context, the path for the adoption of eco-innovation and its wide diffusion may be more complicated compared to other kinds of innovation, as "eco-innovations are, in contrast to such technologies as microelectronics and telecommunications, normally not self-enforcing. Because factors of technology push and market pull alone do not seem to be strong enough, eco-innovations need specific regulatory support" (Rennings 2000:326).

Hence, our analysis of eco-innovation in China includes the demand side, the entrepreneurs and policy, and especially observes how regulation can develop potential windows of opportunities brought on by new rules of the game. In our perspective, eco-innovation has three main pillars: (1) technologies; (2) markets; and (3) regulations – these co-evolve to spur the diffusion of eco-innovation. In brief, technological and market (demand) components are important but not always sufficient as has been shown, since a large part of emission reduction has been enabled by environmental regulations. These regulations change the decision process of consumers and producers and deeply modify the opportunities for the development of alternative solutions (Popp 2001; Jorgenson et al. 2009; Bergh et al. 2011). The research on the lead markets conducted by Beise and Rennings (2005) reveals that consumer demand does not play an important role in all eco-innovation cases, whereas government regulation is always an important factor for eco-innovation.

5.2.2 Diffusion of Innovation, User Innovation and Market Demands

As with all innovations, eco-innovations should not only be developed but also diffused in order to impact consumers and society. Diffusion of innovation is defined as the process by which an innovation is adopted and implemented through certain channels over time among members of the social system (Rogers 1995).

The diffusion of innovation is affected by certain features of technology, including usability, relative advantage, compatibility, observability and complexity, and the channel of communication (Mahajan et al. 1990; Rogers 1995). From a sociological perspective on economics, Witt (2001) suggests that consumers adopt innovative products in basically two ways: either by individual trial-and-error learning or by communicating with and imitating others (social learning process). This latter process may be particularly important for eco-innovation, as it is linked to values.

Sociological studies emphasize the role of social norms and status-seeking behaviour, beginning with Veblen's work on "conspicuous consumption" (Veblen 1994, 1999) and Duesenberry's "relative income hypothesis" (Duesenberry 1949). In stark contrast with the individual maximization of utility according to exogenous preferences, as posited by the neoclassical approach, Veblen developed an evolutionary framework in which preferences are generated and shaped in interaction with the social structure. According to his theories, individuals emulate consumption patterns of other individuals situated at high levels of the social hierarchy (Trigg 2001). Social norms appear to be of such major importance that Veblen coins the famous term "conspicuous consumption" to explain how consumption is a way to gain and signal social status.

His insight has been very much used in economics to discuss inertia in demand and consumption (Duesenberry 1949) and the existence of externalities with the famous "bandwagon effect" depicted by Leibenstein (1950). Increasing returns of adoption are necessary for the development of such externalities. Indeed, as has been shown for innovation, the scale of adoption is largely determined by the success of novelty, while positive externalities are generated as soon as it is diffused beyond restricted communities of users (Arthur 1988).

Von Hippel (1986) proposes that lead users play the role of intermediate between supply and demand, and can be important innovators. Additionally, since users are fundamental to society, user-driven innovation could be viewed as social-technological innovation (Tuomi 2002). However, it is not easy to predict whether these users will have positive or negative effects upon the outcomes in user-driven innovation (Luthje et al. 2005). According to theories of user innovation, if users –particularly lead users – anticipate high benefits from a solution that fits their specific, local needs, producers may find themselves trapped by those local needs. This, in turn, could create small niches but impeding a large scale diffusion. These aspects are particularly important for innovations that may solve ecological and environmental problems.

5.3 THE CHINESE E-BIKE INDUSTRY

5.3.1 Users and Bikes

This chapter is based upon a case study of the Chinese e-bike industry, and of a leading Chinese company. Case studies are suitable methodology for exploratory research and enable researchers to understand a specific context (Yin 1984). In this chapter, the case-study approach helps investigate the role of users in eco-innovation in China. Information is based on on-site observations, semi-structured interviews with the chief executive officer (CEO) of Lvyuan, four phone and email interviews with the assistant of the CEO, and secondary information, including annual reports, press releases, presentation materials to customers and stakeholders, and media material.

Although the technology for e-bikes was patented in the US in the 1890s, it was only successfully adopted and diffused in China in the 1990s and later diffused into Europe and other areas. E-bikes are classified into two categories: scooter-style electric bikes or bicycle-style electric bikes. The latter category uses smaller batteries and a lower-power motor, reaches speeds exceeding 30 km/h and weighs between 40 and 60 kg (Cherry and Cervero 2007). Moreover, when e-bikes coexist with traditional bikes in the bike lanes in China, safety is one of the most important issues associated with growing e-bike use because they are faster and quiet. The worldwide market[2] for e-bikes is expected to increase at a compound annual growth rate of 7.5 per cent between 2012 and 2018, reaching global sales of more than 47 million e-bikes in 2018. The e-bike market generated around $6.9 billion in worldwide revenue in 2012 and should be around $11.9 billion in 2018.

China is the largest global producer of e-bikes and has the largest number of e-bike users in the world. But the historical adoption of bicycles and e-bikes has followed distinct periods of development. The rapidly increasing use of bicycles was begun during the Maoist period, was accompanied by an increase in the improvement of living conditions. This was disrupted during the 1990s with increased competition from automobiles as a mode of transportation (Allaire 2007). Since 2000, the rapid increase in traffic and congestion in big cities has paved the way for a reconsideration of bicycles, and thus has opened up new opportunities for e-bikes according to the specific rules and regulations prevailing in individual provinces and cities. In 1998, 40,000 e-bikes were sold in China. This figure rose up to 10 million in 2005 and 35 million in 2013.[3] Additionally, around 22 million e-bikes were on the road in 2005, but that figure rose to 180 million e-bikes by the end of 2012;[4] by the end of 2013, it will have been about 210 million. Moreover, China is the largest e-bike manufacturing base in the world, making up more than 80 per cent of the global market.

As for the user goals for e-bikes, we can categorize them into three main groups. The first group are individuals who use e-bikes to carry many goods, sometimes heavy ones. This includes the professional carriers for express companies such as postmen and fast-food restaurant employees delivering take-out. The second group are individuals who have to travel for more than half an hour by bus and cannot afford cars. This group includes two distinct sub-groups: blue-collar workers and young white-collar workers. The third group are individuals who use cars but move to e-bikes because of the traffic problems.

5.3.2 Historical Periods

Based upon our extensive case study, this chapter has identified and classified the historical development of the e-bike industry in China into three stages, according to changes in user demand and market growth. In each stage, user behaviour, user demand, the comparative advantage of e-bikes, regulations and industry development are analysed.

The first is the emerging stage, from the mid-1990s to the early 2000s.[5] During this period, a few Chinese companies, like Lvyuan, accessed the e-bike technology. Their perceived market was the needs of individuals to travel over a long distance in a city but without the financial ability to afford automobiles. This was a novel product in a new market. This meant that these early e-bike producers had to create the market. The producers had two main strategies. The first strategy was to introduce e-bikes into the market through the use of field shows, advertising and franchise houses, to let consumers accept the new product. The other strategy, which was the most important for the development of e-bikes, was to convince the administrative offices such as the transportation police to officially allow e-bikes as non-motorized vehicles. The mass production of e-bikes began in 2000.

In order to introduce e-bikes to the market, the producers emphasized not the ecological benefits of e-bikes but rather the cheaper price of e-bikes compared to cars as well as the faster speed of e-bikes compared to bikes. In this stage, e-bike producers did not usually divide the market into niche markets. The first group of users identified above (but excluding professional carriers for express companies) were the main users of e-bikes in China in this period.

During the 1990s, Chinese local governments began to promote the e-bike as a convenient and cheap mode of transportation in response to the congestion within important urban areas in China. E-bikes were welcomed by rural to urban immigrants and citizens. The bikes were gaining an increasing share of two-wheeled transportation in China. In some cities like Chengdu and Suzhou, e-bikes have even exceeded the use of traditional bicycles (Weinert et al. 2007). The national standard for e-bikes was enacted in 1999. Approved in 2004, the Road Transportation Safety Law of China included e-bikes as non-motorized vehicles, giving e-bike users the same rights and duties as bicyclists (Weinert et al. 2007). This law in practice has different applications given the diverse regulations of different cities, as well as the collective willingness of each city to promote, or not, this mode of transport. Beijing's local government is not prone to support them, whereas Shanghai and Chengdu are seen as pro-e-bikes.

The second is the development stage, from the early 2000s to 2010. This was a period of rapid growth, especially after the official recognition of e-bikes as non-motorized vehicles by government regulation. Because of the increasingly serious traffic jams in cities, large numbers of white-collar workers gave up cars to use e-bikes in an effort to escape from terrible rush-hour traffic. Competition in this market was fierce. E-bike providers divided the market into several niches, following different needs, especially one for blue-collar workers, one for white–collar workers and one for ladies. The first and second groups were the main users during this period. Blue-collar users require fast e-bikes that can withstand a load of heavy goods whereas white-collar and ladies require a good lightweight, small design. Producers therefore designed and customized e-bikes to fit the requirement of these different niche markets. They also invested in systems for long-distance travel, safety brakes, design, a cost-reducing production system, and so on. This helped strengthened the comparative convenience of e-bikes to the auto in urban traffic.

In order to meet various customer needs, many e-bikes and e-scooters in China can reach top speeds of around 50 km/h and be loaded with heavy goods. However, this has led many to associate the e-bike with accidents, crime (theft of both the bike and the battery) and even congestion in bike lanes. So some local regulations forbid rights of e-bike users to use certain roads or restrict the types of e-bikes in their cities towards the end of this period (for example, Beijing from 2003 to 2005, Fuzhou since 2007 and Guangzhou since 2006). Speed and weight are central issues concerning the Chinese national standard and used as the basis of the new regulation[6] imposed on e-bike customers, and producers. Nevertheless, due to the great demand of citizens for e-bikes, some cities again allowed e-bikes, and thus stimulated the market. Due to the huge number of e-bike users as well as no speed monitors in the bicycle lanes, the traffic policemen did not check the speeding of e-bikes before 2011. Additional problems were identified, especially environmental burden from the recycling of batteries (in particular, the lead batteries) and the pollution in the battery production, and these are barriers to further development of e-bikes in China.

The third period is the mature stage, after 2010 to the present. With the continuing growth of Chinese cities, the increasing price of gas and oil and the worsening traffic, more and more citizens choose to go to work by e-bike instead of cars, especially for those who need to travel a long distance and would tire on a traditional bike. The third group of users is increasing, using e-bikes as substitutes to cars. Distribution is more possible, through the development of business to consumer (B2C) and consumer to consumer (C2C) in China, and the door-to-door service offer a niche market for those who want to have e-bikes but do not have time to shop for them offline. However, the "Notice on Strengthening the Administration of E-bikes" published by the Ministry of Public Security and other ministries in March 2013 negatively impacted the development of e-bikes.

Despite negative impacts from regulation and administration documents issued by traffic police, the e-bike industry in China continues to develop at a steady pace. Besides the convenience of e-bikes both on the road and for parking, producers and users stress the environmental performance of e-bikes (such as lower carbon emissions) in an effort to score points in the debate with government. Moreover, companies make efforts to sell e-bikes in the rapidly growing global market.

Nevertheless, the gap between consumer demand for e-bikes regarding speed, weight, and battery power on one side and government and city regulations and standards on the other is still a barrier to the developing Chinese e-bike industry. Some cities took action and banned the e-bikes from the road in order to reduce e-bike accidents, which meant millions of e-bikes stayed home. Chinese e-bike manufacturers have to create new strategies to overcome this challenge. See Table 5.1 for a summary of the development of e-bikes in China.

5.4 LVYUAN: A LEADING CHINESE FIRM

In this section, the growth of Lvyuan will be introduced to illustrate the development strategies of a leading firm, developed by an entrepreneur.

Lvyuan (meaning "green energy" in Chinese) is one of the oldest and strongest actors in the China's e-bike industry. Mr Jie Ni identified a potential profitable market for e-bikes, which were faster than traditional bikes, after his visit in 1996 to a Beijing institute working on electric vehicles, and he had the capability to master technologies to develop e-bikes.

The entrepreneur, Mr Ni, thus founded Lvyuan in 1997 in Jinhua, Zhejiang Province on the eastern coast of China. After 15 years of development, Lvyuan has six Chinese manufacturing bases of more than 4,000 employees to offer a whole range of e-bikes – from scooters to bikes – including over 300 parts and accessories in the Chinese market to individual persons through markets and thousands of agencies. Lvyuan has the capacity to annually produce 3 million e-bikes in over 300 models and 5 million batteries for e-bikes. Lvyuan has become one of the top four e-bike manufacturers in China.

This entrepreneurial company has been very active in stimulating an emerging innovation space in China. Lvyuan has set up its own innovation systems, including several research and development (R&D) centres in Jinhua (its headquarters), the Fuzhou subsidiary and elsewhere. Lvyuan does research on electric motors, painting, batteries, processing, and so on. Up until March 2011, Lvyuan held 70 patents in China and had 29 patent applications. In order to improve the quality of process line, Lvyuan operates a Just-In-Time System, adopting robots in their process

Table 5.1 The development of e-bikes in China

Stage	Main adopters of e-bikes	Main adoption reasons	Changes in regulations in China
The emerging stage, from the mid-1990s to early 2000s	Individuals who use e-bikes to carry many and sometimes heavy goods (for example, transplants from rural areas to cities)	The cheaper price of e-bikes compared to cars as well as the faster speed of e-bikes compared to bikes	The Road Transportation Safety Law of China, which included e-bikes as non-motorized vehicles, was issued in 2004 E-bikes had been banned in Beijing since 2002
The development stage, from early 2000s to 2010	Individuals who use e-bikes to carry many and sometimes heavy goods (for example, transplants from rural areas to cities); individuals who have to travel for half an hour or longer by bus and cannot afford cars (including blue-collar workers and young white-collar workers)	The comparative convenience of e-bikes to the auto in urban traffic	Beijing lifted the e-bike ban at the end of 2005. E-bikes have been banned in some cities, for instance, in Guangzhou since 2006 and in Fuzhou since 2007 New regulation about e-bike standards was postponed in 2009
The mature stage, 2011 to the present	Individuals who use e-bikes to carry many and sometimes heavy goods; persons who have to travel for half an hour or longer by bus and cannot afford cars (including blue-collar workers and young white-collar workers); individuals who use cars but use e-bikes because of the traffic problems	No need for gas and oil, and the comparative convenience of e-bikes to the auto in urban traffic Environmental benefits	New e-bike standards are still under discussion. *Notice on Strengthening the Administration of E-bikes* was published by the Ministry of Public Security and other ministries in March 2013 E-bikes have been limited as to approved areas and time periods of in Shenzhen since 2013

lines. In addition, Lvyuan has published some books on technical information about electric vehicles in an effort to diffuse professional information on e-bikes and to train Lvyuan employees. Moreover, Lvyuan has participated in professional committees to formulate a set of e-bike product standards and regulations for the manufacture and use of e-bikes in China (Tyfield et al. 2010).

Lvyuan divides the market into several niches: women, rural–urban immigrants, and so on. For example, Lvyuan develops a type of e-bike for rural–urban immigrants to carry heavy goods, and a type of e-bike for a small cargo for 1 km delivery by logistic service companies emerging from the exploding growth of e-commerce in China. In order to attract women to buying Lvyuan e-bikes, Lvyuan also introduced e-bikes of a small size, which are colourful and have patterns on them.

In addition to providing high-quality and advanced technological e-bikes customized for these niche markets, Lvyuan has agencies and distributors around China to provide comprehensive after-sales services including maintenance, inspection, repair and 24-hour telephone customer service as well as roadside assistance and door-to-door repair service.[7] Lvyuan also advertises on public media.

Lvyuan has experienced strong and stable growth from 2001. Their brand was recognized as one of Zhejiang's famous brands, one of the most reliable brands for low-carbon products in 2010 and one of the top 500 Chinese brands with the power to influence the world in 2008. In 2007 Lvyuan was on the list of top 50 "fast-growing" companies by the magazine *Fast Company* and was awarded the SEE-TNC Enterprise Ecology Award.[8]

Though e-bikes in China provide users with the benefits of great speed and weight-bearing capacity, speed has become a cause of traffic accidents. Therefore, some cities forbid riding e-bikes, as introduced above. In addition, a central government law, issued in 2009 and originally set to be implemented in 2010, limits e-bikes to a maximum speed of 20 km/h and classified e-bikes exceeding these limits as e-motorcycles that require a driver's licence. However, because of the effort of the e-bike industrial community including this company and the large market demand, the law was suspended.

Hence, some uncertainties remain about the future, and no one knows if and when that regulation will finally be implemented. To adapt, Lvyuan is choosing a new strategy. Lvyuan is aiming to develop new technology to govern the maximum speed, improve the security of e-bikes and reduce the possibility of re-equipping e-bikes before the launch of new regulations (Tyfield et al. 2010). Additionally, Lvyuan has published several reports about the standard of e-bikes emphasizing the environmental benefits, development of new products for new niche markets such as e-patrol for organizations like the local police, and plans for the future of the electric vehicle market.

Their future plan is that Lvyuan will enter the electric car industry based on their experience and technology with battery and e-bikes. A successful example exists, namely BYD Company Limited, which upgraded from being a producer of NiMH, NiCd and/or Li-ion batteries to an electric vehicle manufacturer. However, the future of electric-vehicle manufacturing is also uncertain and risky for Lvyuan not only because of technology and market constraints on electric vehicles but also because of the administration system for licences on vehicle models.

In summary, the case of Lvyuan indicates that the technological dimension has been, and remains, critical for developing components necessary for e-bikes and introducing them into the market. The patents give Lvyuan an advantage in competing with other players and helps make them a leader in the Chinese e-bike market. Clearly, though, technology is only one dimension in the diffusion process. The case also illustrates the main role of users in the development and acceptance of the product, and how the company adapts to new uses, develops communities' professional understanding, and tries to impact policy. Users increasingly buy e-bikes as a way to deal with the serious traffic jams in Chinese cities, despite municipal governments banning or strictly regulating e-bikes in urban areas.

5.5 DISCUSSION AND CONCLUSIONS

This chapter has analysed the development of a new innovation space for the e-bike industry in China and how the entrepreneurial company Lvyuan has responded to user needs in niche markets facing government regulation. E-bikes are more environmentally friendly than cars, and they are widely promoted as a case for eco-innovation.

Moreover, the interaction between entrepreneurial venture, user demand and government policy shifts over time, as outlined in Section 5.3. The three stages of development in the Chinese e-bike industry display a co-evolutionary process of institutions, user demand and innovation activities of companies, which is similar to the co-evolution framework of eco-innovation proposed by Foxon (2011). With the market growing and more demands from users, the institutions – especially regarding regulations from local governments – are modified, which affect the strategies and innovation of companies, too. There were no suitable regulations in the first phase, and over time, there have been clear interactions between leading companies and public policy.

The case also describes how users have direct and indirect impacts on the adoption and diffusion of this eco-innovation. The case of Lvyuan shows that diffusion of this eco-innovation in China is very linked to the context and specific needs of consumers. There is no "band wagon" but rather "small niches" which are trying to expand in this very promising market. To benefit from this positive trend, innovative and entrepreneurial firms such as Lvyuan have to take into account the stickiness of local information. This implies that new product development will rely on the needs of local users that are strongly linked to the use environment. For instance, the adoption of e-bikes in the Chinese market in the first development stage mainly constituted a replacement for bikes for daily transportation; e-bikes were adopted because of their affordability compared to cars and their speed compared to bikes. This is quite different from the e-bike's diffusion in Europe, where it replaced the car for sports and leisure. In addition, Lvyuan and other Chinese e-bike producers in the third stage developed smaller-sized e-bikes for the third group of users who use e-bikes as a substitute for cars.

Thus the research indicates that it is crucial to involve users during the development and diffusion of eco-innovation in order to match needs related to the market demand, as well as increase profit and competitiveness in the niche market.

But the nature of these users and their requirements are in this case in seeming opposition to government regulations, and that gap poses a problem for the government. The banning of e-bikes in some cities because of the accidents from high-speed e-bikes seems that sometimes users' behaviours have negative influence on the regulation. Then it will indirectly become a barrier to a company's development path, by hindering further diffusion. Regulation is an important factor, but regulators often are not able to steer effectively against market trends (Beise and Rennings 2005:15). Thus, when the government considers regulations, they need to consider the context, uses and practices in order to have a proper vision of the market and how to solve a perceived problem through policy.

Moreover, the case indicates that users are sources for promoting the development of new products, technologies and solutions, and confirm the theories of user innovation (von Hippel 1986, 2005). The design of e-bikes for niche markets in China, such as colourful e-bikes for career women and high-powered batteries for long-distance usage, confirms sociological aspects found in von Hippel's approach (2005), especially the role of local demand and knowledge of consumers. In fact, similarly to von Hippel's approach (2005), this case of eco-innovation can be understood as triggered by direct and repeated personal experience with a problem that allows an accumulation of local knowledge for designing future products.

There is no "band wagon" but rather "small niches" that are trying to expand in this very promising market. To benefit from this positive trend, innovative firms such as Lvyuan have to take into account the stickiness of local information (von Hippel 2005). This implies that new product development will rely on the needs of local users, which are strongly linked to the use environment.

E-bikes are valued as an environmentally friendly product and benefits in that way from positive social norms (Veblen, 1994, 1999), which possibly creates positive future externalities with potential "bandwagon effects". However, the case does not suggest that values of users as related to the ecological impact of the e-bike was critical at the start stage of the e-bikes and e-scooters in China. Indeed, the rapid expansion of e-bike popularity in China lies not in these values but in being an alternative to the automobile. We argue it is because the e-bike's economic performance and comparative advantage to autos and bikes are quite easily observed, accepted and understood. This is particularly true for the first group of target users in China, especially internal immigrants from rural to urban areas who do not have a good educational background. This confirms aspects of the theories of Rogers on diffusion of innovation (1995).

Historical periods matter also, in this co-evolutionary process. Over time, societal benefits may become important, however, with the environmental concerns of society increasing in China, so that the concept of e-bike as an environmentally positive product is becoming critical to the survival of e-bike industry in China. Therefore Lvyuan has stressed the eco-benefit of e-bikes and e-scooters and their impact on climate change since the end of November 2008, also in order to face challenges raised by regulation. We assume that in the near future, the ecological benefits of e-bikes will be emphasized in the Chinese market, especially after people experienced the terrible air pollution in January of 2014 and winter of 2013.

To sum up, on the one hand, users have a personal experience and needs, accumulated over time and met by the new product; on the other hand, government regulation may be headed in a different direction. In our case, Lvyuan faces the challenge of new national regulation and banning of e-bikes in several cities because of the gap between administrative purposes to solve issues of safety and the demands of different groups of users to solve various transportation problems. In that case, the challenges may become barriers, which stop diffusion of e-bikes.

There are also implications for research questions and policy, in similar cases. This chapter addresses issues about how to expand entrepreneurial ventures into markets that did not previous exist, including highlighting the many different choices for firm strategy. The case of the e-bike industry and of Lvyuan shows that diffusion of eco-innovation in China is very linked to context and the specific needs of consumers. On the one hand, the market expanded rapidly during diffusion, and the research indicates that it is crucial to involve users during the development and diffusion of eco-innovation in order to match the market demand and increase profit and competitiveness in the niche market. It also reveals that the behaviour of users, such as causing accidents or not following regulation, could lead the government to act to restrict further diffusion, thereby being a barrier to the development of companies and the drivers of eco-innovation.

Hence, this chapter ends with a series of questions for further reflection. One aspect is for managing innovation and entrepreneurship. The CEO of Lvyuan faces the challenge from balancing demands from the market and regulation. If you are the manager of an e-bike producer or entrepreneur in a similar situation, what could you do to balance the conflict between user demand and regulation requirements? What should a company do when there are no well-developed industrial standards and rules?

Another implication is for public policy. The Chinese e-bike case suggests that there is a co-evolution of institution (for example, regulation) and industrial development, especially for emerging industries. Perhaps we could assume this is so because: (1) the regulation is always slower than the diffusion of new technology and development of new industry, especially in an emerging country like China without a well-developed regulation system; and (2) China is a large market for many business practices. If there were no such large market as China, would one still expect to find a rapid co-evolution of regulation and industries? Will similar conflicts happen again in the future? What should policy-makers do when a new industry is emerging?

A final area is the implications for other types of eco-innovation. Although there are environmental benefits to e-bikes, why did companies not stress and introduce those benefits before 2008? How do companies and administrations evaluate the performance of an innovation?

NOTES

1. We wish to thank the reviewers for their comments. This research is supported by NSFC project: The Research of Reverse Innovation in China (CNSF: 71172111).
2. This data is from "33.6 million electric bikes production expected for 2013 and start of 48 volt batteries for cars and electric bikes" of Next Big Future: http://nextbigfuture.com/2012/12/336-million-electric-bikes-production.html.
3. Data from "Sales of Lithum e-bikes in China in 2013 increased to 2.5 million sets": http://www.sz.qp110.com/news/type-news.asp?news_id=294287.

4. 2012 data from "China's e-bike market seen by CHR's Roberts starting to plateau": http://www.bloomberg.com/news/2013-03-19/china-s-e-bike-market-seen-by-chr-s-roberts-starting-to-plateau.html.
5. Since target markets overlapped in a period of time, it is not easy to clearly classify years for each stage.
6. In fact, this regulation was planned to be launched in 2010. But it was postponed, and then postponed again in 2012. The new regulation continues to limit e-bikes to a maximum speed of 20 km/h and maximum mass of 40 kg of the national standard in 1999. E-bikes exceeding these limits would be reclassified as e-motorcycles and so would require driver licences and vehicle insurance. Although there is no clear enforcement date for this new regulation, the Ministry of Public Security issued a document (http://www.mps.gov.cn/n16/n85753/n85786/2758399.html; in Chinese) in 2011 to enforce the traffic administration of e-bikes according to the national standard of the e-bike in 1999.
7. Information available from the official website of Lvyuan (in Chinese): http://www.luyuan.cn/page.php?mod=introduction.
8. Data and information was available from the official website of Lvyuan (in Chinese): http://www.luyuan.cn/page.php?mod=introduction.

REFERENCES

Allaire J. (2007), 'China and bicycle: the end of the story?', *Communication à Velocity*. Poster presentation at Velocity, Munich, Germany, 12-15 June.

Arthur, W.B. (1989), 'Competing technologies, increasing returns and lock in by historical events', *Economic Journal*, **99** (March), 116–31.

Beise, M. and K. Rennings (2005), 'Lead markets and regulation: a framework for analyzing the international diffusion of environmental innovations', *Ecological Economics*, **52** (1), 5–17.

Bergha J.C.J.M., Truffer B., Kallis G. (2011), 'Environmental innovation and societal transitions: introduction and overview', *Environmental Innovation and Societal Transitions* **1**, 1–23.

Cherry, C. and R. Cervero (2007), 'Use characteristics and mode choice behavior of electric bike users in China', *Transport Policy*, **14** (3), 247–57.

Kemp, R. and P. Pearson (2007), 'Final report MEI project about measuring eco-innovation', available at: http://www.oecd.org/greengrowth/consumption-innovation/43960830.pdf (accessed 3 October 2012).

Duesenberry, J.S. (1949), *Income, Saving and the Theory of Consumer Behaviour*, Cambridge, MA: Harvard University Press.

Foxon, T.J. (2011), 'A coevolutionary framework for analysing a transition to a sustainable low carbon economy', *Ecological Economics*, **70** (12), 2258–67.

Jorgenson D., R. Goettle, M. Sing Hoc and P. Wilcoxen (2009), 'Cap and trade climate policy and U.S. economic adjustments', *Journal of Policy Modeling*, **31**, 362–81.

Leibenstein, H. (1950), 'Bandwagon, snob, and Veblen effects in the theory of consumers' demand', *The Quarterly Journal of Economics*, **64** (2), 183–207.

Luthje C., C. Herstatt and E. von Hippel (2005), 'User-innovators and "local" information: the case of mountain biking', *Research Policy*, **34**, 951–65.

Mahajan, V., E. Muller and F.M. Bass (1990), 'New product diffusion models in marketing: a review and directions for research', *Journal of Marketing*, **54**, 1–26.

Popp D. (2001), 'The effect of new technology on energy consumption', *Resource and Energy Economics*, **23**, 215–39.
Rennings, K. (2000), 'Redefining innovation: eco-innovation research and the contribution from ecological economics', *Ecological Economics*, **32** (2), 319–32.
Rogers, E. (1995), *Diffusion of Innovations*, fourth edn, New York: Free Press.
Trigg, A.B. (2001), 'Veblen, Bourdieu, and conspicuous consumption', *Journal of Economic Issues*, **35** (1), 99–115.
Tuomi I (2002), Networks of Innovation, Oxford: Oxford University Press.
Tyfield, D., J. Jin and T. Rooker (2010), *Game Changing China: Lessons from China about Disruptive Low Carbon Innovation*, NESTA: London.
Veblen T. (1994), 'The theory of the leisure class', in *The Collected Works of Thorstein Veblen. Vol.1 1899*. Reprint, London: Routledge, pp. 1–404.
von Hippel, E. (1986), 'Lead users: a source of novel product concepts', *Management Science*, **32** (7), 791–805.
von Hippel, E, (2005), 'User-innovators and "local" information: the case of mountain biking', *Research Policy*, **34** (6), 951–65.
Weinert, J., C. Ma and C. Cherry (2007), 'The transition to electric bikes in China: history and key reasons for rapid growth', *Transportation*, **34** (3), 301–18.
Witt, U. (2001), 'Learning to consume – a theory of wants and the growth of demand', *Journal of Evolutionary Economics*, **11** (1), 23–36.
Yin, R.K. (2009). *Case Study Research – Design and Methods*, Thousand Oaks, CA: Sage.

6. Understanding innovation spaces through emerging multinational enterprises in China: an explorative case study of a Chinese wind turbine manufacturer

Dmitrij Slepniov, Astrid Heidemann Lassen, Stine Jessen Haakonsson and Maureen McKelvey

6.1 INTRODUCTION

This chapter analyses the development of innovation space within China – in an inherently global industry of wind turbines – by developing a view of the relationships between entrepreneur and national innovation system, as the company expands internationally to become an emerging-market multinational enterprise.[1] During the past two decades, China has earned a reputation of being the 'manufacturing power house' of the world. Vast numbers of Chinese companies have been very successful in exploiting their access to low-cost labour in order to establish themselves as unbeatable champions of high-volume, low-cost manufacturing. More recently, however, Chinese companies are increasingly seeking to create a foundation for growth and industrial development based on innovation. This new direction requires them to tap into more advanced technologies and knowledge and to upgrade their innovation capabilities. To do so, some Chinese companies have distributed their operations by relocating innovation activities to more innovation-intensive locations, that is the countries in the traditional industrial 'triad', North America, Europe and Japan (Liu et al., 2005; van Wyk, 2009). Other Chinese companies have a different strategy, and they seek to move away from the routine low-tech tasks to more innovation-intensive activities while remaining in China and relying on the domestic knowledge base (Lazonick, 2004; Tzeng et al., 2011). This chapter contributes to this broader phenomenon, by addressing the issue of how Chinese firms exploit and explore national and international innovation spaces as they upgrade to compete through innovation.

This broader phenomenon requires further investigation because this involves a complex process between the firm and the external environment over time. The phenomenon may be influenced by many factors, such as organizational structures, culture, familiarity with innovation and entrepreneurship practices, market access and customer innovation expectations. Moreover, while the objective of the companies is to innovate, their actual strategies of developing into knowledge-intensive segments of their value chains to do so may differ. Two scenarios were introduced above – foreign acquisitions vs. indigenous domestic development – and they provide us with a crude distinction between two strategies through which Chinese companies can access different innovation spaces locally and globally. Naturally, this simple dichotomy does not elucidate the full complexity of the phenomena, and the same company may mix strategies. Still, the dichotomy gives us a starting point to explore how companies move towards a more innovative strategy.

This chapter explores the interconnections between different theoretical approaches, to advance our understanding of firm strategy and dynamics in relation to innovation spaces. There is now a vast body of literature (for example, Freeman, 1987; Archibugi and Michie, 1997; Lundvall, 1999) on national innovation systems. A national innovation system includes educational and government inputs, scientific research, an industry base and collaboration and technology networking across and within industries. However, many of these studies tend to ignore the firm. The physical existence and evolution of the national systems and networks does not mean that they automatically constitute or limit the innovation spaces for companies. Indeed, other literature shows how companies need to have appropriate strategies and management practices enabling them to access these elements. Bunnell and Coe (2001) emphasize the role of firms and individuals in this process.

Furthermore, globalization and especially developments in Asia are changing the perspective of firms on how to engage in innovation activities in order to successfully develop new products, services and organizations. The elements of national innovation systems alone will no longer be sufficient to explain the full spectrum of spaces in which innovation may take place. Firms access innovation spaces increasingly across national boundaries, which can challenge the notion of innovation systems being geographically bound.

This chapter develops these arguments by relating key aspects of a national innovation system to innovation management, focusing specifically on how firms internationalize and how they act in relation to the dimensions of entrepreneurial orientation. This chapter provides detailed information from a case study in China to illustrate these points. The company operates in the renewable energy industry. It was established in 2006 by a 30-year-old Chinese entrepreneur. From its inception, the company has been enjoying 100 per cent annual growth rates, which, to a large extent, are attributed to the innovative nature of the company and its international ambitions. The company can be categorized as a newly created venture, focusing on the innovative exploration and exploitation of a product with a high degree of advanced knowledge. Invoking multiple theoretical perspectives, some researchers suggest that new ventures frequently become active players in the global economy soon after their birth, so-called 'born globals' (for example, Oviatt and McDougall, 1999; Zahra et al., 2000; Fan and Phan, 2007). For such firms to succeed in the global market, a combination of innovativeness, risk taking and entrepreneurship is required. This company has expanded rapidly, so that it can be seen as a type of emerging-market multinational enterprise. In the case study, this chapter addresses the following questions: How did the company manage to access and develop relevant innovation spaces? What were the major drivers and impediments in this process? How does this case inform us about how and why Chinese firms engage with their innovation systems at home and abroad?

The chapter has five parts. The next section presents the theoretical and conceptual background of the study and proposes a framework for analysing the interactions between the firm and the innovation spaces. Section 6.3 describes the methodological approach and the empirical results from the case study. Section 6.4 presents major findings and what we have learnt. The final section includes conclusions and suggestions for how to expand upon our results and interpretations.

6.2 SETTING THE STAGE AND BACKGROUND OF THE STUDY

Let us begin by clarifying our use of the concept of innovation spaces. A key question is what are the relevant characteristics of space? How does space relate to innovation systems and the firm?

The concept of space to be used here is not purely geographical; rather, it can be seen as a multifaceted construct consisting of multiple levels, including institutional, organizational and geographical. Such a complex multi-layered construct is likely to be exposed to temporal volatility and changes over time. On this basis, when relating this conceptualization of space to innovation activity, we can define innovation spaces as dynamic multifaceted systems in which innovation takes place through actions and interactions of focal actors.

The concept of innovation spaces draws upon the concept of innovation systems because they both focus on the role of surroundings and the context, networks and capabilities in the environment for facilitating innovation. Moreover, innovations are seen as products of a context, a system, which existing literature has defined as national, sectorial, technological or regional in its reach.

6.2.1 Innovation Spaces and Internationalization

Research on national innovation systems sees innovation policy and absorptive capabilities as prerequisites for successful transformation into a knowledge-intensive economy. Hence, the approach is concerned with the functioning of national policy and economy and the performance of a country in terms of innovative capacity. A national innovation system consists of structures and institutions embedded in a national context, which directly and indirectly constitutes a particular country's innovation capacity as well as its absorptive capacity and attractiveness to technology and innovation activities from abroad. However, according to Lundvall (1999) most innovations reflect an interactive process where the institutional setup of the innovation system interacts and looks for feedback from knowledge creation and entrepreneurial initiatives at the firm level.

From the sectorial and technological innovation systems perspectives, the process of innovation cuts across distinct national geographies (Edquist and Johnson, 1997; Bergek et al., 2008). Malerba (2002) argues that sectors have particular patterns of institutions, networks and knowledge structures that are specific per sector globally, with Malerba (2009) providing conceptual tools and analyses of six major sectors. Dicken (2011) argues that technological systems and characteristics of technological innovation vary greatly by country in that companies may distribute their innovation activities in order to access diversified sources of knowledge and different types of capabilities across national innovation systems (Ernst, 2006). Still there are interactions between them and the current dynamics of increasing integration and interdependence between different localities that cannot be ignored. Actors in a given location interact in and actively construct innovation networks involving actors that are not only geographically but also cognitively distant.

In this chapter, we are particularly interested in how these innovation processes occur in a global setting and their relationship to the entrepreneurial orientation of the firm. We follow existing literature to argue that international interactions form global pipelines or channels that are particularly important for advancing our understanding of the innovation spaces as not exclusively locally or nationally bound. Private actors are very active in establishing such global channels through their engagement in global innovation networks (Ernst, 2002; 2006). Moreover, firms engage increasingly in the globalization of innovative activities (Archibughi and Michie, 1995) in order to exploit new markets and tap into complementary knowledge sources (Kuemmerle, 1999). This is also true in the context of emerging markets, such as China. Research has shown that emerging-market multinationals are very likely to move beyond their national innovation systems (Barnard and Chaminade, 2011). In order to construct and operate global innovation networks, companies need to coordinate innovation efforts across space and often across nationally bound traditions.

In order to succeed in the transition to global innovation, firms in emerging markets will need to apply new communication and coordination methods across innovation activities that may engage a number of independent actors across national innovation systems. Having a home base in a transition economy that is moving towards a knowledge economy, as is the case with China, implies that the globalization of the innovation space may be necessary to access foreign technology as well as high-end markets.

Table 6.1 summarizes the conceptualization of innovation spaces discussed here. It links the elements of the national innovation system (A), the firm both in the national context (B) and the international context (D) and international linkages outside the national innovation system (C).

Table 6.1 Conceptualization of innovation spaces from the firm's perspective

Innovation spaces			
National innovation system and firm-level characteristics		Linkages outside the national innovation system and firm-level characteristics	
A: Key elements of NIS (for example, availability of skills, scientific research, industry collaboration)	B: Intra-firm (for example, organizational culture, learning processes)	C: Global channels and external collaboration (for example, capabilities in host location)	D: Intra-firm (for example, foreign subsidiary capabilities, network resources)

6.2.2 Entrepreneurial Orientation and Strategies for Accessing Innovation Spaces

Hence, this chapter proposes that innovation spaces may be constructed by firms' actions and can be operated differently by different companies depending on how well they can access and exploit the national innovation systems. For this, the individual firm must develop appropriate capabilities and create networks to access external capacities. To understand this process, several elements inspired by entrepreneurship literature help to operationalize the key factors influencing the behaviour of firms navigating these innovation spaces. For example, McDougall and Oviatt (2000) list the following elements: cooperative alliances, economic development activities, entrepreneur characteristics and motivations, exporting and other market entry modes and venture financing.

The advantage of focusing on the innovativeness and entrepreneurial orientation of a firm is a better understanding about how firms engage with processes and actors in their innovation spaces. This also provides us with insights into the dynamic interplay between firm and system levels, which in general, and in the Chinese context in particular, is relatively underexplored.

Furthermore, focusing upon innovativeness and entrepreneurship is relevant, as this literature can provide a highly relevant framework, which emphasizes not merely the formation of new firms, but rather the process of exploring and exploiting opportunities demonstrated by the entrepreneurial organization (Quinn, 1979; Burgelman, 1983; Stevenson and Jarillo, 1990; Stopford and Baden-Fuller, 1994). This entrepreneurial orientation may be enacted by a new start-up company as an entrepreneurial venture as well as by multinational enterprises that are large. Orientation is a concept that includes innovative activities and strategies aimed at the development of new products, services, technologies, administrative techniques, strategies and competitive postures (Miller, 1987; Lumpkin and Dess, 1996).

The entrepreneurial orientation (EO) construct includes innovativeness, pro-activeness, risk-taking, autonomy and competitive aggressiveness, which are each discussed below.

Innovativeness is in a broad sense seen in relation to entrepreneurial orientation. It includes a firm's tendency to engage in and support new ideas, novelty, experimentation and creative processes that may result in new products, services or technological processes. It is recognized that the degree of innovativeness can vary in terms of radicalism, but it essentially refers to a willingness to depart from existing technologies or practices and venture beyond the current state of the art (Kimberly, 1981).

Pro-activeness is related to the act of taking the initiative in the entrepreneurial process and to acting in anticipation of future problems, needs or changes. For example, Lieberman and Montgomery (1988) emphasized the importance of the first-mover advantage as the best strategy for capitalizing on a market opportunity. Thus, pro-activeness by anticipating and pursuing new opportunities and by participating in emerging markets has become associated with entrepreneurship. The pro-activeness dimension of entrepreneurial orientation most closely resembles the prospector-type ideas suggested by Miles and Snow (1978).

Risk taking is a quality that is often used to describe entrepreneurship. Risk can have various meanings, depending on the context in which it is applied. In this chapter, risk is seen in the context of strategy, which according to Baird and Thomas (1985: 231–2) can take three forms: (1) venturing into the unknown; (2) committing a relatively large portion of assets; and (3) borrowing heavily. Venturing into the unknown involves a sense of uncertainty that may apply generally to different types of risk. Financial risk refers to a risk-return trade-off – a term used by Miller and Friesen (1978) when they defined risk taking as the degree to which managers are willing to make large and risky resource commitments (that is, those that have a reasonable chance of costly failures).

Autonomy is also an essential element of entrepreneurship, and it is traditionally defined as the formation of new and independent ventures. In an organizational context, autonomy may also refer to the independent actions of an individual or a team in bringing forth an idea or a vision and carrying it through to completion (Lumpkin and Dess, 1996). As such, autonomy means that actions are taken free of rigid restrictions, where the individual or team throughout the process remains free to act independently, make key decisions and proceed. Fostering autonomy in an organizational context may involve flattening hierarchies and delegating authority to operating units. This literature draws on the idea of, for example, Burns and Stalker's (1961) organic management systems, promoting decentralization, low formalization, dynamic behaviour, learning and flexibility in structures and processes of the organization (Burns and Stalker, 1961; Drucker, 1988; Nonaka, 1991; Senge, 1994).

Competitive aggressiveness refers to a firm's predisposition to directly and intensely challenge its competitors to achieve entry into a new market or improve its position in an existing market. In other words, this reflects a strong focus on outperforming competitors. According to Lumpkin and Dess (1996), competitive aggressiveness is characterized by responsiveness and a willingness to be unconventional rather than relying on traditional methods of competition. Other dimensions of competitive aggressiveness include the propensity of firms to analyse and target a competitor's weaknesses (Macmillan and Jones, 1984), as well as its tendency to be first to market with new product offerings on a frequent basis (Porter, 1985).

The five dimensions outlined above contribute to our analysis of how innovation spaces are accessed both domestically and internationally as well as how innovation activities are organized by companies around the world. Schmitz and Strambach (2009) analyse firms from developed countries, and they highlight a fundamental shift from centralizing innovation activities at or near headquarters to having a much more decentralized, outsourced and dispersed innovation set-up. Doz and Wilson (2012) argue that reliance solely on local resources and on a more traditional composition of the innovation process is shortsighted. They emphasize the necessity of new approaches to innovation spaces. These works illustrate that there is an emerging body of literature concerned with Western companies.

Moreover, some recent studies also acknowledge the increasing significance of emerging-market 'challengers' (for example, Ramamurti, 2012; Pananond, 2013). How these emerging-market firms construct, access, develop, enlarge and engage in innovation remains largely uncharted territory, although several chapters in this book address emerging-market multinational enterprises (see Chapters 10, 11 and 12).

Table 6.2 presents the framework used to structure the case study, and the framework may be of relevance to similar studies conducted in the

Table 6.2 Analytical framework for the case study

	Innovation spaces as linking national innovation systems and firms			
	National innovation system and firm-level characteristics		Linkages outside the national innovation system and firm-level characteristics	
Dimension of entrepreneurial orientation	A: Key elements of NIS (for example, availability of skills, scientific research, industry collaboration)	B: Intra-firm (for example, organizational culture, learning processes)	C: Global channels and external collaboration (for example, capabilities in host location)	D: Intra-firm (for example, foreign subsidiary capabilities, network resources)
Innovativeness Pro-activeness Risk-taking Autonomy Competitive aggressiveness				

future. To comprehend the key dimensions of innovation space in relation to the dynamics of an entrepreneurial venture that internationalizes, Table 6.2 relates each of the location types mentioned in Table 6.1 (A, B, C and D) to the five elements of the EO construct.

Table 6.2 summarizes the elements of the framework, as discussed in the literature review. By structuring the framework in the matrix form, we seek to relate the elements of strategy and entrepreneurial orientation (vertical axis) to the innovation spaces (horizontal axis) conceptualized in this chapter as dynamic multi-layered (institutional, organizational, geographic) systems. This matrix helps us more precisely to analyse how and where innovation takes place by virtue of the actions and interactions of focal actors.

6.3 CASE STUDY OF WIND TURBINE COMPANY IN CHINA

6.3.1 Methodology

The chapter employs a qualitative methodology, which combines theories to propose an analytical framework, and then it applies this framework to an explorative case study of a Chinese company in the renewable energy industry. The dynamic perspective of knowledge-intensive entrepreneurship also draws upon the phases, as developed in the theoretical model and case studies detailed in (McKelvey and Lassen 2013a, 2013b). The company was followed by two of the authors independently from the spring of 2012 to the spring of 2013.

The case-study strategy, one of several strategies of qualitative enquiry, has been chosen for this investigation for several reasons. First, case studies can describe, enlighten and explain real-life phenomena that are too complex for other approaches that require tightly structured designs or pre-specified data sets (Voss, 2009; Yin, 2009). Second, the case-study strategy is well equipped instrumentally for advancing an understanding of issues or concepts that have not been deeply investigated so far (Eisenhardt, 1989; Yin, 2009).

Despite having many advantages, case-study research also poses significant challenges, such as the observer's perceptual and cognitive limitation, a high probability of overlooking some key events and lack of generalizability (for example, Meredith, 1998).

To help address these challenges, we followed practical guidelines and steps discussed in the qualitative methodology literature (for example, Voss, 2009; Yin, 2009). The current research therefore relied on the use of triangulation. Multiple sources of evidence (semi-structured interviews, documents and on-site observations) as well as triangulation of multiple data-points within each source of evidence (for example, multiple respondents at various management levels both in China and in Europe) were used. Interviews constituted a major source of evidence for the case study. In total, seven interviews were conducted, two of which took place at the companies' research and development (R&D) departments in China, and the remaining five took place in Denmark. The duration of interviews varied from one hour and two and a half hours. Interview data combined with secondary material such as media material, presentation material to customers and stakeholders, and these were used to build the database for the case and the case narrative, which is presented below. The descriptive case narrative has also been interpreted in terms of the framework in order to answer our research questions.

6.3.2 Venture Creation and Rapid Growth

The company, called here 'Energy Company' (or Energy Co.), was established in 2006 by a 30-year-old Chinese entrepreneur who, prior to this venture, spent six years in London's financial sector working as an energy trade strategy analyst. By returning to China shortly before establishing his company, he benefited from the promotion of returnee policies, since home-comers are given higher levels of protection of intellectual property rights (IPR), greater freedom to run private companies and special high-tech zones and parks.[2]

After returning to China, the entrepreneur wanted to start a business. However, with no technical background, it was almost a coincidence that the company was established in the wind energy industry. He noted in an interview: 'My ambition was to found an international company [using] Chinese wisdom to rival high-tech innovation giants like IBM and Microsoft with creativity, or the power of the mind, as its competitive edge, rather than intense exploitation of natural resources and cheap labour.'

At the time, sustainable energy technologies seemed to be an emerging growth industry in China due to the passing of the first Renewable Energy Law (REL) of China in 2005. The provincial government was sponsoring projects on the production of blades for wind turbines. That was how the Energy Co. started in 2006; since then, the company has gained a prominent position in the highly competitive wind energy market in China.

From its inception, the Energy Co. enjoyed a 100 per cent annual growth rate and aimed at becoming a competitive and smart energy management expert. In 2012, just six years after the company was established, the founder proudly announced that the company was on target to erect 1000-megawatt capacity turbines in China. The revenue figures for 2013 exceeded US$1 billion. The largest part of the revenue stream was generated in the Chinese onshore wind market by wind turbine sales. One of the company's employees described the founder: 'Our CEO is into innovation and new things. We do not do only traditional designs, but focus on smart and new solutions.' Therefore, it was not surprising that the Energy Co. product-offering portfolio gradually increased. In 2013, it included the development of software for monitoring and controlling wind farms, wind turbine services, energy storage solutions, integrated project management services, asset management and energy investment services related to wind energy.

The size of the company has also increased since it was founded. In 2008, it employed 45 employees. In 2013, the number of employees worldwide reached 600 with 60 per cent of them in R&D functions. In 2012, the company's R&D efforts were boosted by 150 new R&D specialists, who were hired by the company's Shanghai office. Fifty more were hired in 2013. The specialists in Shanghai worked closely with the company's engineering and manufacturing base in Jiangyin and its sales office in Beijing.

6.3.3 The Role of the Chinese National Innovation System

Since the millennium, China has developed into a core location for the wind turbine industry and is today the largest market for wind turbines. In 2012, despite the slow-down, China maintained its position as a global wind leader in cumulative terms with a total of 75.32 GW. The Chinese market represented about 27 per cent of the global wind energy market (Global Wind Energy Council, 2012). Furthermore, during the past two decades, the Chinese industry gradually developed from relying on production capabilities and technology imported from Europe to having national technological capabilities and a large supplier base. The development of the industry could, to a large extent, be attributed to China's wind energy policy transformation, including the REL and the Clean Development Mechanisms (CDM).

Being a private company in the energy sector in China is a challenge and the Energy Co. was by no means exempt from the challenges of this institutional context. However, having the status of a relatively small private actor in the industry also carries certain benefits in that it was not seen as a threat to the state-owned enterprises (SOEs) dominating the wind turbine industry in China. As a private actor with large R&D investments, the Energy Co. quickly developed into one of the most high-tech wind turbine manufacturers in China.

The founder of the company acknowledged the role of the government programmes in the successful development of its business in China:

> A national program to recruit professionals globally recognized my work and gave me funding and policy support, and the municipal government of Jiangyin built infrastructures worth RMB 10 million or more. Capital and preferential policies from the provincial government also enabled my company to swiftly transform research results into tangible commercial products.

However, with 60 per cent of the employees in R&D-related functions, the internal innovation drive of the entrepreneur and venture also played a very important role. In 2008, the company became the first wind-power generator supplier in China to independently develop wind-power core control technologies and load optimization technologies. The Energy Co. also became a pioneer of the ultra-low-wind-speed wind turbine generator technology. In 2010, the Energy Co. designed and launched the world's first wind turbine generator moving assembly line. This showed the company's ambition to be at the forefront of product and process innovation.

The company adopted the philosophy that talents were the core assets of the enterprise and that employees played a crucial part in building an innovative corporate culture. Operating in an industry that is highly dependent on continuous technological innovation, the company needed to build a highly stimulating innovative internal environment. In addition to the in-house capabilities, the company also engaged with Chinese universities to access the best students and tap into specialized knowledge. In China, partnerships were formed with two leading universities: Peking University and Tsinghua University. Specific examples of their collaboration with these universities included an initiative with Peking University to establish a Wind Power Technology Research Institute, which focused on research within aerodynamics. At Tsinghua University, the company established several research centres on smart wind farms, smart grids and energy storage.

6.3.4 Accessing Knowledge in Global Innovation Spaces

While benefiting from interactions with the Chinese national innovation system, the Energy Co.'s quest for innovation was not limited to its domestic context, as internal and external sources of ideas and knowledge. In 2009, the company acquired its first research subsidiary outside China. The subsidiary was in Denmark, at the heart of a large wind cluster. It employed 29 engineers in 2012 and was dedicated to designing the next generation of wind turbines based on partial pitch and dual-blade modular design. The unit in Denmark had a greater focus on research and experimentation with new concepts, while the R&D function in China mainly focused on short-term development and well-proven technologies. The Energy Co. was very eager to get this innovation subsidiary up and running in Denmark, or as an R&D manager described Denmark, in the 'The Wind Valley of the World'.

In 2012, an R&D manager in Denmark recalled the history of the acquisition in Denmark:

> When the company in China started, it was clear that looking long-term, they needed a greater technical depth. They dreamed of getting an innovation centre in Europe. When the founder was visiting, it seemed that there was a good chemistry between him and the engineers here, and we could talk. So the opportunity was realized. As time went by, we gradually expanding, and we hired some people. Now we are 29.

The Energy Co. managed to attract the world-class engineers by providing them with a competitive salary package and, more importantly, giving them the mandate and freedom to experiment with new concepts and technologies.

Having a local presence in Denmark allowed the company to engage with research centres specializing in wind energy at two Danish universities. These engagements proved beneficial for recruiting the best competencies and for accessing the best test facilities; hence, there was an international engagement with the Danish national innovation system that was similar to the Chinese. Moreover, the company actively engaged in the Danish national innovation system in order to gain the advantages that come with being an active part of this global hub for wind energy.

By 2012, all these initiatives helped the Energy Co. establish a strong position in the domestic market. Its wind turbines were erected in 11 provinces in China. However, the company experienced multiple challenges when expanding outside China. In 2013, the company started operating its first farm in Europe. The European market was saturated and known for its loyalty to well-established brands.

The company's R&D manager explained:

> It is tough to explain to Europeans that [our product is] good quality [. . .] especially if we source components from Western companies. That is one of the major barriers [to] entering the European market because they have a preconception that if it comes from China then it is bad quality. We hope to get a track record here and prove the sceptics wrong.

Testing new turbines in Europe in collaboration with the large certification firms and complementing turbines with service packages required by the European market were the core strategies used to overcome this barrier.

The company saw collaboration with suppliers as another important element of this strategy. As the company's R&D manager stated: 'When we develop and assemble our turbines, we engage in this work [with] the best suppliers from around the world, for example, a German company for gearboxes and a Danish company for blades. It helps us to achieve high quality and product performance.'

In an attempt to internationalize its innovation efforts further, the company added an R&D centre in Osaka, Japan, and a research centre in Houston, TX, to its network. Like the Danish subsidiary, these research centres were owned by the Energy Co. The three overseas research units work independently, and the Energy Co. faces a big challenge to integrate and align their operations. One tool they used to increase integration was the increased mobility of employees. Chinese employees were expatriated to Denmark to ensure better integration with the Danish unit, which enjoyed a high degree of autonomy. However, the challenge of integration remains in Japan and US.

The company also adopted a number of open innovation mechanisms. The company's R&D manager in Shanghai stated, 'We use open innovation.' Primarily, this allows the Energy Co. to team up with world's leading R&D centres, think tanks and other innovative companies. One example of a more recent involvement with external capabilities was the company's involvement with the China Thousand Talents Program (recruitment program of global experts) in order to discover new solutions for the global smart energy industry.

6.4 ANALYSIS AND WHAT WE HAVE LEARNED

This section focuses upon analysis, in terms of the conceptual framework. The case study identifies a set of practices and principles that contributed to the Energy Company's success with accessing innovation spaces in China and abroad. Figure 6.1 maps the elements of the company's innovation space and its web of internal and external linkages.

Figure 6.1 indicates the dynamics of the company, as through domestic contacts and internationalization, they accessed national innovation systems and exhibited entrepreneurial orientation. The development of the entrepreneurial company shows that some drivers and impediments to how the Energy Co. accessed and navigated various innovation spaces have country-specific idiosyncrasies, which may be related to respective national innovation systems. At the same time, the fundamental characteristics of the venture creation and innovation process within China still resemble results previously found in the literature dealing with the similar

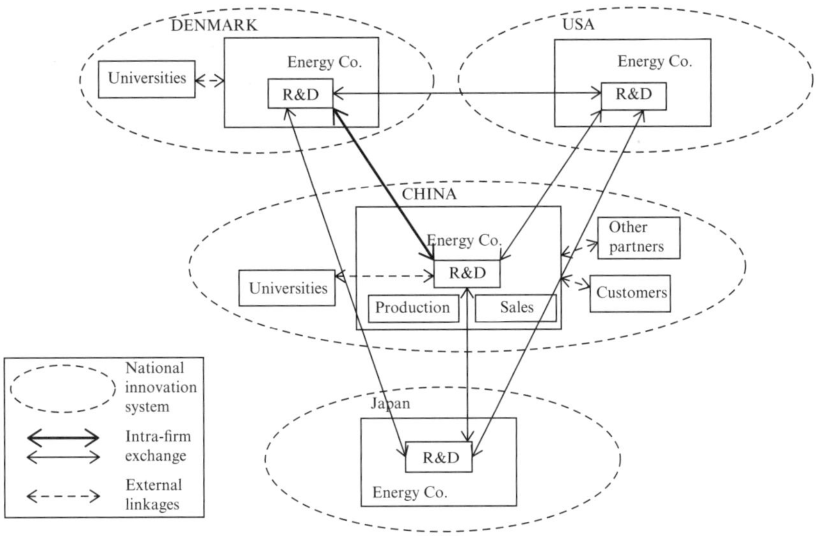

Figure 6.1 Dynamics of the energy company accessing innovation space

challenges in Western companies. Its strategy for the internationalization of innovation has been knowledge augmenting, specifically, its strategy to gain a position at the cutting-edge of technology by locating itself at the most promising places for wind turbine development.

Table 6.3 summarizes the case study, and when doing so, it illustrates that the framework is useful for explaining how the company could move across national innovation spaces in order to develop its own linkages and characteristics. Specifically, it summarizes how the various dimensions of entrepreneurial orientation are used in this process of accessing innovation spaces.

Table 6.3 structures the descriptive case, in terms of the variables in the analytical framework, as found in Table 6.2. The analysis reflects the fact that from the inception, the entrepreneur and the Energy Co. venture has been driven by the values of innovativeness and willingness to focus on novel solutions and designs.

The Energy Co. cannot be classified as a first mover when it comes to more traditional products that are often inspired by Western analogues. Currently a fast follower, the company is pro-actively seeking new solutions and technologies through its constructed innovation space; hence, we can say that the venture acts in anticipation of future changes in the industry.

The Energy Co. strategy aims at acquiring knowledge and technology

Table 6.3 Analysis of this Chinese company in terms of innovation spaces and entrepreneurial orientation

	Innovation spaces			
	National innovation system and firm-level characteristics		Linkages outside the national innovation system and firm-level characteristics	
Dimension of Entrepreneurial Orientation	A: Key elements of NIS (for example, availability of skills, scientific research, industry collaboration)	B: Intra-firm (for example, organizational culture, learning processes)	C: Global channels and external collaboration (for example, capabilities in host location)	D: Intra-firm (for example, foreign subsidiary capabilities, network resources)
Innovativeness	Ability to identify opportunities provided by NIS in China and utilize them in building a commercial success	'Our CEO is into innovation and new things. We do not do only traditional designs, but focus on smart and new solutions' Development focus and continuous improvement of the traditional product line	Teaming up with world's leading R&D centres, think tanks, universities and companies	Research focus and prototypes
Pro-activeness	Pro-active search for collaboration with universities and research institutes. Partnerships with two leading universities: Peking University and Tsinghua University	The Energy Co. is proactively maintaining continuous technological innovation	Research centres in the 'triad' countries (for example, Denmark) engage in external collaboration initiatives	Proactive search of the next breakthrough solutions and creating change in the industry through owned research

				centres in the 'triad' countries (Denmark, Japan and the USA)
Risk-taking	Being a relatively small private actor means continuous risk taking the industry populated by foreign MNEs and local SOEs	Going beyond traditional and standardized product portfolio: diverse product portfolio including complex smart energy solutions	Opening wind farms in Europe	Wholly-owned research centres established in Denmark, Japan and the US
Autonomy	Being a relatively small private actor means independence from hierarchies and delegating authorities	Development base is highly integrated with the company's production setup in Jiangyin	Independent and actor in joint international collaboration initiatives	High degree of freedom and autonomy
Competitive aggressiveness	Strong positions on the domestic market and turbines in the 11 provinces in China	Recruitment of 200 new R&D specialists in 2012–2013 Building a highly stimulating innovative environment internally The first wind power generator supplier in China that independently developed wind power core control technologies and load optimization technologies	Being among the first Chinese players in the 'Wind Valley of the world'	Foreign knowledge base is highly dispersed Competitive salary packages for engineers outside China

abroad. As such, the Energy Co. contributes to the stream of Chinese knowledge-seeking outward foreign direct investment (OFDI). The company took a high-risk OFDI strategy, with first overseas investments made just three years after the start of its operations. This targeted resource commitment by the company is also influenced by the entrepreneurial spirit of the founder.

Autonomy plays a big role in the company's behaviour. On the one hand, autonomy is associated with independence; on the other hand, being a relatively small private actor means continuous risk-taking in the industry populated by foreign MNEs and local SOEs in China and globally. Competitive aggressiveness is visible in the high propensity of the company to act unconventionally and have a strong focus on outperforming competitors.

The results from this case study suggest many avenues for future research to advance our understanding of innovation spaces, and it may be useful in informing our expectations about how Chinese companies can tailor their strategies to develop sustainable innovation spaces domestically and abroad. We would like to conclude with a few questions for reflection about the implications to other contexts:

- Traditional international business theories tend to argue that firms evolve into international actors over time as they mature. This company has done the opposite. What are the advantages and disadvantages of this strategy?
- As a Chinese company with its core research departments located in Northern Europe, what would be the main barriers for innovation? How can these be overcome?
- How does this case advance our understanding of innovation in Chinese companies? Are there aspects of innovation that are specific to companies located within the Chinese national innovation system?

NOTES

1. We would like to acknowledge our respective universities, including: Copenhagen Business School; the University of Aalborg and the Center for Industrial Production; and the University of Gothenburg. Moreover, Maureen McKelvey would like to acknowledge and thank Volvo AB and the Sten A. Olsson Foundation for Research and Culture for supporting this research. This chapter has been written in context of the research programme *Radical Innovations for the Enhancement of the Swedish Economy*, running at the Institute of Innovation and Entrepreneurship, with Professor Maureen McKelvey as principal investigator.
2. The returnees were offered cheap office space, tax exemptions, interest-free loans, access

to international schools, a special waiver of the one-child policy and other facilities (Cao, 2004). Since the 1980s, more than 110 science parks and 6000 enterprises have been the result of more than 15,000 returnees (Jonkers, 2008).

REFERENCES

Archibugi, D. and Michie, J. (1997). 'Technological globalisation and national systems of innovation: an introduction.' In Archibugi, D. and Michie, J. (eds), *Technology, Globalisation and Economic Performance*, Cambridge: Cambridge University Press, pp. 1–23.

Baird, I.S. and Thomas, H. (1985). 'Toward a contingency model of strategic risk taking'. *Academy of Management Review*, **10** (2), 230–43.

Barnard, H. and C. Chaminade (2011). 'Global innovation networks: what are they and where can we find them?' CIRCLE Electronic Working Papers.

Bergek, A., Jacobsson, S., Carlsson, B., Lindmark, S. and Rickne, A. (2008). 'Analyzing the functional dynamics of technological innovation systems: a scheme of analysis.' *Research Policy*, 37, 3, 407–29.

Bunnell, T.G. and Coe, N.M. (2001). 'Spaces and scales of innovation.' *Progress in Human Geography*, 25, 4, 569–89.

Burgelman, R.A. (1983). 'Corporate entrepreneurship and strategic management: insights from a process study.' *Management Science*, 29, 12, 1349–64.

Burns, T. and G. Stalker (1961). *The Management of Innovation*. London: Tavistock.

Cao, C. (2004). 'China's efforts at turning "brain drain" into "brain gain".' *EAI Background Briefs*, Singapore: East Asian Institute, National University of Singapore.

Dicken, P. (2011), *Global Shift: Mapping the Changing Contours of the World Economy*. London: Sage.

Doz, Y.L. and Wilson, K. (2012). *Managing Global Innovation: Frameworks for Integrating Capabilities Around the World*. Boston, MA: Harvard Business Review Press.

Drucker, P. (1981). 'The coming of the new organization.' *Harvard Business Review*, **66**, 45–53.

Edquist, C. and Johnson, B. (1997). 'Institutions and organizations in systems of innovation.' In Edquist, C. (ed.), *Systems of Innovation: Technologies, Institutions, and Organizations*, London: Routledge, pp. 41–64.

Eisenhardt, K.M. (1989). 'Building theories from case study research.' *Academy of Management Review*, 14, 4, pp. 532–50.

Ernst, D. (2002). 'Global production networks and the changing geography of innovation systems: implications for developing countries.' *Economics of Innovation and New Technology*, 11, 6, pp. 497–523.

Ernst, D. (2006). 'International offshoring: Asia's emerging role in global innovation networks.' East-West Center, Honolulu, HI.

Fan, T. and Phan, P. (2007). 'International new ventures: revisiting the influences behind the "born-global" firm.' *Journal of International Business Studies*, **38**, pp. 1113–31.

Freeman, C. (1987). *Technology Policy and Economic Performance: Lessons from Japan*. London: Pinter.

Global Wind Energy Council (2012). *Global Wind Report: Annual Market Update 2012*. Brussels: GWEC.
Jonkers, K. (2005), 'A comparative study of return migration policies targeting the highly skilled in four major sending countries.' *Analytical Report, MIREM-AR 2008/05*. Robert Schuman Centre for Advanced Studies, European University Institute.
Kimberly, J.R. (1981). 'Managerial innovation.' In P.C. Nystrom and W.H. Starbuck (eds) *Handbook of Organizational Design*, vol. 1, New York: Oxford University Press, pp. 84–104.
Kuemmerle, W. (1999). 'Foreign direct investment in industrial research in the pharmaceutical and electronics industries – results from a survey of multinational firms.' *Research Policy*, **28**, 179–93.
Lazonick, W. (2004). 'Indigenous innovation and economic development: lessons from China's leap into the information age.' *Industry and Innovation*, 11, 4, 273–97.
Lieberman, M., and D. Montgomery (1988). 'First-mover advantages.' *Strategic Management Journal*, **9**, 41–58.
Liu, X., Buck, T. and Shu, C. (2005). 'Chinese economic development, the next stage: outward FDI?' *International Business Review*, 14, 1, 97–115.
Lundvall, B.-A. (1999). 'National business systems and national systems of innovation.' *International Studies of Management & Organization*, 29, 2, 60–77.
Lumpkin, G.T. and Dess, G.G. (1996). 'Clarifying the entrepreneurial orientation construct and linking it to performance.' *Academy of Management Review*, 12, 1, 135–72.
MacMillan, I. C. and P.E. Jones (1984). 'Designing organizations to compete.' *Journal of Business Strategy*, **4**, 11–26.
Malerba, F. (2002). 'Sectoral systems of innovation and production.' *Research Policy*, **31**, 247–64.
Malerba, F. (2009). *Sectoral Systems of Innovation: Concepts, Issues and Analyses of Six Major Sectors*. Cambridge: Cambridge University Press.
McDougall, P. and B. Oviatt, B. (2000). 'International entrepreneurship: the intersection of two research paths.' *Academy of Management Journal*, **43**, 902–8.
McKelvey, M. and Lassen, A.H. (2013a). *Managing Knowledge-Intensive Entrepreneurship*. Northampton, MA, USA and Cheltenham, UK: Edward Elgar.
McKelvey, M. and Lassen, A.H. (2013b). *How Entrepreneurs Do What They Do: Case Studies of Knowledge Intensive Entrepreneurship*. Northampton, MA, USA and Cheltenham, UK: Edward Elgar.
Meredith, J. (1998). 'Building operations management theory through case and field research.' *Journal of Operation Management*, 16, 4, 441–54.
Miles, R.E. and Snow, C.C. (1978). *Organizational Strategy, Structure, and Process*. New York: McGraw-Hill.
Miller, D. (1987). 'Strategy making and structure: analysis and implications for performance.' *Academy of Management Journal*, 30, 1, 7–32.
Miller, D. and P. Friesen (1978). 'Archetypes of strategy formulation.' *Management Science*, **24**, 921–33.
Nonaka, I. (1991), 'The knowledge-creating company.' *Harvard Business Review*, **69**, 96–104.
Oviatt, B.M and McDougall, P.P (1999). 'A framework for understanding

accelerated international entrepreneurship.' In *Research in Global Strategic Management (Volume 7)*, Emerald Group Publishing Limited, pp. 23–40

Pananond, P. (2013). 'Moving along the value chain: emerging Thai multinationals in globally integrated industries.' *Asia Business & Management*, 12, 1, 85–114.

Porter, M. (1985). *Competitive Advantage*. New York: Free Press.

Quinn, J.B. (1979). 'Technological innovation, entrepreneurship, and strategy.' *Sloan Management Review*, 20, 3, 19–30.

Schmitz, H. and Strambach, S. (2009). 'The organisational decomposition of innovation and global distribution of innovative activities: insights and research agenda.' *International Journal of Technological Learning, Innovation and Development*, 2, 4, pp. 231–49.

Senge, P. M. (1994). *The Fifth Discipline Fieldbook: Strategies and Tools for Building a Learning Organization*. New York: Currency, Doubleday.

Stopford, J.M. and Baden-Fuller, C.W.F. (1994). 'Creating corporate entrepreneurship.' *Strategic Management Journal*, 15, 7, 521–36.

Stevenson, H.H. and Jarillo, J.C. (1990). 'A paradigm of entrepreneurship: entrepreneurial management.' *Strategic Management Journal*, **11**, 17–27.

Ramamurti, R. (2012). 'Competing with emerging market multinationals.' *Business Horizons*, 55, 241–9.

Tzeng, C., Beamish, P.W. and Chen, S. (2011). 'Institutions and entrepreneurship development: high-technology indigenous firms in China and Taiwan.' *Asia Pacific Journal of Management*, 28, 3, 453–81.

Van Wyk, B. (2009). 'Why, who and where? Chinese OFDI on the world stage!' INSEAD articles, Fontainebleau, France: INSEAD.

von Hippel, E. (1988). *The Sources of Innovation*. New York: Oxford University Press.

Voss, C. (2009). 'Case research in operations management.' In Karlsson, C. (ed.), *Researching Operations Management*, New York: Routledge, pp. 162–96.

Yin, R.K. (2009). *Case Study Research – Design and Methods*. Thousand Oaks, CA: Sage.

Zahra, S.A., B. Matherne and J.M. Carleton (2000). 'Technological resource leveraging and the internationalisation of new ventures.' *Journal of International Entrepreneurship*, **1**(2), 163–85.

7. Manoeuvring global innovation spaces: an explorative case study of a Korean entrepreneurial venture in nanotechnology

Astrid Heidemann Lassen and Dmitrij Slepniov

7.1 INTRODUCTION

This chapter analyses the development of innovation space for nanotechnology within Korea, by applying the framework developed in Chapter 6 about the relationship between entrepreneur and national innovation system.[1] In recent decades, South Korea has earned a reputation as one of the fastest-growing economies in the world (Lee, 2013). Korean companies, such as LG, Samsung, Daewoo, Hyundai and Kia Motors, have established themselves as world-dominating conglomerates, highly successful in developing and introducing advanced technology solutions as well as exploiting their access to low-cost labour in the Asian regions. The success of such companies has been driven by a strategically prioritized industrial policy in Korea, which has enabled very dedicated and resourceful efforts in prioritized areas.

While undoubtedly a highly successful approach, it has also had very particular effects on the national innovation system in Korea. Today, the top-ten largest Korean companies [*chaebols*] account for more than 80 per cent of Korean gross domestic profit (GDP), and this figure is growing. Yet, the industry structure is still dominated by small and medium-sized enterprises (SMEs), which account for 99.9 per cent of all companies in Korea (Yang, 2009). These figures indicate that Korea has developed a highly skewed economy, where a minority of companies are very large and highly successful, and the vast majority of companies are small and relatively unproductive in comparison.

As Korean industry growth has slowed down over the past years, attention has gradually turned to the problems related to a skewed economy. Among such problems is the lack of experience with successful entrepreneurship. In 2013, the Korean government initiated a large-scale industrial program labelled "Creative Economy" to foster innovation and start-up companies. The government plans to spend 200 billion won (approximately €135 million) on start-ups directly and 300 billion won (approximately €202.5 million) on mergers and acquisitions (M&As). Angel investors will receive generous tax incentives. Highly skilled foreigners will be offered entrepreneur visas to start businesses. The government estimates that its measures will ultimately unlock 4.3 trillion won (approx. €29 billion) in funding (Bloomberg, 2013).

However, many questions remain about which strategies Korean knowledge-intensive ventures should develop and follow in order to manoeuvre within the innovation spaces developing under these particular circumstances. Such strategies may be influenced by factors such as business structures and culture, familiarity with innovation and entrepreneurship practices and market access and customer innovation expectations. Therefore, the phenomenon in Korea requires further investigation, which links firm creation with the external environment of the innovation system (McKelvey and Lassen, 2013). The principal objective of this chapter is to contribute to the debate about innovation spaces by discussing how Korean entrepreneurial firms develop their business strategies in order to exploit and explore global innovation spaces.

7.2 MANOEUVRING GLOBAL INNOVATION SPACES

In pursuing the objective of this chapter we draw on the conceptual framework developed by Slepniov et al. (2014), found in Chapter 6. This framework defines an innovation space as a multifaceted organism not restricted geographically, but rather seen as consisting of multiple levels, including institutional, organizational and geographical aspects (Table 7.1).

This definition sets the global innovation space apart from the concept of national innovation systems, although it clearly relies on similar lines of reasoning. The literature on innovation systems is concerned with the functioning of the national economy as a whole and its performance in terms of innovative capacity (McKelvey, 1991). A national innovation system therefore consists of structures and institutions that both directly and indirectly facilitate a country's innovation capacity (Lundvall, 1999).

A global innovation space, on the other hand, is from an organizational perspective, found in the national as well as the global linkages that directly and indirectly affect the possibilities of the organization to explore and exploit new knowledge. This space emerges from the interaction

Table 7.1 Conceptualization of innovation spaces from the firm's perspective

Innovation spaces			
National innovation system and firm-level characteristics		Linkages outside the national innovation system and firm-level characteristics	
A: Key elements of NIS (for example, availability of skills, scientific research, industry collaboration)	B: Intra-firm (for example, organizational culture, learning processes)	C: Global channels and external collaboration (for example, capabilities in host location)	D: Intra-firm (for example, foreign subsidiary capabilities, network resources)

Source: Slepniov et al 2014

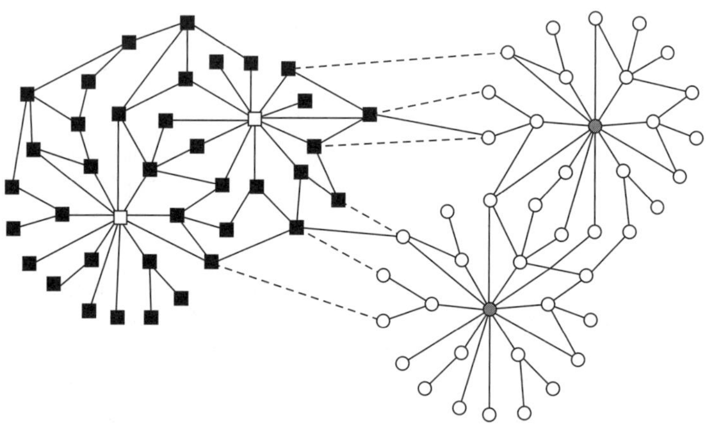

Figure 7.1 Visual illustration of global innovation spaces

between different national innovation systems, and takes shape after the activity conducted by actors in these systems. In this sense, the innovation space is to be considered more than the sum of interacting national innovation systems; the interaction creates a synergetic new space – the global innovation space.

Figure 7.1 provides us with a visually intuitive understanding of how the concept of innovation spaces builds on the national innovation system thinking, but surpasses the systemic and geographically confined definition.

The metaphor of an organism is suitable, as it emphasizes the ever-changing nature of the global innovation space. A more detailed discussion of the mutuality and diversity of the two terms is also found in Chapter 6 (Slepniov et al. 2014).

7.2.1 Operationalizing the Conceptualization

The proposed dynamic definition of a global innovation space encourages us to understand how the individual firm operates within this space as well as the specific organizational strategies and processes supporting the innovativeness of the firm. Focusing on the innovativeness of a firm has numerous advantages for understanding how firms engage with their innovation spaces. This focus compels us to think about the processes by which innovative firms explore and exploit opportunities in the global context. It also provides us with insights into the dynamic interplay between firm-level and system-level dimensions, which is relatively underexplored in the Korean context in particular.

To comprehend the key dimensions of innovation space we have, in addition to geographical location, developed four dimensions of organizational innovativeness: process, strategy, organization and technology. These dimensions parallel key areas of organizational models (for example, Leavitt, 1965), and, in this chapter, they are operationalized into aspects that have a particular impact on the innovativeness of the organization. The framework is a further development of the work presented in Slepniov et al. (2014), where the entrepreneurial dimension is particularly in focus.

By expanding the framework, we capture more of the organizational factors, thus allowing us to analyse innovative behaviour. Table 7.2 presents the framework that is to be used in pursuing the principal objective of this study. The elements of the framework are distilled from the review presented below.

First, the process dimension refers to the approach to creating innovation, and it is operationalized into knowledge absorption and knowledge diffusion. Knowledge is considered an essential factor of innovation (von Hippel, 1988; Cockburn and Henderson, 1998; McKelvey and Lassen, 2013). For example, according to Chesbrough (2003), openness to using external sources of information and ideas in the firm's innovation process, as well as interaction among different partners, is of high importance when creating value through innovation activities. The influence the absorptive capacity of knowledge has on innovation has also been studied extensively by, for example, Cohen and Levinthal (1990) and Zahra and George (2002), and the absorption of new ideas and knowledge is considered a key input for organizational innovativeness. In terms of new

Table 7.2 Analytical framework for the case study

		Innovation spaces			
		National innovation system and firm-level characteristics		Linkages outside the national innovation system and firm-level characteristics	
		A: Key elements of NIS (for example, availability of skills, scientific research, industry collaboration)	B: Intra-firm (for example, organizational culture, learning processes)	C: Global channels and external collaboration (for example, capabilities in host location)	D: Intra-firm (for example, foreign subsidiary capabilities, network resources)
Process	Knowledge absorption				
	Knowledge Diffusion				
Strategy	Pro-activeness				
	Risk-taking				
	Competitive aggressiveness				
Organization	Knowledge sharing				
	Customer insight				
	Network				
	Competence development				
	Team structure				
	Role of founder				
Technology	Product				
	Process				

product development, Asian firms in general are often associated with cost-efficient imitations of Western products. However, we may consider imitation to be a type of knowledge absorption with innovation potential, as imitation and reengineering generates in-depth product-specific knowledge, which in turn could harbour innovation potential and learning. In the Korean context, the initial imitation of Western products has sometimes led to the subsequent development of superior competences (for example, mobile phones, industrial robotics, and so on). Therefore, in this chapter we apply the term "knowledge absorption" to focus on how global knowledge is utilized as an input for creation of innovative output.

We also propose that knowledge diffusion is a relevant consideration from an organizational perspective. One often tends to think of the absorption of knowledge as conceptually distinct from its diffusion, with the latter concept having to do with how widely a given stock of knowledge is used (Jovanovich and Rob, 1989). This distinction, made by Schumpeter (1934), between the creation of new knowledge on the one hand and its diffusion on the other, has proved useful in research on the spreading of particular technologies. However, studies show that most advances in knowledge are not attained at once and then slowly diffused. Rather, innovation is often the outcome of many knowledge-building blocks, and it embodies many ideas. To assemble a lot of ideas one generally needs to involve a lot of different actors to share, discuss, evaluate and rethink knowledge in order to make advances through iterative development. If so, the absorption of knowledge is essentially an interactive and iterative process, and therefore one that is largely inseparable from its diffusion. In a Korean context, we find clear examples of how rapid diffusion into the marketplace of relatively early versions of products allows for specific feedback of features, functionalities, design, price and so on, which in turn allows for further optimization and advancement of the product leading to increased market potential. Hence, absorption and diffusion must be considered parts of a joint product that are equally important for the understanding of how innovation is created through the dynamics of global innovation spaces.

Second, we operationalize the strategic dimension by applying the three original and central aspects of the entrepreneurial orientation (EO) concept: proactiveness, risk-taking and competitive aggressiveness (Covin and Slevin, 1991). These aspects are widely associated with entrepreneurial strategies pursued by innovative firms.

Proactiveness is an act of taking the initiative in the entrepreneurial process and acting in anticipation of future problems, needs or changes. Risk-taking can have various meanings, depending on the context in which it is applied. In this chapter, risk is seen in the context of strategy, which according to Baird and Thomas (1985: 231–2) can take three forms: (1) venturing into the unknown; (2) committing a relatively large portion of assets; and (3) borrowing heavily. Competitive aggressiveness refers to a firm's predisposition to directly and intensely challenge its competitors to achieve entry into a new market or improve the present position in an existing market; that is, a strong focus on outperforming competitors (MacMillan and Jones, 1984; Porter, 1985). These dimensions essentially parallel the focus of the framework by Slepniov et al. (2014).

Third, the dimension of organization captures elements related to the influences of organizational structures, knowledge-intensity and managerial practices on innovation. Theories on innovation management widely recognize that the interplay between these aspects is central to innovation. Jensen et al. (2007), for example, argue that organization and communication structures that encourage and make use of experience-based learning, knowledge sharing and interaction – such as project teams, problem-solving groups and task rotation – contributes positively to the performance of innovative activities. Brown and Eisenhardt (1997) and Dutta and Evrard (1999) emphasize that fluid structures, low formality and high flexibility are significant enablers of innovation. Cohen and Levinthal (1990) argue that innovation requires a diversified learning process, involving learning-by-using, learning-by-doing and learning-by-sharing internal and external sources of knowledge. Ahjua (2000) argues that the success of such processes is largely determined by the ability to manage and integrate the web of formal and informal relationships inside and outside the firm. This in turn highlights the importance of structures, knowledge and management as elements central to the understanding of innovativeness. In this chapter, we further operationalize the abovementioned elements into: network of customer insight, knowledge sharing, competence development, team structure and the role of the founder.

Fourth, the final dimension included in the framework is technology. We here apply technology as a proxy for the object of innovation. Technological innovations are those innovations that embody inventions from, for example, the industrial arts, engineering, applied sciences and/or pure sciences (Garcia and Calanone, 2002). Examples include innovations from the electronics, aerospace, pharmaceuticals, robotics and information systems industries. This obviously creates a limitation in scope of innovation. However, we argue that the limitation is relevant for two reasons: technological innovation is without comparison the most influential type of innovation, and it is where most innovative breakthroughs are created. This definition is relevant here, because the focus of the current study is on an entrepreneurial and high-tech innovative organization in which innovation in most forms and shapes will evolve around technology exploration and exploitation. In addition to the purely technological innovation, we also argue that technology and knowledge related to this domain is in fact a prerequisite for much product and process innovation in other areas. Classic examples of process innovations, which are highly reliant on technological expertise, are the float glass process for flat-sheet glass manufacturing and the Bessemer process for converting iron to steel, which revolutionized the steel industry. This means that our focus on technology will be discussed both in terms of technology as an innovative product and as a catalyst for process innovation.

The four dimensions combined contribute to understanding and discussing the way in which firms access global innovation spaces and how innovation activities can be organized by companies.

7.3 CASE STUDY OF A KOREAN ENTREPRENEURIAL VENTURE

7.3.1 Methodology

This chapter employs the same qualitative methodology as in Chapter 6. The methodology and rationale for using it are to develop and analyse an explorative case study of the development of a knowledge-intensive entrepreneurial venture Korean company in the nanotechnology industry.

The company was followed by the authors from the spring of 2013 to the autumn of 2013. Two visits to Korea were made, involving ten interviews. Interviews at the company were conducted with the chief executive officer (CEO), head of R&D, project manager for R&D, head of global sales and project manager for production. Moreover, visits and observations were made to the R&D department as well as the departments of production and planning, marketing, sales and packaging.

Our explanation of the case-study strategy, one of several strategies of qualitative enquiry, is identical to the explanation that was originally developed in Slepniov et al. (2014). It has been chosen for this investigation for several reasons. First, case studies can describe, enlighten and explain real-life phenomena that are too complex for other approaches requiring tightly structured designs or pre-specified data sets (Voss, 2009; Yin, 2009). Second, the case-study strategy is well equipped instrumentally for advancing understanding of particular issues or concepts that have not been deeply investigated so far (Eisenhardt, 1989; Yin, 2009).

Despite having many advantages, case-study research also has several pitfalls and poses significant challenges, such as the observer's perceptual and cognitive limitation, a high probability of overlooking some key events, as well as the challenge of generalizability (for example, Meredith, 1998). To address these challenges, we followed practical guidelines and steps discussed in qualitative methodology literature (for example, Yin, 2009; Voss, 2009). The current research relied on the use of triangulation. Multiple sources of evidence (semi-structured interviews, documents and on-site observations) as well as triangulation of multiple data-points within each source of evidence (for example, multiple respondents at various management levels in South Korea) were used. These data combined with secondary material (media material, presentation material to customers and stakeholders) were used to build the database for the case. The case narrative is presented below.

7.3.2 Venture Creation and Market Segments

Park Systems is a developer and manufacturer of atomic force microscopes (AFM) used for metrology. The company was founded as a PSIA venture in 1997 by Dr Sang-il Park, one of the pioneers in developing commercialized AFM equipment. Prior to the PSIA venture, Park founded Park Scientific Instruments (PSI) in California, where he served as the chairman and CEO for nine years (1988–97). Before starting his own business, Park worked with Professor C.F. Quate at Stanford University, the birthplace of the AFM. Since returning to Korea in 1997, Park has been actively involved in academic and public activities, earning recognition in the industry as well as multiple awards, including the Iron Tower Order of Industrial Service Merit and the Industrial Technology Innovation Award.

PSIA changed its name to Park Systems because the word "systems" better reflected the company's focus on total metrological solutions. In addition, the company also started offering an industrial product line that extended its innovative technology to a variety of metrological products and applications, including hard-disk inspection, next-generation sliders, sidewall/overhang imaging, and profiling and semiconductors.

In 2013, within the field of AFM technology, Park Systems was one of the three largest players worldwide with a market share of approximately 10 per cent. The industry was characterized as highly concentrated, with the largest player in the market, Bruker, having around a 50 per cent market share, which it gained by acquiring smaller companies.

By looking at various market segments, Park Systems was the largest player in the industrial segment. In fact, Park Systems was the only company actively applying AFM technology to industry use. The market in this area had not yet been well developed and Park Systems proactively took first-mover risks. This decision was taken by the founder of the company. Over time, the results of this move started to reveal themselves when the new industrial customers started using the Park Systems' AFM solutions to substitute for older technologies, such as critical dimension scanning electron microscopy (CDSEM). The superiority of the AFM technology and the gradual success of new products developed on its basis strengthened the company's commitment to this business area. The feedback the company got from the customers suggested that the AFM technology better met new demands of an industry working with smaller products and new materials.

The competition in the industrial segment was not as intense as the research segment of the market. Yet, this field had not yet been developed, meaning that the market at that time was very limited. Although the future looked promising for the AFM technology, Park Systems' strategy and organization had to be developed in the way that could enable the company to overcome extant challenges.

In both market segments – research and industry – Park Systems had approximately 1000 customers worldwide. In the industrial segment, these included big companies, such as IBM, Rolex and Samsung. In the research segment, customers included universities, research institutes and laboratories. The level of customization and the price in the research segment was much lower than in the industrial segment. According to estimates by the company's R&D director, the products sold to research customers cost five to ten times less than those developed and produced for industrial customers.

7.3.3 Manoeuvring the Innovation Space

The research base of the company was largely concentrated in Korea. The research activities were often supported by government grants. Park Systems were rather successful in acquiring industry project-related grants. More than 60 per cent of such grants from the governments programs were given to "teams". These teams often consisted of competing companies. This meant that in order to receive the grant, Park Systems had to collaborate with competitors. As this could be problematic, the competitor and Park Systems divided the objective between them. A third party supervised the technical development done in each of the companies.

However, collaborative work with other companies in the industry was not the only way for the company to build its research network. The development based on the grants from the Science and Education Department was often done in collaboration with universities in Korea and overseas; this meant that the company had to rely heavily on experts from around the world. These partners were also used when new products had to be tested. The company divided the research partners into three categories: (1) the so-called brainy testers who were used to get access to their expertise; (2) the friendly testers, who through years of collaboration were shown to give good and constructive feedback to Park Systems; and (3) big laboratories where a lot of people would have access to the instruments, which meant the company's products were exposed to potential customers.

Park Systems strove to be not only a manufacturer of nanotechnology, but also a solution partner for their customers. This meant that the collaboration with industry customers was also extensive in the area of new product development and testing. Testing could be done at the customers' manufacturing sites, resulting in large amounts of feedback. The testing was done over a period of 6–12 months, whereupon the customers had to buy the product if they wanted to continue using it.

The R&D department of Park Systems was divided into mechanical, electrical, software and systems engineering teams, employing 30 engineers altogether. As the largest team, systems engineering consisted of 7–10 people. In addition, there was a technical management team to handle matters, such as intellectual property and ISO certification. The company was ISO 9001 and ISO 14001 certified.

With regards to internal cross-functional collaboration, the R&D department collaborated with the sales and marketing department and more extensively with the manufacturing department. The sales department was in charge of customer contact; therefore, the link to customers had to go through this department. The link to the manufacturing department was important because the products developed were going through different stages. The usual process consisted of three stages. First, an alpha development stage took approximately 8–12 months, during which the product was tested internally for three months and input from the manufacturing department was received. Then a beta version was developed, which was exposed to industrial application and testing by customers. This took around nine months. Input was again received from customers and the manufacturer. After this, a final product was released.

Normally, three to four projects were running at the same time in the R&D department. Three of these normally turned into final products.

The biggest success of the R&D department in 2013 was the launch of the next-generation AFM and getting a big customer. The biggest challenge was to continually meet market requirements. This was a big challenge, especially in the industry segment where standards had not yet been established. Customers therefore often had unrealistic requirements for specifications, with which Park Systems had to deal. Another challenge was that the products had to be very advanced in order to meet these requirements. Suppliers could not deliver on these specifications, which was a barrier to meeting their requirements. For instance, only 10 per cent of the probes received from the supplier could be used due to poor quality. The same problem with market requirements was expected to be present in 2014.

Upstream in the supply chain, Park Systems had approximately 20 to 30 first-tier suppliers from all over the world delivering probes, vibration isolation systems, acoustic enclosures, optic vision, big stages and so on. All of these were unique and advanced components, which made the sourcing of components very hard. When Park Systems was established, the suppliers were smaller than the company. However, they gradually outgrew Park Systems. Even though the balance had shifted, the relationship with the suppliers remained very good.

Recruitment in Korea can be a hassle for a company like Park Systems, as many talented people often prefer to work for the large chaebols (multinational enterprises). However, Park Systems found ways of dealing with high employee turnover. The average worker stayed in the company for six to seven years. Park Systems was also becoming an increasingly decentralized company with an empowered middle management level. Before 2013, the CEO and founder made 100 per cent of all decisions; after 2013, in many cases, the CEO disengaged from unilateral decision-making.

Its future as an independent manufacturer of metrology instruments is unknown, as many companies in this industry are acquired by larger players. In order to retain independence, Park Systems could be more aggressive. In 2013, the company was putting a lot of emphasis on technological development of both product and processes. However, not enough was done in the areas of sales and marketing. If the company is to become not only an expert in the industry, but also set the trends, this lack of sales and marketing will have to change. To realize this, different solutions can be proposed. First, the brand identity and reputation has to be improved. Moreover, there is a need for an increased number of application areas for AFM and technologies at Park Systems besides the AFM. The input from different universities could be one way of solving this. The company could then be in charge of the application aspects, as this was its original field of expertise, when it was one of their core competences.

7.4 ANALYSIS AND CONCLUSIONS

Our description and analysis of the case study of Park Systems identifies a set of practices and principles that contributed to the company's success in accessing innovation spaces in South Korea and globally. As in the Chinese case reported in Slepinov et al. (2014), we found that some drivers and impediments in this process have country-specific idiosyncrasies. For example, the challenges related to being an entrepreneurial venture in South Korea could exemplify some of the impediments unique in the Korean context. Still, the fundamental characteristics of the process of venture creation do resemble those previously found in the literature dealing with similar challenges in Western companies. The strategy for internationalization of innovation at Park Systems is based on collaboration with customers and locating its own outposts in the most promising places for nanotechnology development in Europe and the US. Table 7.3 illustrates how the company deconstructed and targeted its innovation spaces throughout its history.

By applying the analytical framework, we can illustrate how Park Systems could develop global knowledge and marketing, and in so doing, manoeuvre their innovation space as influenced by local, national and global aspects. The ability to be innovative is determined by the continuous interplay between national and global knowledge, market demands and possibilities and business partners and competitors. The particular industrial and system characteristics of the Korean context highlight the necessity of this interplay between global and national levels. In this setting, high-tech entrepreneurial ventures would have difficulties, if they had to depend exclusively on the national innovation system. Nationally in Korea, the arena for support of small venture growth is relatively underdeveloped; instead, the all-dominating presence of large chaebols helps influence which types of technologies are supported and technological specializations of the country. The technologies applicable in business areas of interest for the chaebols are prioritized and developed either through their own corporate R&D resources or by the acquisition of small ventures working with such technology. The market possibilities for independent ventures outside this are rather limited due to, for example, the lack of industrial prioritization, lack of resource allocation possibilities (venture capital, skilled human resources, and so on) and lack of possibilities for a tie-in with established technological trajectories. Such system dynamics restrict independent ventures to niche markets with low national growth potential.

For these reasons, the tie-in with global markets becomes essential for the creation and development of knowledge-intensive entrepreneurial ventures. Due to the clear challenges to entrepreneurial ventures presented by the Korean context, this case study provides interesting insights for studying and understanding the particularities of what, in spite of the challenges, enables such entrepreneurial ventures to prosper. The approach applied by Park Systems to manoeuvring the national and

Table 7.3 Analysis of how this Korean company could manoeuvre global innovation spaces

		Innovation spaces			
		National innovation system and firm-level characteristics		Linkages outside the national innovation system and firm-level characteristics	
		A: Key elements of NIS (for example, availability of skills, scientific research, industry collaboration)	B: Intra-firm (for example, organizational culture, learning processes)	C: Global channels and external collaboration (for example, capabilities in host location)	D: Intra-firm (for example, foreign subsidiary capabilities, network resources)
Process	Knowledge absorption	Industrial market segment	Lack of standards and random requests from customers	High requirements from global customers driving the continuous ideation	Establishment of global offices in close vicinity to central global knowledge hubs
	Knowledge diffusion	Research market segment	Test-panels	Strong ties to former partners in Silicon Valley	Test-panels of key partners
Strategy	Pro-activeness	Continuous pursuit of research funding	Frequency of NPD	Continuous development of new customers relationships and markets	Global network of trusted retailers established

Table 7.3 (continued)

		Innovation spaces			
		National innovation system and firm-level characteristics		Linkages outside the national innovation system and firm-level characteristics	
		A: Key elements of NIS (for example, availability of skills, scientific research, industry collaboration)	B: Intra-firm (for example, organizational culture, learning processes)	C: Global channels and external collaboration (for example, capabilities in host location)	D: Intra-firm (for example, foreign subsidiary capabilities, network resources)
Strategy	Risk-taking	Entrepreneurial start-up in a chaebol-dominated environment	Penetration of the new market segments	High customer demands accepted	Establishment of new global offices to create proximity to central knowledge partners and customers
	Competitive aggressiveness	Using empowerment and delegation of responsibility as a competitive response to the challenge of human capital	Focus on excellence rather than competitiveness	Focus on differentiation from other providers of AFM	Supplier relationships with competitors
Organization	Knowledge sharing	Collaboration with competitors, universities	Dedication to team work and	Collaboration with	Training of trusted retailers

		communication across groups	international universities	
Customer insight	Collaboration with customers, testing and development	Integration of customer insights into both NPD and services	Collaboration with customers, testing and development	
Network	Universities, suppliers, partners and customers	Employee participation in knowledge networks	Universities, suppliers, partners and customers	Managing R&D units and outposts abroad, for example, Germany
Competence development	Driver of change in attitude towards entrepreneurship in Korea. for example, awarded the P. F. Drucker Award for Innovation	Employee profile, low employee turnover, gradual delegation of power and responsibilities	Most management level employees have PhDs from the US	Enhancing competences of suppliers
Team structure	R&D collaboration with network partners	Cross-functional teams	Collaboration style highly inspired by the Silicon Valley working style	R&D collaboration with foreign network partners
Role of founder	Bringing Silicon Valley culture to Korea	Pioneer in the AFM	Strong ties to former research partners in Silicon Valley	Strong ties to former research partners in Silicon Valley

Table 7.3 (continued)

		Innovation spaces			
		National innovation system and firm-level characteristics		Linkages outside the national innovation system and firm-level characteristics	
		A: Key elements of NIS (for example, availability of skills, scientific research, industry collaboration)	B: Intra-firm (for example, organizational culture, learning processes)	C: Global channels and external collaboration (for example, capabilities in host location)	D: Intra-firm (for example, foreign subsidiary capabilities, network resources)
Technology	Product	Fast follower in the research segment, first mover in the industry segment	NPD process based on close collaboration between management, R&D and Market groups	Inspiration of industrial application areas found in global markets	Collaboration on industrial application areas with global partners
	Process	Fast follower in the research segment, first mover in the industry segment	Focus on service innovation and development of new industrial benefits of using the AFM technology		

global innovation space thus provides rare insight into how interrelated practices for manoeuvring innovation spaces are developed and managed, as well as the distinct advantages and challenges related to this.

The results of the analysis demonstrate a refined approach to both knowledge absorption and diffusion, where the global arena plays as large a role as the national arena, and possibly even a larger role. Reliance on and integration of research knowledge, market knowledge and applied knowledge is characteristic of the approach to the global innovation space. In this sense, knowledge from the US, the European Union and South Korea are continuously in focus for the development of this company.

A similar approach is apparent when focusing on the strategic approach of the company, as we have identified how proactiveness and risk-taking are in this case built-in aspects of gaining first-movers' advantages, attracting funding and moving into new market segments. This is evident by the fact that Park Systems is the only AFM provider breaking into the industrial application segments.

The internal organization has also been categorized as open and proactive. Here we find an unusually – apparently in the South Korean context – high degree of delegation of power to middle management, knowledge sharing across different layers and responsibilities assumed by employees and teams of employees. These aspects of human resources and desired employee characteristic were strongly emphasized as a decisive factor for individuals to choose employment with Park Systems in a national system where employment in SMEs is otherwise associated with low skills, poor career prospects, low social recognition and low desirability. The culture of the organization is highly affected by the experiences of the founder from Silicon Valley, and, in this sense, we also see the influence of the global arena.

The technological development taking place in the organization is cutting-edge, not only in a South Korean context, but also globally. The translation of this technology into value offerings in the marketplace therefore also requires innovative thinking in relation to product categories, usability areas and integration with other technologies. This confirms theoretical arguments made about knowledge-intensive entrepreneurs, about the need to continually redefine the technologies and markets (McKelvey and Lassen, 2013). In Park Systems, we therefore find a continuous focus on not only the development of actual product lines, but also on the process offerings and systems associated with the product. This particularly relates to key customers, who are typically located in Europe and the US. In-depth understanding of global customers' needs and requirements is a key factor in being able to provide technological product and process innovation.

Let us conclude with some questions for reflection:

- Traditional international business theories tend to focus on sales as the main driver of internationalization. This company has a significant proportion of their sales internationally. However, the global presence of the company also has many other essential benefits that affect their innovativeness. What are these, and how does the identification of such benefits enrich our understanding of global innovation spaces?
- As a small South Korean company with core research and operations located in South Korea, what are the main barriers for innovation? How can these be overcome?
- How does this case advance our understanding of sources of innovation in South Korean companies?

The generalization of these findings is a subject for further research. This specific case does offer insights into how companies from emerging economies can tailor their strategies in order to develop sustainable innovation spaces.

NOTE

1. We would like to acknowledge and thank Aalborg University and the Center for Industrial Production.

REFERENCES

Ahuja, G. (2000). 'Collaboration networks, structural holes, and innovation: a longitudinal study'. *Administrative Science Quarterly*, 45(3), 425–55.

Baird, I.S. and Thomas, H. (1985). 'Toward a contingency model of strategic risk taking'. *Academy of Management Review*', 10(2), 230–43.

Bloomberg (2013). 'South Korea can't just order up creative economy'. 30 May.

Brown, S. and Eisenhardt, K. (1997). 'The art of continuous change: linking complexity theory and time-paced evolution in relentlessly shifting organizations'. *Administrative Science Quarterly*, 42, 1–34.

Chesbourgh, H. (2003). *Open Innovation: The New Imperative for Creating and Profiting from Technology*. Boston, MA: Harvard Business School Press.

Cockburn, I. and Henderson, R. (1998). 'Absorptive capacity, coauthoring behavior, and the organization of research in drug discovery'. *Journal of Industrial Economics*, XLVI(2), 157–82.

Cohen, W.M. and Levinthal, D.A. (1990). 'Absorptive capacity: a new perspective on learning and innovation'. *Administrative Science Quarterly*, 35, 128–52.

Covin, J.G. and Slevin, D.P. (1991). 'A conceptual model of entrepreneurship as firm behavior'. *Entrepreneurship: Theory and Practice*, 16(1), 7–24.
Dutta, S. and Evrard, P. (1999). 'Information technology and organization within European small enterprises'. *European Management Journal*, 17(3), 239–51.
Eisenhardt, K.M. (1989), 'Building theories from case study research'. *Academy of Management Review*, 14(4), 532–550.
Garcia, R. and Calantone, R. (2002). 'A critical look at technological innovation typology and innovativeness terminology: a literature review'. *The Journal of Product Innovation Management*, 19, 110–32.
Jensen, M.B, Johnson, B., Lorenz, E. and Lundvall, B.A. (2007). 'Forms of knowledge and modes of innovation'. *Research Policy*, 36(5), 680–93.
Jovanovich, B. and Rob, R. (1989) 'The growth and diffusion of knowledge'. *Review of Economic Studies*, 56, 569–82.
Leavitt, H.J. (1965). 'Applied organizational change in industry: structural, technological and humanistic approaches'. In J.G. March (ed.), *Handbook of Organizations*, Chicago: Rand McNally, pp. 1144–70.
Lee, K. (2013). *Schumpeterian Analysis of Economic Catch-Up: Knowledge, Path-Creation and the Middle Income Trap*. Cambridge: Cambridge University Press.
Lundvall, B.-A. (1999). 'National business systems and national systems of innovation'. *International Studies of Management & Organization*, 29(2), 60–77.
MacMillan, C. and Jones, P.E. (1984). 'Designing organizations to compete'. *Journal of Business Strategy*, 4, 11–26.
McKelvey, M. (1991). 'How do national systems of innovation differ? A critical analysis of Porter, Freeman, Lundvall and Nelson'. In G.M. Hodgson and E. Screpanti (eds), *Rethinking Economics: Markets, Technology and Economic Evolution*, Aldershot, UK Brookfield, VT, USA: Edward Elgar, pp. 117–37.
McKelvey, M. and Lassen, A.H. (2013). *Knowledge Intensive Entrepreneurship*. Northampton, MA, USA and Cheltenham, UK: Edward Elgar.
Meredith, J. (1998). 'Building operations management theory through case and field research'. *Journal of Operation Management*, 16(4), 441–54.
Porter, M.E. (1985). *Competitive Advantage*. New York: Free Press.
Schumpeter, J. (1934). *The Theory of Economic Development*. Cambridge, MA: Harvard University Press.
Slepniov, D., Lassen, A.H., Haakanson, S.J. and McKelvey, M. (2014). 'Understanding innovation spaces of emerging MNEs: evidence from a Chinese wind turbine manufacturer'. In M. McKelvey and S. Bagchi-Sen (eds) (2014). *Innovation Spaces in Asia*, Northampton, MA, USA and Cheltenham, UK: Edward Elgar, Chapter 6.
von Hippel, E. (1988). *The Sources of Innovation*. New York: Oxford University Press.
Voss, C. (2009). 'Case research in operations management'. In C. Karlsson (ed.), *Researching Operations Management*, New York: Routledge, pp. 162–96.
Yang, J. (2009). 'Small and medium enterprises (SME) adjustments to information technology (IT) in trade facilitation: the South Korean experience'. *Asia-Pacific Research*, Working Series No. 61.
Yin, R.K. (2009). *Case Study Research – Design and Methods*. Thousand Oaks, CA: Sage.
Zahra, S.A. and George, G. (2002). 'Absorptive capacity: a review, reconceptualization, and extension', *Academy of Management Review*, 27(2), 185–203.

8. The emergence of knowledge-intensive entrepreneurship in China: four start-up companies in nanotechnology in Suzhou

Jun Jin, Zhengyi Zhang and Maureen McKelvey

8.1 INTRODUCTION

This chapter addresses the development of innovation spaces within China, through an analysis of the development of knowledge-intensive entrepreneurship in nanotechnology, including global linkages to access knowledge and resources. Recent research results suggest that China offers many opportunities, driven by aspects such as expanding markets, high number of graduates in science and engineering, and consumers with new demands. These factors suggest that we need to further understand and conceptualize the process of entrepreneurship within Asia.[1] We need to get beyond the view that China and other emerging economies are simply destinations with low-cost labour and materials, and instead explore them as innovation spaces where national innovation systems are emerging as enablers of science and technology development.

A key aspect of focusing upon innovation spaces in Asia implies that we must further our understanding of how and why emerging markets may present radically new opportunities for innovation, opening up for new global actors, and at the same time posing new challenges to established multinational corporations (MNCs). This chapter explores these issues. Indeed, China is seen in recent research as an important arena for the generation of business opportunities and thereby moves away from low-cost manufacturing towards innovation-led growth. Shapira et al. (2011) argue that nano is an area that is at the intersection of national innovation systems and global linkages, which supports our idea that research must think dynamically about the emergence of specific innovation spaces at particular times and places, which have local and global dimensions.

By focusing upon innovation, we can think of competition through innovation, in terms of types of knowledge, and competition for potential markets. In emerging economies, innovation could focus on the needs of customers, as epitomized by bottom-of-the-pyramid innovations. A set of related concepts like 'chinovation', 'base of the pyramid', 'jugaad' and 'frugal innovation' stress that emerging-market countries may well change the nature of innovation by focusing on new low-cost services and goods that are sustainable and prudent (see Chapter 13, this book). Related concepts like social entrepreneurship stress the need to solve larger societal challenges. However, innovation is traditionally concerned with scientifically and technologically advanced knowledge. This requires entrepreneurial firms to develop extensive in-house capabilities as well as extensive networks. This chapter addresses the latter type of innovation on-going in China – the focus is nanotechnology developed in start-up firms in Suzhou.

Using knowledge to innovate has become a key goal of public policy in China (see Chapter 2, this book). Chinese government encourages university–industry cooperation, and commercialization of patents from universities or institutes. Moreover, nanotech is one of priority innovation areas in the Chinese 12th Five-Year Development Plan and the Chinese Science and Technology (S&T) plan. China aims to become the leader. The research on nano start-ups and commercialization in China is limited, and the empirical discussion in this chapter provides the reader with an understanding of the reality of technology development.

Nanotechnology and nanoscience are emerging technologies, and these technologies are heavily science-based. They tend to be interdisciplinary and involve a distributed form of knowledge amongst many different actors such as universities, existing large firms, and entrepreneurial firms. Knowledge-intensive entrepreneurship in this emerging innovation space is relevant, because opportunities can be created between the entrepreneur (person), the entrepreneurial venture (organization) and the national innovation system. The theoretical framing derives from Malerba et al. (forthcoming) and McKelvey and Lassen (2013a, 2013b). This chapter will focus especially upon the phases of start-up and managing this type of start-up, which is heavily dependent upon advanced knowledge.

This chapter first provides some insights into the particular pattern of knowledge development within the emerging nanotechnology/nanoscience field (Section 8.2). Section 8.3 describes four case studies of nanotech firms in one Chinese city. Section 8.4 compares and contrasts the cases, while

Section 8.5 analyses main factors in the establishment and development of four cases. Section 8.6 provides conclusions and implications.

8.2 CONCEPTUAL FRAMEWORK OF ENTREPRENEURSHIP IN NANO

8.2.1 Emergence of Nanotechnology and Nanoscience

Studying entrepreneurship in China through firms in nanotechnology is interesting, due to the fast pace of development of knowledge in a set of related fields.

Nanotechnology works in nanometer scale, from 1–100 nanometers. This is seen as revolutionary because nano involves research at the atomic and molecular levels to provide the exploitation of novel phenomena and properties of matter (Salamanca-Buentello, 2005; Wonglimpiyarat, 2005; Fiedler and Welpe, 2010). Nanotechnology was first introduced in 1959, technically by Nobel Prize winner Richard Feynman, and commercially by Drexler in 1986, and some researchers identify it as the resource of the next Schumpeterian wave (Drexler, 2004; Linton and Walsh, 2008; Allarakhia and Walsh, 2012; Mangematin and Walsh, 2012). Nano is a multidisciplinary field in that there is multidiscipline networked research (Islam and Miyazaki, 2010). There are nano devices on the market, start-ups being created and large firms making investment in production capacities (Mangematin and Walsh, 2012). According to the estimation by the National Science Foundation (US), this technology will have a market of more than US$1 trillion after 2010 (BCC Research, 2003; PCAST, 2005; Cheng, 2013).

Existing literature about the field of nanotechnology specifically stresses certain characteristics. These emerging technologies are usually interdisciplinary and involve a distributed form of knowledge among many different actors such as universities, existing large firms and entrepreneurial firms (Meyer, 2000a, 2000b; Rothaermel and Thursby, 2007; Mangematin and Walsh, 2012; Bourelos et al., 2013).

Huang et al. (2011) provide a comprehensive review of the social science debates about nanotechnology and nanoscience. Four debates about nano are used in their literature review and reflections for methodology: (1) Is nanotechnology an interdisciplinary field? (2) Are nanoscience and nanotechnology closely interlinked? (3) Is nanotechnology development path dependent? and (4) Who is winning the global nano race?

The relationship between science and technology has been conceptualized as both far away and close. David and Dasguptha (1994) argue that the differing institutional arrangements of firms and universities will focus different researchers on different outcomes. Nightingale (1998) argues that science and technology have different cognitive bases, while Murray (2002) argues that while scientific and technological networks in tissue engineering may co-evolve, the networks are still separate.

However, the empirical work on nano suggests that there are close relationships between basic research and applied research. Basic research is usually analysed in terms of scientific excellence related to publications while applied research is analysed in terms of industrial invention related to patents. A series of papers about nano confirm these close links (Meyer, 2000a, 2000b; Rothaermel and Thursby, 2007; Mangematin and Walsh, 2012; Bourelos et al., 2013). Research also indicates close ties between the firms and universities seen as networks and organizations (Nikulainen and Palmberg, 2010; Thursby and Thursby, 2011; Genet, et al. 2011). Specifically in nano, the same individuals who do scientific research – and are successful at it – tend to also be successful in doing patents. Most research is on the attributes and performance of individuals active in research in nano, and this microdata can be aggregated to the level of the organization to provide an integrated perspective of nano development pathways.

Based upon this brief overview of the attributes of knowledge within nanotechnology and nanoscience, we would expect that firms in nano are focused on collaboration among scientists in universities and firms. Aspects of collaboration may include mobility between organizations, co-publication and so on. Moreover, given that the knowledge base of the firm is dependent upon advanced science and technology, the nano firms should access capital and financing in order to develop internal capabilities through human resource development as well as the creation of external networks of scholars.

8.2.2 Nano in China

The existing research on nanotechnology in China focuses on policies and basic research at institutes or universities based on analyses of patent data and publication data (Huang et al., 2004; Li et al., 2008). Nano is heavily dependent upon the national innovation system as well as global sectoral innovation systems for various industries that apply the advances of nano to develop new products.

China now is considered to be the latecomer nation, seeking to catch up with globalization and with the advanced countries, industry by industry, technology by technology, firm by firm. China wants to exploit its late arrival to tap into advanced technologies, rather than having to replicate the entire previous technological trajectory (Mathews, 2007). By examining international patent and firm foundation trends in the nanotechnology industry, Niosi and Reid (2007) suggest that China as well as India and Brazil have the possibility to catch up in nano. An emerging industry for latecomers could be very important, because the changing industrial structure can create a mechanism in place for producing viable, competitive firms for the future (Mathews, 2007). As the economy grows, new funds may be invested, intellectual property (IP) and regulatory frameworks will be upgraded, and human capital will flow into the new area (*The Economist*, 2002; Niosi and Reid, 2006).

As in many countries, programmes and initiatives on nanotechnology in China were started in the 1980s (Zhao et al., 2008). Nanotech is one of priority innovation areas in the Chinese 12th Five-Year Development Plan and Chinese S&T plan. China is planning to become the leader in nanotechnology. In 2008, China's nanotechnology research and development (R&D) investment was estimated to be about US$250 million (Liu et al., 2009). The current analysis is that China has a good performance in terms of patents and research outcomes, and represents a potentially large market. However, the Chinese nanotechnology industry is still relatively weak although the Chinese government has several incentive policies to encourage nanotechnology commercialization (Guan and Ma, 2007).

Interestingly, China now has many research centres that focus on nanotechnology. These specialized research centres have been established in the Chinese Academy Sciences (CAS), Peking University, Tsinghua University, as well as two national nanotech centres, namely the National Center for Nanoscience and Technology (NCNST funded with 250 million RMB) located in Beijing and the National Center for Nano Engineering and Technology (NCNET; funded with 200 million RMB) located in Shanghai. More than 120 research organizations have engaged in the R&D of nanoscience and nanotechnology. Most of the 300 companies and 7000 scientists engaged in nanotechnology development in China are concentrating solely on nanomaterial, considered by some to be an area which may quickly become a commodity (Niosi and Reid, 2006).

Since 2002, China has produced the second largest number of nano publications (Kostoff, 2006; Guan and Ma, 2007). Moreover, a 2009 comparison of China/US research publication outputs showed that China is about to overtake the US in nano research output (Kostoff, 2012). China is thus predicted to do very well in nanotechnology, especially in scientific research, with this fast speed of publications attributed to support from government and policies traced back to the early 1980s (Liu and Zhang, 2007).

The current view towards commercialization and industrialization reveals other challenges. The Chinese nanotech industry is relatively weak in commercializing basic research and in the production of nano devices. Nanotech products (nanomaterials) are primarily used in the microelectronics, textile, cosmetic, pharmaceutical, construction material and food industries. In recent years, China has focused on the application of nanotechnology in advanced materials, information technology (IT) and communications, energy, environmental protection, medical and health care, and agriculture (Aydogan-Duda, 2012). Shapira and Wang (2009) argue that for China the pathways from laboratory research to successful commercialization remain problematic. China's performance in nanotechnology patenting and product development is weak compared to its research strength, suggesting a significant gap between the research base and industrial development.

As yet little is known about the innovation practice of firms on nanotechnology in China, and therefore this chapter provides interesting and novel case studies of nano firms within the Chinese innovation system. The case studies include four firms active in a particular institutional setting. The government of China has established a BioBay, which gives incentive policies and eco-environment to bio-industry and nano-industry in the Suzhou Industrial Park. This Suzhou Industrial Park provides much support for nanotechnology's industrialization and commercialization, and provides novel insights into the development of one of China's emerging industries in a highly science-based field with a predicted rapidly expanding market.

8.2.3 Knowledge-Intensive Entrepreneurship

Entrepreneurship has been analysed in many dimensions. Research on entrepreneurship in general provides a variety of inputs and models – from business planning to design processes. However, in the context of nano – as developed above – specific types of scientific and technological knowledge are crucial to the entrepreneurial venture, and that knowledge is evolving rapidly (Malerba et al., 2015).

Hence, this chapter draws its analytical variables from a particularly relevant approach, McKelvey and Lassen (2013a), which develops a process model of managing knowledge-intensive entrepreneurship (KIE) at the level of the entrepreneurial firm. While the variables and processes are known from general entrepreneurship, and indeed the model is based upon an extensive literature review, the KIE model only focuses upon the most important variables for each phase, when the KIE venture is dependent upon advanced knowledge as a key competitive asset. This line

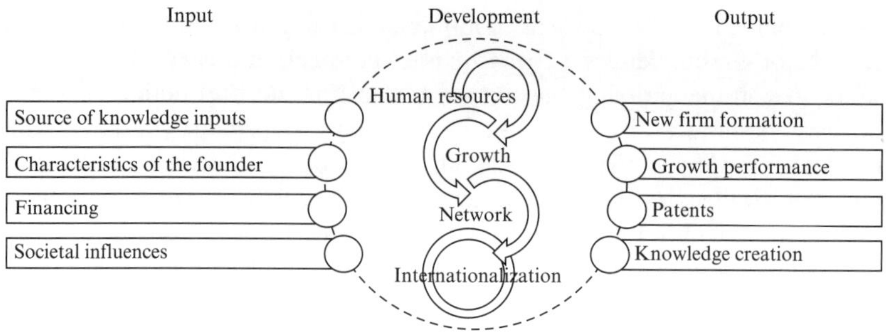

Source: McKelvey and Lassen (2013a)

Figure 8.1 Model of managing a knowledge-intensive entrepreneurial venture

of research provides a synthetic model, which identifies the key variables and phenomena that drive KIE ventures, as seen in Figure 8.1.

A key aspect is how these variables are interacting over time in a specific process, and especially in relation to knowledge and developing the company: how does the individual firm manage them during start-up and development phases? McKelvey and Lassen (2013b) also provide extensive case studies. Existing case studies suggest that at the firm level, the processes identified below then to have extensive feedback loops, dependent upon the type of technology (nano) and the institutional context (China). The key advantage of using case studies is that the interactions between the variables over time can be analysed, in relation to the development of the KIE venture.

This chapter focuses primarily upon the variables that are an input into the development of a KIE venture: source of knowledge inputs, characteristics of founder, financing and societal influences. The reason for this focus is the context of nano in China. Existing research suggests that founders impact firms, and that firms need highly specialized knowledge and human resources as well as likely high levels of capital to support the R&D and the extended network structure needed to develop basic research results into industrial applications in nano. Moreover, societal influences are necessary for the discussion, given that existing research on both nano and China in specific has emphasized the role of the state, and the extensive involvement of public actors in financing both science and firms in this area. Given that we are describing a process of development of a KIE venture from this perspective, we also describe the human resources as well as local and global networks that are directly related to the development of nano. This model thus provides us a structure to describe the empirical findings of the case studies at the firm level, and focus upon variables in relation to the development of science and technology.

8.3 METHODOLOGY

Case study is widely recognized as a method for explorative research (Yin, 2009). The methodology can help answer the research questions of our study, based on the principles of engagement with practice (Voss, 2009; Yin, 2009). Through case studies, we can enlighten and explain real-life phenomena that are too complex for tightly structured designs or pre-specified data sets (Voss, 2009; Yin, 2009). In addition, case studies are suitable for unravelling concepts (Yin, 2009). This chapter focuses upon KIE in China, specifically within nanotechnology, and uses the variables defined above to describe the development of these entrepreneurial ventures.

The method of case study is divided into two categories: single case and multi-case. The former is suitable when the case we observe can be categorized as the extreme case, or else the only one which is typical, and which can be studied in-depth across a span of time. However, a multi-case study can be considered more robust (Herriott and Firestone, 1983). Multiple case studies follow the replication logic, which emphasizes similar or comparable results, rather than sample logic of the averages in a population. Therefore, this chapter employs a multi-case study approach as a method to explain the commercialization and industrialization of nanotechnology in China.

Most of the case studies are founders, chief executive officers (CEOs) and chief financial officers (CFOs). In general, each was interviewed for one to two hours. The interviews were conducted in Chinese, then transcribed. This chapter does not code the interviews, as our purpose is to provide insights and illustrations into their emergence and their development, and try to describe the reasons.

The case description presented below is based on in-depth interviews. In the summer of 2013, one Chinese-speaking author with colleagues visited companies' offices and research centres, and did interviews with seven senior managers of the four companies. More specifically, the founders of these four enterprises as well as either the CEO or CFO constituted the interviewees. Moreover, we interviewed the administrative officer of Suzhou Industrial Park (SIP) to confirm the incentives to locate there for the nanotechnology enterprises. Public information from websites and companies' introduction documents were also used as background information and to triangulate the data.

8.4 CASE STUDIES OF FOUR NANO COMPANIES IN CHINA

8.4.1 Beaver Nano-Technologies Co., Ltd

In terms of founding date, founder and CEO, Dr Arthur Ren is the CEO of Beaver Nano-Technologies. In 2010, Ren established Beaver Biosciences Inc. when he returned to Guangzhou from Massachusetts Institute of Technology (MIT) because he believed the technological advance of Prof. Zhang's[2] patents would be succeeded in a large growing market of nano-biotechnologies in China. Ren served as chief technology officer (CTO) of Beaver Biosciences Inc., and his PhD supervisor at MIT (US), Zhang, served as the president of the company. In 2011, Ren moved to Suzhou and founded Beaver Nano-Technologies at BioBay in SIP, due to the science park dedicated to biotechnologies and nanotechnologies. Beaver Nano-Technologies has operated since 2012.

In terms of capital for founding the venture, the capital invested was around 16 million RMB, and attracting capital was also linked to recognition awards. Ren was awarded the title of 'science and technology leading talent of Suzhou Industrial Park' in 2012; and he was selected as one of the Innovation Leader Talents of Suzhou and Jiangsu Province. Because of this, Ren received millions of renminbi from the Suzhou and Jiangsu governments as research funds to support his R&D. In addition to these research funds, there was also his individual investment of RMB1 million as well as RMB5 million venture capital monies when Beaver Nano-Technologies was founded. In three years, Ren has, as an individual, also invested into the company several million RMB.

In terms of main technology, product and industry, Beaver Nano-Technologies is an enterprise focusing on research and development, and leading to the commercialization in the field of biomedical engineering, specifically nanomaterials science. Currently, six product lines have been introduced into research and in vitro diagnostics commercial markets. Beaver Nano-Technologies provides four groups of product: cell culture surface, BeaverBeads magnetic, ELISA plate and magnetic separator. These products are to be used in the diagnosis and health-care industries in China, specifically in the fields of protein purification, antibody purification, immunoprecipitation, co-immunoprecipitation, ELISA test, cell sorting and nucleic acid isolation, and so on. The company owns around ten Chinese intellectual property rights (IPRs) related to peptide bio-nanotechnology.

Although the potential market of nano biomedical materials in China is growing and large, there are few Chinese companies working to develop such kinds of products at this time. The competitive advantage of Beaver is the price of its products, which is much lower than those provided by MNCs.

In terms of human capital, expansion has been swift, and depends upon skilled labour. Beaver has grown from ten persons at its foundation to 50 in 2013. With the exception of one person who is working for information systems, all the other employees have a background within biotechnology and materials science. Many of the employees graduated from Suzhou University, which is located in SIP. Moreover, Beaver has licensed patents of Zhang and other scientists, who are stakeholders of the company because of their technical investment. At SIP, some companies are partners of Beaver, in order to reduce production costs. Beaver Nano-Technologies also cooperates with some universities and institutes on R&D. The development of the Beaver Nano-Technologies is summarized according to our main variables in Table 8.1.

Table 8.1 Development of Beaver Nano-Technologies Co, Ltd

Founding date	Founded Beaver Nano-sciences in 2010 in Guangzhou; moved to Suzhou in 2011; operating since 2012
International background of CEO and/or of founder	The founder, Mr Ren, got his PhD degree from MIT, US
Foundation capital ($) and intellectual property rights	The founder, Ren, invested millions into the company in three years. VC 5 million, and an investment subsidiary of Suzhou Industrial Park 10 million The company has rights to commercialize patents of Prof Zhang and other scientists The company applied for ten Chinese patents
Main technology and main product or industry	Nanomaterials used in biotech, for diagnosis and health care
Human resources	Employees increased from 10 persons to 50 persons, all biotech and materials science background, one IT Also time and technology investment from the foreign scientists (who are working as part-time, no salary, provide technical advisory, own shares of the company)

8.4.2 Polynova Materials Suzhou Ltd

In terms of founding date, founder and CEO, Polynova Materials Suzhou Ltd was founded in 2008 by Mr Zhejun Huang and his partners. Attracted by the incentive policies and local nano-industry environment of SIP, Polynova Technology was set up in 2010 at SIP as a subsidiary of Polynova Materials, and in order to focus on research and new product development of nanosized antistatic materials.

The founders' group, including Huang and his partners, have more than ten years' work experience with materials at leading MNCs, such as BSF and General Electric. In 2008, they left their respective MNCs to set up a company to research, produce and sell nano-sized conductive polymer materials based on their patents and a dream of entrepreneurship.

In terms of foundation capital, Polynova Technology's raised RMB2 million as the registered capital. With the development of the company, the founders received some venture capital from a venture capital investment subsidiary of SIP in 2011, and also raised their individual investments. The registered capital increased to RMB15 million in 2013.

Because Polynova Technology is the first Chinese company to own patents on antistatic materials in China and finalizes the first industrial standard on antistatic materials in China, Polynova is recognized as a leading high-tech company in nanotechnology. Thus the company receives research grants from Jiangsu government and investment from SIP.

In terms of technology, products and industry, Polynova concentrates its R&D and production on nano-sized antistatic materials. Moreover, the company provides service and antistatic solutions for electronic products, liquid crystal display, semiconductor products, and so on. In the field of antistatic materials, the competition that Polynova faces is only from MNCs, and not from other Chinese enterprises. As with other nanotechnological enterprises, their main competitive advantage is from their price and technology. The high profit of this emerging industry and the low competition enables Polynova to keep growing and making a high profit. For instance, even in 2008, Polynova received enquiries from Nokia and Apple, and passed qualification to become their suppliers. The sale revenue reached 10 million RMB in 2009, and Samsung and Apple are their key customers right now.

In terms of human resources, in 2013 there were around 30 persons employed by Polynova. The majority hold at least bachelor degrees. The case study of Polynova is summarized according to our main variables in Table 8.2.

Table 8.2 Development of Polynova Materials Suzhou Ltd

Founding date	In 2010, the parent company founded a subsidiary at Suzhou Industrial Park The parent company was founded in 2008
International background of CEO and/or of founder	Founders have working experience in foreign nanotechnology company, education background on nanotechnology relative sciences
Foundation capital (¥) and patents	2 million registered capitals in 2008 from four founders. The registered capitals increase to 15 million. A small part of VC was attracted in 2011 Patents on antistatic materials
Main technology and main product or industry	Nano-size conductive polymer materials for semiconductor industry
Human resources	From four employees in 2008 to 30 employees in 2013 Only a few employees (for example, the driver) do not hold a bachelor degree

8.4.3 Hanano Materials Co., Ltd, Suzhou

In terms of founding date, founder and CEO, Hanano Materials Co., Ltd, Suzhou was founded by Dr Xinjiang Chen, together with Mr Yang and Ms Zhou, in November 2011 at SIP after one year of preparation. The three internationally experienced founders played different roles. Yang invested in the enterprise as a venture capitalist, and became the CEO; Chen invested with his patents and became the general manager; and Zhou (who was overseas studying in Japan) was in charge of all administration. Chen had studied in Japan for his PhD for six years. He then joined the Suzhou Institute of Nano-Tech and Nano-Bionics of Chinese Academy of Science (SINANO) as a researcher when he returned. He kept the research position at SINANO when he helped to set up Hanano, and the industrialization of technologies is one of the criteria by which researchers in SINANO are evaluated.

In terms of foundation capital and patents, when the company was set up, they obtained an RMB7 million investment from the investment subsidiary of SIP, as well as RMB6 million venture capital from Yang and others. Most of the investments were used to pay for R&D and commercialization of patents.

In terms of technology, products and industry, Hanano is a high-tech company specialized in the R&D, manufacturing and application of carbon nanomaterials, especially the single-walled carbon nanotube (SWNT) transparent conductive films (TCFs). With more than seven years' research, Hanano has achieved a great breakthrough on the development of SWNT computer to plates (CTPs). Hanano has also developed carbon nanotube transparent conductive films (for example, touchscreens) based on their patents, and these films are ultra-thin, flexible, environmentally friendly and give high performance.

Hanano designed their production process and equipment in 2012, and has built several production lines to produce SWNT TCFs since 2013. The self-developed SWNT TCFs can be applied to CTPs and their properties have reached the level of internationally leading technology. Moreover, Hanano completely owns the independent IPs on the entire producing process, including but not limited to the mass synthesis of SWNTs, the fabrication of TCFs, and so on. Furthermore, Hanano can give the solution on CTP with high performance and low cost to module producers and mobile phone manufacturers.

The competitive advantages of Hanano can be described as threefold. First, Hanano uses recycled materials to produce environmentally friendly transparent conductive films. Being more environmentally friendly makes their products more easily accepted by the market, as more and more customer companies pay attention to the environment. Second, since the company does not rely on natural resources, the company can reduce their production costs, leading to a low price. Finally, the company makes TCFs with the single-walled carbon nanotube material structure. In fact, there is only one other company in the world that currently produces TCFs using similar technology to Hanano. Chen estimates that Hanano was the first company in the world to commercialize the technology of carbon nanotube transparent conductive films. However, scale-up is difficult. Since the company is new, and there are high technological requirements of equipment in the production line, it is not easy for Hanano to build its first production line for mass scale manufacturing.

In terms of human resources, the company has grown from three persons in 2011 to 15 persons in 2013. Half of them have master's or doctorates. The development of our case-study firm Hanano is summarized according to our main variables in Table 8.3.

8.4.4 Nano-Micro Technology Company

In terms of founding date, founder and CEO, Nano-Micro Technology Company[3] was founded in 2006 in Shenzhen by Dr Biwang Jiang, based

Table 8.3 Development of Hanano

Founding date	2011 founded
International background of CEO and/or of founder	General manager (one of three founders) had six years in Japan, the HR manager (one of three founders) came back from Japan too
Foundation capital (¥)	13 million when founded, in which 7 million from Suzhou Industrial Park as the leading scientist, other from individual investment (CEO of Hanano, Mr Yang)
Main technology and main product or industry	Nanomaterials for touch screen
Human resources	Primarily researchers expanded from three persons to 15 persons, including two PhD, two overseas, more than 80 per cent have bachelor or above degrees, 50 per cent have master's

on his six US patents and European patents. Jiang has more than ten years of study and research experience in the US. He worked as a project leader at Rohm and Haas (Rohm and Haas was merged by Dow in 2009) from 2000 to 2005. He is a professor of a Chinese graduate school in Shenzhen and has received awards from the Recruitment Program of Global Experts of China.

In terms of foundation capital and patents, these are linked in this case study. Most of the RMB10 million investment in 2006 was from individual investors who believed in the potential market of Jiang's technology. In 2007, Jiang moved the company to SIP, attracted by the research and industrial environment. Since 2007, Nano-Micro Tech has been granted six invention patents.

In terms of technology, product and industry, Nano-Micro Tech is devoted to developing a world-leading technology for providing the highest quality and largest selection of uniformly sized nano- and micro-spheres (microparticles). Nano-Micro Tech started to introduce products to market in 2009. After that, due to their quality and cheaper price than MNCs, their sales reached more than RMB30 million in 2012, with around 20 per cent from overseas markets. This is a substantial market growth from their RMB5 million in 2009. In addition, Nano-Micro Tech became the second company other than GE in 2012 to provide a series of ion exchange chromatography media for highly efficient separation and

Table 8.4 Development of Nano-Micro Technology Company

Founding date	2007 founded
International background of CEO and/or of founder	The founder (US nationality) had 10 years' study and work experience in US.
Foundation capital (¥)	More than 10 million RMB investment from founders and several individual investors
Main technology and main product or industry	Highest quality and the largest selection of uniformly sized nano- and microspheres (microparticles) for pharmaceuticals, foods, biotechnology, and other industries
Human resources	In 2013, 104 employees. Of which 40 work for R&D and eight hold PhD degrees

purification of a wide range of biomolecules such as proteins, oligonucleotides and peptides.

Now Nano-Micro Tech is capable of supplying monodispersed particles of almost any size ranging from five nanometres to 1000 microns of different materials, structures and functionalities. At the same time, the company has developed from one floor of a building for R&D and administration to a 5000-square-metre research and manufacturing plant in SIP and a 1000-square-metre R&D lab in Shenzhen High Tech Industrial Park. In terms of their local environment, more than ten enterprises at the SIP are customers of Nano-Micro Tech.

In terms of human capital, nearly 40 per cent of the 104 employees working in 2013 at Nano-Micro Tech are in R&D. Nano-Micro Tech cooperates with universities for research as well as customers from the same park for co-development. The development of our case-study firm Nano-Micro Technology Company is summarized according to our main variables in Table 8.4.

8.5 SYNTHESIS AND ANALYSIS OF VENTURES AND INNOVATION SPACES

The overview of the four cases suggests that all four have highly specialized skills and access to relevant networks, attracted significant foundation capital in millions of RMB, and expanded rapidly in terms of employees. The oldest company is seven at the time of writing, with over 100 employees, while the two youngest companies had been founded three years previously, with 50 and 15 employees, respectively. The companies have impressive discussions on their websites about their technologies, products and industrial applications, suggesting an understanding of very specific knowledge and its application to products for their firms. Interestingly, the firms have identified that they can use their specialized knowledge within nanotechnology towards specific uses in quite different industries, from food to diagnostics to electronics.

In this section, the synthesis and analysis goes further into two questions. The first is whether, and how, the scientific and industrial background of the company founders seems to influence the subsequent emergence and development of these knowledge-intensive entrepreneurial ventures. The second is to go into more detail about societal influences in terms of the firms' local environment related to public policy, specifically the support as well as the S&T environment of SIP.

8.5.1 Scientific and Industrial Background of Case-Study Firms' Founders

To answer the first question posed above, a first aspect is the characteristics of the company founders, and their relation to creating nanotechnology companies. The four case studies all indicate that the founders (or at least one of the founder-group members) has a high-level, scientific background in nanotechnology, and this linkage to science appears necessary to start up nanotechnology enterprises. Founders within each of these four cases have worked in companies, or conducted research at universities, within nanoscience and technology industry for several years prior to starting their firm. In the interviews, they explain that this industrial and scientific experience has helped them understand the market trends and the trends of technological development, in ways relevant to identifying opportunities for this firm. Additionally, all of the companies hold patents on nanoscience. Moreover, these factors interact over time. The individual founders – and their entrepreneurial ventures – are recognized as industrial leaders by the science park, so they all then received capital to attract technical leading talents (up to millions RMB) from the park and/or from the central Chinese government to support their research.

These cases support previous research, which found that nanoscience and nanotechnology have very close links between industry and science. This means that if an entrepreneurial venture is to succeed in nano, they need specialized knowledge of technology. In these cases, it appears important, moreover, that at least one of the founders has these qualifications, rather than hiring in someone later on. Quite probably, if the founder does not have a strong scientific background from either industry or university, they will not know the potential market nor create networks nor be able to control the complex industrial processes needed in production. Moreover, in the Chinese context, the case studies indicated that personal reputation was important in able to attract the investment necessary for the venture. Hence, the variables of the characteristics and traits of founders had a profound influence on both financing, attracting additional human capital, and the later development of the companies.

8.5.2 Suzhou Industrial Park and the Local Environment

To answer the second question posed in Section 8.5, an interesting aspect is the role of societal influences, and specifically how and why public policy in the local environment impacts KIE ventures. The four case-study companies are all located in the SIP, specifically in the Biobay and factories in the Nanopolis Suzhou, which since 2013 has been dedicated to nanotechnology. This so-called Nanopolis has been developed to provide access to human resources and equipment for the particular needs of these types of companies (www.nanopolis.cn). The website (in 2014) states:

> The Nanopolis is a 1sq km 'mini-city' designed to drive nanotech commercialization by providing an ideal ecosystem for nanotech R&D, incubation of venture companies, and a smooth and rapid ramp-up to pilot production. Areas of focus include functional materials, energy and environment, display technologies, MEMS, power electronics and nano medicine.

As of 2014, there were 15 companies located in the Nanopolis, as well as two European government agencies interested in the development of nano in China. As seen in the four case studies, but also visible for the other firms, the companies represent a wide range of applications into different industrial areas, and many firms are growing rapidly.

Let us explain a bit more about the benefits of this specific science park to the companies, as found in the case studies. Suzhou Industrial Park has investment in foundation capital or operational capital from two aspects. One is venture capital from an investment company of SIP, and the other is the research funds given to founders as leading talents from SIP and/or other government agencies. The second kind of investment is like a kind of government-financed angel capital to support R&D. The SIP helps company founders to get these two kinds of funds. In addition, SIP provides support to start-ups to find venture capital outside of the park.

Another benefit of SIP is the S&T environment. The SIP builds an eco-business system to enable industrialization of advance technologies. When the companies grow in the future, the park will supply scarce resources in society, that are seen as facility services, such as more office space or factory land. None of the four case-study enterprises thinks it will be difficult to obtain enough office space in SIP during future development of their entrepreneurial ventures. All of them already have offices at Nanopolis, and their network relationships are enhanced by being located there. There are some biotech and materials companies in the park, who could be the suppliers and customers of the four case-study companies. As mentioned in the case studies, Nano-Micro Tech has more than ten customers from SIP, who are high-tech companies. In addition, the companies cooperate to find solutions to new technological problems. This helps to stimulate the co-evolution of companies in an industrial network in the park.

There are additional benefits to being located in SIP, which can positively affect returnee founders from abroad, as well as attract additional human resources. The SIP, in fact, has all the functions of a city, even including a customs office as well as education affiliates from kindergartens to universities. The park is famous for its facilities, park plan and service to Chinese start-ups because the administration committee of SIP adopt experiences from their Singaporean partners, and it was established as a joint park established by the Chinese and Singapore government. Attracted by the production facilities (including the possibility to find land for industrialization), three of the four case-study firms were attracted by the eco-system of SIP and moved to there for the industrialization of nanotechnologies.

Suzhou Industrial Park also offers low-rent apartments for employees in these companies, and salary subsidiaries to talents with master's degrees or doctorates who work there. These policies help start-ups in the park to attract talent. Some employees of the interviewed firms have been eligible for this assistance. The SIP also provides different kinds of information to companies. For instance, during the interviews we observed that one officer told the CEO about an opportunity they could pursue to reduce their costs, thereby an example of information exchange through close networks.

Finally, SIP attracts universities from around the world to set up their campuses or research centres there. The universities in the park provide an education and an intellectual environment to the families of employees, and also cooperate on R&D with companies in the park. The Suzhou Institute of Nano-Tech and Nano-Bionics of Chinese Academy of Science (SINANO), the leading research centre on nano-tech and nano-bionics in China, is located in SIP, and is the close neighbour of the four case-study companies. These four companies cooperate with SINANO on research projects. In addition, one founder works at SINANO. Nanotechnology companies cooperate with other universities in the park, such as Suzhou University. Two patents of Nano-Micro Tech were cooperative research achievements with Suzhou University.

Thus, the KIE entrepreneurial ventures in nano can find important partners for network relationships – such as institutes, suppliers and even customers – in the park, as well as access to scarce resources such as new office space, and access to special policy measures to stimulate business based upon advanced knowledge.

8.6 CONCLUSIONS AND IMPLICATIONS

This chapter explores how an emerging innovation space has developed in China, in nanotechnology, through entrepreneurial ventures, and through global and local linkages to access knowledge and resources. The discussion focuses upon key variables and processes enabling the start-up of four nano KIE entrepreneurial ventures located in Suzhou Industrial Park in China. Much emphasis in this chapter has been on the development of specialized knowledge assets of the company – specifically sources of knowledge input, characteristics of the company founders, and human resources – as well as the ability to attract additional resources through financing, societal influences and networks.

While the knowledge-intensive entrepreneurship model outlines phases and processes, an emerging innovation space suggests that there are dynamics within the innovation system that will affect the KIE venture. Which one factor is most important? How do they interact? Are there other important factors?

From a policy perspective, it is clear that characteristics of the founder as well as societal influences have a large impact upon these KIE ventures. A key issue is therefore how a local government may build conditions to support an eco-system for emerging industries. In this case, the scientific background of the founders is vital, as are societal influences in terms of the incentive policies and the eco-industrial environment. As an emerging knowledge-intensive industry, knowledge and networks about nanoscience and nanotechnology are vital, and hence, founders and their companies need to understand the technologies as well as potential applications, including trying to forecast the potential markets.

The case studies suggest that the development of research capabilities of Chinese scientists on nanotechnology is the first step and base of industrialization. In other words, although emerging technologies like nano may provide many opportunities to emerging countries to catch up, these cases suggest that for that to happen, the scientists in the emerging countries must develop relevant technological capabilities and invest in basic research. In all four case studies, the founders and many co-founders have taken advanced degrees and/or worked at skilled jobs abroad, specifically in US and Japan. Clearly then, the global diffusion of knowledge through labour mobility, in these cases, helps the development of a new innovation space in nanotechnology in China as well as providing global network relationships.

Another aspect is the specific role of societal influences, and specifically the support through financing and policies of the science and technology environment of Suzhou Industrial Park. The SIP is a joint park between the Chinese and Singaporean governments, and the Chinese administrative officers learn management experience from their Singaporean partners. This is a different type of flow of knowledge, related to public policy, but of clear importance here to make the science park more effective. Suzhou Industrial Park has more autonomy than others in China, such, for example, that SIP has its own customs office, tax bureau, and so on, which facilitate business as well as providing special support for skilled labour.

Thus, a question for future discussion is whether it is possible to reproduce SIP in other industrial parks or science parks in China. In the four case studies of this chapter, the concentration of specialized nano firms and these societal influences seem to have developed a local innovation system with strong feedback loops – and which creates a new innovation space in China highly connected to other Asian countries (Japan, Singapore) and to the US. Moreover, the case studies suggest that both the incentive policies and the local eco-industrial environment are important in attracting founders with a relevant scientific background, as well as additional human capital. Perhaps these factors do not play the same effective roles in providing services and access to scarce resources and networks in other places in China as those in SIP, because of the autonomy of different science parks and regions.

NOTES

1. Professor Maureen McKelvey would like to acknowledge and thank the Sten A. Olsson Foundation for Research and Culture and Volvo AB for supporting this research. This chapter has been written in context of the research programme Radical Innovations for the Enhancement of the Swedish Economy at the Institute of Innovation and Entrepreneurship, with McKelvey as principal investigator. Jin and Zhang wish to acknowledge that their research is supported by NSFC projects (CNSF 71172111 and CNSF71232013). McKelvey would also like to thank Berna Beyhan and Evangelos

Bourelos, who are co-authors on other papers about the emergence of nanotechnology in Sweden, as part of the Sten A. Olsson research programme and as Broman Post-Doctoral Scholars at the Institute of Innovation and Entrepreneurship.

2. Professor Zhang is the laboratory head at the Center for Bits and Atoms, Massachusetts Institute of Technology (MIT), Cambridge, MA, US. He was the supervisor of Dr Ren when Ren pursued his PhD degree at MIT.
3. Nano-Micro Technology Company was named Nano-Micro Biotechnology Company before 27 January, 2014. In the following research, the company is called Nano-Micro Tech.

REFERENCES

Allarakhia M. and Walsh S. (2012). 'Analyzing and organizing nanotechnology development: application of the institutional analysis development framework to nanotechnology consortia'. *Technovation* **32**: 216–26.

Aydogan-Duda, N. (ed.) (2005). *Making It to the Forefront: Nanotechnology – A Developing 77 Country Perspective, Innovation, Technology, and Knowledge Management Vol. 14*. New York: Springer.

BCC Research (2003). 'Advanced ceramic powders and nanoceramic powders'. Available at: http://www.bccresearch.com/market-research/nanotechnology/advanced-ceramics-nanoceramic-powders-nan015g.html (accessed 20 October 2014).

Bourelos, E., Beyhan, B. and McKelvey, M. (2013). 'Academic inventors and knowledge technology transfer in nanoscience in Sweden'. In Bourelos, E. (2013), *Creation and transfer of academic knowledge in Sweden*, PhD thesis, University of Gothenburg, Sweden. Available at: http://hdl.handle.net/2077/34457 (accessed 20 October 2014).

Cheng, A.C. (2013). 'A fuzzy multiple criteria comparison of technology valuation methods for the new materials development'. *Technological and Economic Development of Economy* **19**(3): 397–408.

Daspugta P. and P. David (1994). 'Toward a new economics of science'. *Research Policy* **23**: 487–521.

Drexler, K.E. (1986). *Engines of Creation: The Coming Era of Nanotechnology*. New York: Anchor Doubleday.

Drexler, K.E. (2004). 'Nanotechnology: from Feynman to funding'. *Bulletin of Science, Technology Society* **24**: 21–7.

The Economist (2002). 'Biotech's yin & yang'. 12 December.

Fiedler, M. and Welpe, I.M. (2010). 'Antecedents of cooperative commercialization strategies of nanotechnology firms'. *Research Policy* **39**: 400–10.

Genet C., Errabi K., Gauthier C. (2011). 'Which model of technology transfer for nanotechnology? A comparison with biotech and microelectronics'. *Technovation* **32**(3-4): 205–15.

Guan, J. and Ma, N. (2007). 'China's emerging presence in nanoscience and nanotechnology: a comparative bibliometric study of several nanoscience giants'. *Research Policy* **36**(6): 880–6.

Herriott, R.E. and Firestone, W.A. (1983). 'Multisite qualitative policy research: optimizing description and generalizability'. *Educational Researcher* **12**(2): 14–19.

Huang, Z., Chen, H., Chen, Z. and Roco M.C. (2004). 'International nanotechnology development in 2003: country, institution, and technology field analysis based on USPTO patent database'. *Journal of Nanoparticle Research* **6**(4): 325–54.

Huang C., Notten A., Rasters N. (2011). 'Nanoscience and technology publications and patents: a review of social science studies and search strategies'. *The Journal of Technology Transfer* **36**(2): 145–72.

Islam, N. and Miyazaki, K. (2010). 'An empirical analysis of nanotechnology research domains'. *Technovation* **30**: 229–37.

Kostoff, R.N. (2012). 'China/USA nanotechnology research output comparison – 2011 update'. *Technological Forecasting & Social Change* **79**: 986–90.

Kostoff, R.N., Stump, J.A., Johnson, D., Murday, J.S., Lau, C.G.Y. and Tolles, W.M. (2006). 'The structure and infrastructure of the global nanotechnology literature'. *Journal of Nanoparticle Research* **8**: 301–21.

Li, X., Chen, H., Dang, Y., Lin, Y., Larson, C.A. and Roco M.C. (2008). 'A longitudinal analysis of nanotechnology literature: 1976–2004'. *Journal of Nanopartical Research* **10**: 3–22.

Linton, J.D. and Walsh, S.T. (2008). 'A theory of innovation for process-based innovations such as nanotechnology'. *Technological Forecasting and Social Change* 75: 583–94.

Liu, L. and Zhang, J.J. (2007). 'Characterising nanotechnology research in China'. *Science Technology Society* **12**: 201–16.

Liu, X., Zhang, P.Z. and Li, X. (2009). 'Trends for nanotechnology development in China, Russia, and India'. *Journal of Nanoparticle* **11**: 1845–66.

Mangematin, V. and Walsh, S. (2012). 'The future of nanotechnologies'. *Technovation* **32**(3–4): 157–60.

Mathews, J.A. (2007). 'Catch-up strategies and the latecomer effect in industrial development'. *New Political Economy* **11**(3): 313–35.

Meyer, M. (2000a). 'Does science push technology? Patents citing scientific literature'. *Research Policy* **29**: 409–34.

Meyer, M. (2000b). 'Patent citations in a novel field of technology – What can they tell about interactions between emerging communities of science and technology?' *Scientometrics* 48: 151–78.

Malerba, F., Caloghirou, Y., McKelvey, M. and Radosevic, S. (2015). *Dynamics of Knowledge Intensive Entrepreneurship: Business Strategy and Public Policy*. London: Routledge.

McKelvey, M. and Lassen, A.H. (2013a). *Managing Knowledge Intensive Entrepreneurship*. Northampton, MA, USA and Cheltenham, UK: Edward Elgar.

McKelvey, M. and Lassen, A.H. (2013b). *How Entrepreneurs Do What They Do: Case Studies of Knowledge Intensive Entrepreneurship*. Northampton, MA, USA and Cheltenham, UK: Edward Elgar.

Murray F. (2002). 'Innovation as co-evolution of scientific and technological networks: exploring tissue engineering'. *Research Policy* **31**: 1389–403.

Nightingale P. (1008). 'A cognitive model of innovation'. *Research Policy* **27**: 689–709.

Nikulainen T. and C. Palmberg (2010). 'Transferring science-based technologies to industry—does nanotechnology make a difference?'. *Technovation* **30**: 3–11.

Niosi, J. and S.E. Reid (2007). 'Biotechnology and nanotechnology: science-based enabling technologies as windows of opportunity for LDCs?'. *World Development* **35**(3): 426–38.

President's Council of Advisors on Science and Technology (PCAST) (2005). 'The national nanotechnology initiative at five years: assessment and recommendations of the national nanotechnology advisory panel'. Available at: http://www.whitehouse.gov/sites/default/files/microsites/ostp/pcast-nni-five-years.pdf (accessed 20 October 2014).

Rothaermel, F.T. and Thursby, M. (2007). 'The nanotech versus the biotech revolution: sources of productivity in incumbent firm research'. *Research Policy* **36**: 832–49.

Salamanca-Buentello, F., Persad, D.L., Court, E.B., Martin, D.K., Daar, A.S. and Singer P.E. (2005). 'Nanotechnology and the developing world'. *PLoS Medicine* **2**(5): e97.

Shapira, P. and Wang, J. (2009). 'From lab to market? Strategies and issues in the commercialization of nanotechnology in China'. *Asian Business and Management* **8**(4): 461–89.

Shapira, P. Youtie, J. and Kay, L. (2011). 'National innovation systems and the globalization of nanotechnology innovation'. *Journal of Technology Transfer* **36**: 587–604.

Thursbya J.G. and M.C. Thursby (2011). 'Has the Bayh-Dole act compromised basic research?'. *Research Policy* **40**: 1077–83.

Voss, C. (2009). 'Case research in operations management'. In C. Karlsson (ed.), *Researching Operations Management*, New York: Routledge, pp. 162–96.

Wonglimpiyarat, J. (2005). 'The nano-revolution of Schumpeter's Kondratieff cycle'. *Technovation* **25**: 1349–54.

Yin, R.K. (2009). *Case Study Research – Design and Methods*. Thousand Oaks, CA: Sage.

Zhao, F., Zhao, Y.L. and Wang, C. (2008). 'Activities related to health, environmental and societal aspects of nanotechnology in China'. *Journal of Cleaner Production* **16**: 1000–1002.

9. The role of returnees in developing entrepreneurial ventures in high-tech sectors in China

Peilei Fan

9.1 INTRODUCTION

This chapter explores the global linkages and internationalization of the Chinese innovation space, by focusing upon the role of returnees in Chinese high-tech sectors of internet infrastructure and services, health biotechnology, and telecommunication equipment. China has a long history of internationalization, through students. In 1854, China sent its first overseas student, Wing Yung (容闳), to study law.

Graduating from Yale, Yung later devoted himself to China's revolution and modernization, including supporting Emperor Guanxu's Hundred Days' Reform. Since then, China has experienced several waves of sending overseas students. They have followed the footsteps of Yung and devoted themselves to revolutionizing and modernizing the ancient "Middle Kingdom" after they have returned to China, many becoming important political, economic, and scientific figures in modern Chinese history (Wang, 2010). The economic reform and the opening of China orchestrated by Deng Xiaoping in 1978 marked an unprecedented wave of sending students overseas, mainly to North America, Japan, and Western Europe. From 1978 to 2013, China has sent more than one million students and scholars to study and conduct research abroad (MOE, 2013). However, in recent years, looming on the horizon has been the increasing reverse wave of returnees from overseas, starting what many consider as the "brain circulation" phenomenon. Aside from contributing to the scientific community, many returnees have active involvement in high-tech entrepreneur activities, encouraged by policies set by the central and local governments. For instance, by 2010, 150 entrepreneurial parks and over 10000 enterprises were set up in China by returnees (Wang, 2010). The success of high-tech returnee companies is closely related to China's rising position in the world economy and the expansion and upgrading of

China's domestic market, which has rendered tremendous opportunities for returnee entrepreneurs who are skillful in integrating China with the globalized knowledge economy.

In recent years, a growing number of overseas returnees have established start-up companies in their home countries. This phenomenon has received increasing attentions from policy-makers, entrepreneurs, and academics in emerging economies such as China and India. In reference to Dai and Liu (2009) and Kenney et al. (2013), I here define a returnee as a foreign national who left their home country to be educated abroad, especially in OECD, and/or worked abroad for a period of time and then returned to their native country to set up a business. The returnees have played an important role in economic development, especially high-tech industries' development in China, India, South Korea, and Taiwan (Chang, 1992; Choi, 1995; Dai and Liu, 2009; Saxenian, 2002, 2006).

This chapter examines this phenomena by studying six cases of returnee firms in China's high-tech sectors: Asia Info, Sohu, and Baidu in Internet infrastructure and services, Sibiono and Sandia Medi Tech in health biotech, and UT Starcom in telecom-equipment. These companies set up by the returnees have played an important role in facilitatating the access of Chinese firms to global financial and technology resources that otherwise may have been unavailable to domestic firms in China. Specific topics addressed are the sources of financing of these returnee firms, sectoral variations, and the role of government intervention. This provides us with insights on the contribution the returnee firms have made to the development of China's high-tech industries, financing and location pattern of returnee companies, and the government intervention.

The chapter is organized as follows. After the introduction, Section 9.2 reviews literature on returnees and industrial and economic development of host countries, introduces the general background of returnees in China's high-tech sectors in China, and raises research questions of this chapter. Section 9.3 describes methodology and the six case firms for the research. Section 9.4 illustrates the findings of these returnee firms. Section 9.5 offers discussion on the impact of financing, sectorial variation, location, and the role of the government intervention on returnee high-tech firms. Section 9.6 concludes.

9.2 LITERATURE REVIEW AND RESEARCH BACKGROUND

It is widely acknowledged that globalization and the knowledge economy have made faster movement of technology-related capital and human talent across geographic boundaries. The recent work of UNU-WIDER (Solimano, 2008) and Saxenian (2006) provide both a good grasp of the international movement of talent and a critical perspective of the opportunities and challenges faced by the returnee entrepreneurs. Three types of movement of talents have dominated the scene: industrial related (entrepreneurs, and technical talent); campus related, that is, scientific and academic talent; and international students, health professionals, and cultural workers (Solimano, 2008).

The impact of international movement of talent has been analyzed both at local and global perspectives based on experiences of various countries (Chang, 1992; Choi, 1999; Saxenian, 2006; Solimano, 2008; Yoon, 1992; Zhou and Hsu, 2011). From a local perspective, the source countries have been portrayed to have both loss and gain. Here, loss of talents and waste of investment characterize the so-called brain drain. In contrast, the gain refers to the increasing return on investment in human capital and potential brain circulation as a result of returnees, illustrated by the experiences of Taiwan, South Korea, and Israel. For destination countries, most literature has agreed on the gain. For instance, Argentina, US, Canada, and Australia at the end of the 19th century, and recently the US and Europe, have all benefited tremendously from scientific discovery, technology development, and industrial upgrading as a result of receiving countries for international talents.

From a global perspective, the international movement creates global efficiency gain, that is, marginal productivity (MP) of human capital increases when talents move from countries with low MP to countries with higher MP. Further, it has the distributional effect. There is an increasing disparity between the sending and receiving countries as a result of talent moving "for good," causing (possibly) a dearth of talent for the development of innovation systems in the sending countries and a virtuous cycle of human capital development in the receiving countries. Nevertheless, return migration can bring remittance flows, capital, technology, managerial skills, and business contracts that can alleviate the distribution effect.

Various push and pull factors affect the mobility of international talents (Solimano, 2008), including earning and development gaps, non-pecuniary motivations such as attachment and excitement to a particular destination, the demand for capital, talent, and technology. Further, agglomeration and concentration of talents can create a gravity effect, that is, talents are usually attracted by the concentration of other talents. Moreover, linguistic compatibility, networks, and sociocultural affinity also play a role for the movement of international talents (Fan, 2011a). Finally, policy regimes and immigration regulations have significant effects on the mobility of international talents. Special visas for attracting talents in Australia, Canada, Germany, UK, and the US have made them ideal locations for international talents. In recent years, Ireland, Singapore, and Scotland have implemented this kind of policy to attract talent, whereas in the attempt to transform brain drain to brain circulation, developing countries also initiated similar policies to attract returnees. When the high-tech returnee entrepreneurs seek their fortune in distant homelands by launching companies far from established centers of skill and technology, how have they fared *vis-à-vis* multinational corporations and domestic companies in their home countries? Have they profoundly transformed the development of high-tech industries of their home countries or are they just marginal additions to the already rising high-tech industries of their home countries? These are questions that not only attract academic research but also policy-makers.

In the context of international movement of talent, a few researchers have studied the scale of brain drain and returnees in China in academics and industries, analyzing causes for going abroad and returning and comparing performances of returnees and their companies *vis-à-vis* local professionals and firms (Cao, 2008; Chang and Deng, 1992; Chen, 2008; Dai and Liu, 2009; DeVoretz and Zweig, 2008; Filatotchev et al., 2011; Huang, 2008; Kenney et al., 2013; Tian, 2011; Zweig, 1997, 2006; Zweig et al., 2004, 2007, 2008). Since the economic reform, there has been a continuous increasing trend of Chinese returnees, along with the increasing Chinese students and scholars who go abroad. While political stability has been a significant factor affecting the returning decision of the returnees, like other Asian economies such as Taiwan and South Korea (Chang, 1992; Choi, 1999), the main pull for returnees is the rising economic power and market opportunities in China (Zweig, 2006). Returnee companies have been found to be more inclined to import global technology and capital than other types of firms, including local or multinational corporations (Filatotchev et al., 2011; Zweig et al., 2004), thus becoming an effective vehicle for technology development (Chen, 2008).

Overall, the "brain circulation" literature has largely concentrated on "theoretical conceptualizations or descriptions of the national or global environment" (Devoretz and Zweig, 2008: 171) without empirical examination at the micro or even meso levels, especially for specific industrial sectors. Many scholars have argued that more studies on returnee entrepreneurs and companies that are founded or managed by returnees (returnee companies) are needed (Saxenian, 2002; Saxenian et al., 2002; Solimano, 2008).

While the literature on China's returnees has reviewed the policy dynamics of government effort (Cao, 2008; Kshetri, 2007; Zweig, 2006), they mostly focus on policies related to academic returnees (Jonkers and

Tijssen, 2008). Although surveys on returnee companies in Zhongguancun (Chen, 2008; Dai and Liu, 2009; Filatotchev et al., 2011) have provided classification and the profile of returnee companies *vis-à-vis* local and multinational corporations (MNCs), they lack the deep insight that case studies can provide. While case studies of Huang (2008) have addressed this issue, his cases focus on returnee companies in only one sector, the internet service, which has its unique market orientation, therefore lacking a comparative perspective that a study on cases from multiple sectors can provide.

Within this context, I seek to understand the contribution of the returnee companies to the development of China's high-tech sectors as well as critical factors affecting the growth of returnee companies, such as financing, location choice, and government intervention. Specifically, I will address the following research questions through case analysis of six returnee companies in three different high-tech sectors:

- What are the characteristics of returnee companies in different high-tech sectors, for instance, in terms of financing, technology strategies, market orientation, and government involvement?
- How have companies set up by returnees contributed to China's catching-up in these sectors?

While this chapter does not focus on what causes have motivated returnees to return, I agree with Zweig (2006) that market opportunity presented in the rising China provides the ultimate pull for returnees to decide to go back to China and set up their own companies. Further, despite the inter-city competition for attracting returnees, from tax breaks for returnee firms, to subsidized housing for returnee entrepreneurs, to schooling for children of returnees and job opportunities for spouses of returnees, the fact that returnees' companies concentrated only in a few locations in China, such as Beijing, Shanghai, and Shenzhen, indicates that places with the most promising opportunities for high-tech industries will attract the most returnees.

9.3 RESEARCH STRATEGY: METHODOLOGY, DATA AND COMPANIES STUDIED

For this research, I used a multi-case analysis approach to conduct in-depth analysis of the phenomena and build theory (Ragin, 1994), that is, a longitudinal assessment of business environment changes and corresponding changes in companies, to understand the returnee companies. Multi-case analysis rather than single case allows us to "identify similarities with subsets of cases that distinguish them from other subjects" (Ragin, 1994: 124). I collected data from various resources including government statistics, reports, news, and articles on the companies and their founders, and interviews with firms, industrial experts, and government agencies. I selected six returnee companies in three high-tech sectors: internet, health biotech, and telecom equipment, with four of them listed in NASDAQ.

9.3.1 Details About the Entrepreneurial Companies

These companies are considered significant in their respective sectors in the sense that they are or were recognized as successful and profitable high-tech companies with good reputations (Filatotchev et al., 2011; Frew et al., 2008, Huang, 2008). Even though UTStarcom has had unsatisfying performance in recent years, due to its significant market share in the 2000s and impact on wireless technology development in China, I select UTStarcom[1] as one of the cases and hope to bring more insight on important factors that may obstruct the success of returnee companies. Table 9.1 provides an overview of the profiles of the six entrepreneurial ventures in different sectors.

All three internet companies, that is, Asiainfo, Sohu, and Baidu, are listed in NASDAQ. Although a less well-known company to the general households, Asiainfo is a top provider of software and information technology (IT) services to the telecommunication industry, listed as number one in custom billing management in Asia-Pacific and independent software vendor (ISV) in Asia-Pacific Wireless Market in 2013. Sohu was ranked as the world's 3rd and 12th fastest-growing company by Fortune in 2009 and 2010 respectively (Fortune, 2009, 2010), and as of 2013, Sohu was listed as the 44th globally and 10th in China in terms of its internet traffic by Alexa, an internet-traffic-ranking branch of Amazon.[3] Baidu, as the leading Chinese language internet search provider, had over 63 percent market share in 2010.[4]

A quick glance at the founders of these three companies reveals several common traits: they all have post-graduate education and rich working experiences in the US; they returned to China for its market opportunity; and they have used international financing or government funding to support the growth of their firms. Asiainfo was founded by Edward Shuoning Tian, who obtained a BS in Biology from Liaoning University and his MS in Resource Management from the Chinese Academy of Science before going abroad in 1987 and obtaining a PhD in Resource Management from Texas State Technical College (TSTC) in 1992. After working at ICF and TSTC, he established Asiainfo in Dallas in 1993 with an initial investment capital of $0.5 million. His close partner and Asiainfo's co-founder, James Jian Ding, also had his undergraduate education in China (Beijing University) and postgraduate degrees in the

Table 9.1 Profile of the six entrepreneurial ventures for case studies

Company Name	Establishment year and HQ	Main business
Asiainfo (NASDAQ: ASIA, since 2000)	1993, Dallas (moved major operation to China in 1995)	Construction of national and provincial networks in China, telecom software, IT security.
Sohu (NASDAQ: SOHU, since 2000)	1996, Beijing	Chinese language web services (alumni, game, real estate, mapping, wireless-value added) portal, search engine.
Baidu (NASDAQ: BIDU, since 2005)	1998, Beijing	The leading Chinese search engine, collaborative encyclopedia, first company in China to be included in NASDAQ-100 index (Dec. 2007).
Sibiono Gene Tech	1998, Shenzhen	A leading company in gene therapy, developed first gene therapy of the world. Signature product: Gendicine (recombinant adenovirus type 5 gene therapy carry human p53 gene) for head-and-neck cancer squamous-cell carcinoma.
Sandia MediTech	2004, Shanghai,	Development of new medicine, provide contract research for pharmaceutical and medical biotechnology companies.
UTStarcom (NASDAQ: UTSI, since 2000)	1996, US (moved global operation center to Beijing in 2010)	PHS (in the late 1990s and early 2000s), IPTV (already divested), broadband, and Next Generation Networks (NGN).

US (master's degree in information science and EMBA from UCLA). After three years' work as a system engineer at the University Texas in Arlington, Ding joined Tian to found Asiainfo and became its vice president from 1993 to 1999 until he became the president and chief executive officer (CEO) of the company in 1999. Tian returned to China in 1995 and in the same year, his company, Asiainfo, built China's first test internet for the largest telecom service provider in China, China Telecom. From 1995 to 1998, Asiainfo built national backbone networks and provincial networks in 19 provinces. Since 1998, Asiainfo has entered into other fields such as telecom software, email, and IT security.

Sohu was founded by Charles Chaoyang Zhang, a graduate of Tsinghua University who later obtained a PhD in physics from Massachusetts Institute of Technology (MIT) in 1993. After two years' work as a postdoctoral researcher and later as an MIT Asia-Pacific Region Liaison Officer, he returned to China in 1995 as a chief representative of an American company, ISI, in China. In 1996, Zhang established his company, ITC, later renamed as Sohu, with two MIT professors, Dr Nicolas Negroponte and Dr Edward Roberts. As the first internet company with venture capital investment in China, Sohu received four rounds of financing before its initial public offering (IPO) in NASDAQ in 2000.

Baidu, the leading Chinese search engine, collaborative encyclopedia, and the first company in China to be included in NASDAQ-100 Index (December 2007), was founded by Robin Yanhong Li, a graduate of Beijing University who later obtained an MS in information management from SUNY Buffalo in 1994. Li worked in Dow Jones and Company and Infoseek from 1994 to 1998 and returned to Beijing to establish Baidu in 1998 with an initial capital of $1.2 million. In 2001, $10 million was raised during its second-round financing and the company started to directly provide a searching service to individual users. In 2004, Google chipped in $10 million to Baidu, equal to 2.6 percent of Baidu's equity. On August 5, 2005, the IPO day of Baidu in NASDAQ, its stock price increased by 354 percent, becoming one of the ten stocks that had the most increase in share prices in history.

The development of China's health biotech sector is said to be driven by government policies and national programs that support technology development of the sector, the highly qualified returnees who brought with them scientific talent and international credibility, and the large and diverse demand from the Chinese market (Frew et al., 2008). I offer two cases to represent returnee companies in this sector. Founded in 1998 in Shenzhen, Sibiono GeneTech is a well-known health biotech company as it was the world's first company that developed and marketed a gene therapy product, Gendicine, a recombinant adenoviral-p53 agent primarily used for head and neck cancer that gained the approval by the China's State Food and Drug Administration (SFDA) in October 2003. Its founder, Zhaohui Peng, is a returnee entrepreneur. He not only studied biochemistry and learned gene therapy techniques at Chiba University in Japan and University of California at Los Angeles but also worked for a US biotech company in San Diego before returning to China in 1998. Furthermore, about 10 percent of Sibiono GeneTech's staff is overseas returnees who were attracted by the competitive salary and career opportunity offered by the company (Frew et al., 2008). From 1998 to 2003, Sibiono received $6.25 million seed funding from the Chinese government.

Founded by a group of returnees who have experience in the US biopharmaceutical industry with an investment of RMB50 million (RMB10 million angel investment) in 2004, Sandia Meditech is a leading pharmaceutical and biotech research and development (R&D) outsourcing company, or contract research organization (CRO), located in Shanghai. In contrast to most returnee companies that focus on the domestic market, Sandia Meditech provides drug discovery and development to about 100 global clients, mainly in North America, Europe, and Japan. Although its initial focus is custom synthesis, it has expanded to process research, application programming interface (API), and intermediate manufacturing, and services in preclinical trials. Its first-round financing resulted in $2.5 million capital provided by IDG Capital Partners (IDGVC). The second-round financing in 2008 totaled tens of million dollars of funding from venture capitalists in China, the US, and Japan. Sandia experienced fast growth due to the needs of global pharmaceutical and biotech companies for contract research and its unique advantages of resources for R&D and patients. Sandia was named as Deloitte Technology Fast 500 Asia Pacific from 2008 to 2012. Chen Chen, its CEO, was a post-doctoral scientist at Texas A&M and University of Illinois at Chicago, and has 15 years of work experience in Neurocrine, a California-based biopharmaceutical company. Out of 600 research staff, about 8 percent are overseas returnees.

China's telecom-equipment industry is probably the most exciting high-tech industry, as two leading domestic companies, Huawei and ZTE, have grown into MNCs with impressive innovation capability that compete on equal footing with other MNCs in the sector (Fan, 2011a). UTStarcom was founded in 1995 by merging California-based Unitech Telecom, Inc., founded by Honglu Liang in 1991, with New Jersey-based Starcom Networks Systems Inc, founded by Ying Wu. Both key founders were born in China and educated in the US. The company started with its market orientation in China and gained popularity and a substantial market share by launching Personal Handy-phone System (PHS), a scaled-down version of the cellular network, in 1997. After divesting the IPTV business in 2012, it currently focuses on two core areas: broadband and next-generation networks (NGN). While its headquarters is located in Silicon Valley, China serves as its main manufacturing base and hosts most of its employees; it has other business and R&D centers located in Japan, South Korea, India, the Middle East, and Europe. Although the company is not as well known as Huawei and ZTE and has not performed well since the middle of the 2000s, its rise and fall in the Chinese telecom market reflects the opportunities and challenges facing returnee companies in industries that have been heavily regulated by the government. Table 9.2 provides an overview

Table 9.2 Characteristics of the returnee entrepreneurs and access to resources

Company name	Overseas experience of founder(s)	Source of financing; government policy; major milestones of the companies
Asiainfo (NASDAQ: ASIA, since 2000)	Edward Shuoning Tian: Ph.D. from Texas Tech. Univ., three years US working experience Jian Ding (MS and EMBA from UCLA): three years' US working experience	Initial business in China were from telecom service providers at the central and provincial levels
Sohu (NASDAQ: SOHU, since 2000)	Charles Chaoyang Zhang, Ph.D. from MIT, two years' US working experience	First Chinese Internet company with VC capital investment
Baidu (NASDAQ: BIDU, since 2005)	Robin Yanhong Li, MS from SUNY Buffalo, four years' US working experience	Initial and pre-IPO funding mainly raised in the US
Sibiono Gene Tech	Zhaohui Peng, studied in Japan (Chiba Univ.) and US (UCLA); worked in US before returning to China	A quarter of its investment in Gendicine is from the Chinese government; government support: approval of Gendicine commercialization; biotech VCs mainly created by government, international VCs discouraged by the government
Sandia MediTech	Chen Chen, post-doctor at Texas A&M and UIC, 15 years' US working experience Xiaochuan Yang, Ph.D. from U. of Chicago, >20 years' US study and working experience	Initial capital of RMB10 million in 2004 from family and friends RMB10 million in 2004 from Anhui Guotou $2.5 million from IDGVC in 2005 Over $10 million in 2007 from both domestic and international investors
UTStarcom (NASDAQ: UTSI, since 2000)	Wu Ying, master's from NJIT, four years' US working experience (Bell Lab) before he return to China in 1992	Financing mostly relied on overseas financing, especially from Japan's Softbank (by 2000, a total of $160 million), but also include Intel, Matsushita, and Mitsubishi; Government policies on PHS, 3G, and IPTV had a huge impact on the company

of the characteristics of the returnee entrepreneurs and access to resources, especially financing.

9.4 ANALYSIS OF FINDINGS ABOUT THE ROLE OF RETURNEES

Based on a database of over 700 firms (half returnee companies and half domestic companies), Dai and Liu (2009) found that returnee companies perform better than those owned by local entrepreneurs due to their technological and commercial knowledge and due to the positive impact from their international networks. Based on comparison of Taiwan, China, and India, Kenney et al. (2013) argue that returnees only played an active role in the secondary developmental phase after indigenous entrepreneurs and policy-makers had laid the groundwork for the industry, and that they were not critical in the initial stage of the development of the information and communication technology (ICT) industries of these countries. Will this generalization be applicable to the case studies of this chapter? While sectorial variation exists, cases in this chapter indicate that these leading returned companies have made various impacts in their respective industries, and some of them were critical in the initial stage of the development of the high-tech industry in the sense that they have led other domestic companies to create the market and business, especially the internet services.

9.4.1 Ventures in the Internet Sector

Returnee companies in the internet sector have made a profound impact to facilitate China's catching-up in the internet infrastructure and services. Asiainfo has played a major role in building China's internet infrastructure whereas Sohu and Baidu have provided Chinese internet users with services of comparable quality that Western countries have. In the late 1990s, Asiainfo not only initiated the building of China's Internet, that is, the test internet in Beijing and Shanghai, but also successfully finished ChinaNet (the backbone internet in China that connected all provincial capitals and went into operation in January 1996) as well as internet in 19 provinces.[2]

The success of all these companies relied on developing technology appropriate to China's market. For instance, in the late 1990s, Asiainfo developed payment software that suited China's context and could compute fees that customers need to pay to Internet Service Providers (ISPs). The existing software for consumers' payment to ISPs in the US have been based on the wide usage of credit cards in the US and the dominant model of fixed monthly charges. As it was not appropriate to use credit cards to pay the fee and the fee is usually calculated based on minutes used for the internet, the software used in the US was not appropriate to be directly used in China. Asiainfo successfully developed payment software specifically for Chinese market, which was adopted quickly.

Sohu and Baidu developed Chinese search engines and Chinese content web services, although the technology base and business models actually come from the US. Baidu's founder Robin Yanhong Li had worked for Dow Jones and Company to help develop the online version of *The Wall Street Journal* and developed the RankDex site-scoring algorithm, which resulted in a US patent for the technology in 1996. Li later used this technology to build the Baidu search engine, specially tailored with Chinese-language search terms. While Sohu provides Chinese users with internet services popular in the US and other developed countries, such as news, search, email, games, and wireless messaging, in China, two core pillars, online games and brand advertising, generate most of its sales and profit.

The three case companies all highlight the importance of foreign, especially US-based financing from initial seed capital to the second and even fourth-round financing to IPO at NASDAQ. For instance, all early financing of Asiainfo came from the US. Similarly, for Sohu and Baidu, most initial funding and pre-IPO funding came from US venture capital. Finally, it is interesting to note that even within this sector, three companies had different experiences with the government. While Asiainfo has worked with China Telecom and Ministry of Information Industry (MII) very closely, Sohu and Baidu have had little involvement with the government.

9.4.2 Ventures in the Health Biotech Sector

Even though this sector occupies a small percentage of the pharmaceutical industry, China's health biotech sector was growing at 30 percent annually from 2000 to 2005, with the majority firms (>90 percent) producing non-novel products (biogenerics), a small portion of firms generating innovative products, and another small group providing contract research services. Sibiono is one of the successful examples of Chinese firms in the fields of innovative gene therapies and stem cells, whereas Sandia represents the experiences of companies in contract research services. It is worthwhile to highlight the role of the government in facilitating the development of companies with innovative products and contract services, especially in terms of financing and regulation, as illustrated by Sibiono's experience.

First, the most important government support for Sibiono has been the approval of Gendicine by the SFDA, which has surprised many Western scientists. Gene therapies worldwide had made little progress in the 1990s and early 2000s and suffered a major setback due to the death of Jesse Gelsinger, who was participating in a gene therapy trial and died from multiple organ failures four days after starting the treatment in 1999. The US Food and Drug Administration (FDA) put a temporary halt on all gene therapy trials using retroviral vectors in blood stem cells in January 2003. Although SFDA in China routinely does not approve a new drug if it has not been approved in the US, it made an exception for Gendicine due to the large proportion of skin cancer patients in China and the significance of Gendicine in saving many lives (*People's Daily*, 2004). Nevertheless, given the controversial results from Phase I and II and the small number in Phase III trials, which involved only 120 patients, some Western scientists, such as James Wilson, editor of *Human Gene Therapy*, where Peng published the result of Gendicine trials in English, were surprised to see Gendicine approved for commercialization in China (Rusconi, 2003).

Like other domestic companies, Sibiono also depended heavily on government financing for its initial development. Although China set up technology policies to facilitate funding for R&D activities due to its lack of appropriate mechanism to fund high-tech ventures, government support remains the main avenue for financing domestic biotech firms, especially start-up firms (Jia, 2006a). Sibiono invested a total of $9.66 million into Gendicine, and a quarter of the investment came from various government resources, such as Ministry of Science and Technology, the 863 Program,[3] and Shenzhen municipal government (*People's Daily*, 2004). Other health biotech domestic firms, such as Medgenn, who developed an angiogenic inhibitor, and Shanghai Sunway, who developed Endostar, the first commercialized oncolytic virus, H101, had similar experiences (Fan, 2011b). Further, biotech venture capital (VC) firms have been mainly created by governments and strict regulations have discouraged international VCs to enter the market, making their contribution insignificant in China's biotech sector (Li et al., 2004).

The contract service companies have grown rapidly in China's health biotech sector, benefiting from the low-cost scientific talent (as low as 10 percent of the cost of similar expertise in the United States), low cost and large amount of patients for clinical trials,[4] and raw materials available in China (Frew et al., 2008). The contract services provided by Chinese companies can range from early stage research, pre-clinical development, clinical service, and manufacturing. For instance, Sandia provides drug discovery and research, pharmaceutical development, and new drug pre-clinical development. After esarning a PhD from the University of Chicago, Dr Xiaochuan Wang, the founder of Sandia, started to work in the US pharmaceutical industry and had been in charge of the whole process of drug discovery and R&D. At the time, the CRO started to boom globally and while Wang was working for a US company, she was asked by the company to look for a CRO company in China. However, after exhaustive research in Beijing, she could not find a qualified company as most CROs at the time conducted only simple processing of synthesized chemicals. This type of low-end CRO sector then had low entry barrier and faced strong competition from the Indian companies who have ten years lead-time over Chinese companies. However, at the high end of the CRO sector, which mainly serves clients for drug discovery and development, there was great opportunity for Chinese firms to grow without strong competition from the Indian companies. This experience finally made Wang return to China to establish a high-end CRO company to take the advantage of China's human resources as well as the large consumer market. The returnee companies such as Sundia have great advantage in linking global clients and the Chinese advantage in human resources, leading to the explosive growth of the export-oriented CRO in China (Baike, 2013). While China is only at its initial stage of CRO industry and its CRO market size had only RMB22–28 billion, 7–10 percent of the global CRO market, it had rapid growth and had about 500 domestic CROs in 2012 (Reuters, 2013).

9.4.3 Ventures in the Telecommunication Sector

The experiences of returnee companies in the telecom sector, illustrated by the case of UTStarcom, share common traits as some of the returnee companies in the internet and biotech sectors in terms of financing relied on overseas sources, working closely with the Chinese government, and developed technologies mainly for the Chinese market.

UTStarcom has mostly relied on overseas financing, especially from Japan's Softbank. In 1995, Softbank invested US$30 million and by 2000, financing from Softbank reached a total of US$160 million. Investments from others include Intel, Matsushita, and Mitsubishi. In addition, working closely with Ministry of Information Industry, UTStarcom took off due to technology it developed specifically for the Chinese market: the Personal Handy-phone System (PHS). The PHS, dubbed as local wireless and a scaled-down version of cellular network that was launched by UTStarcom in 1997, has limited range and roaming abilities, but fit the needs of Chinese consumers at the time due to its low cost – it had only 20 percent cost of GSM (Global System for Mobile Communications) phones. However, it should be noted that the government policies, such as on PHS, third-generation cell phone technology (3G), and Internet Protocol Television (IPTV) have had a huge impact on UTStarcom's development.

As a telecom equipment producer for PHS, UTStarcom took off quickly after China Telecom launched PHS to use its extra capacity of fixed telecom networks in 1997. When the Ministry of Information Industry (MII) announced the halt of the development of PHS in June 2000, UTStarcom's stock price plunged by 46 percent in one day. Within a month, when MII explained that PHS was the extension of fixed phone networks and it would continue its support to PHS, UTStarcom's stock started to rise again. Knowing that PHS was a transitional technology that would be replaced sooner or later, UTStarcom invested heavily in 3G R&D and new technologies for mobile telecommunication, by acquiring technologies, human talent, and related companies.

While other domestic companies such as Huawei and ZTE have accumulated rich experience through R&D on first and second generations of mobile telecommunication technologies, UTStarcom was a newcomer in the mobile technology field, despite the experiences of PHS system, its main business that brought huge profit for the company. By 2005, UTStarcom had invested over RMB1 billion in 3G and started to have difficulty surviving the market competition as a 3G license was still not issued by MII. The long-delayed issuance of 3G licenses forced UTStarcom to adopt another strategy to survive in 2005: betting on the deployment of IPTV. This time, government policies and regulations had played unfavorably for UTStarcom again.

The IPTV business is governed by two government bodies in China: the MII and the State Administration of Press, Publication, Radio, Film and Television (SAPPRFT). Due to the governance separation between these two governing bodies, even though MII has been very interested in deploying IPTV, local bureaus of SAPPRFT showed little interest in developing IPTV as this would have made their existing broadcasting networks outdated, thus affecting their current profit. Despite some successful cases such as IPTV in Harbin, there had been more cases of the local bureaus of SAPPRFT intentionally delaying and disrupting the deployment of the IPTVs, such as the case in Quanzhou. The difficult situation forced UTStarcom to divest its IPTV business at the end of August 2012 (China Entrepreneurs, 2010).

9.5 DISCUSSION: ENTREPRENEURS, FINANCING AND POLICY

The case analysis of six returnee companies in three different high-tech sectors in China provides us insight on several issues, such as the contribution they made to China's high-tech industries, financing and location patterns of returnee companies, and the government intervention. Despite sectorial variations, returnee companies studied in this chapter have made important rather than marginal contributions to the development of their respective high-tech industries in China. Returnees' started ventures leading to companies that have relied heavily on foreign financing, but also government funding, especially when it is highly risky and long-term oriented. Further, also following a gravity effect, returnee companies are attracted by other high-tech companies and congregate in only certain locations, that is, in Beijing, Shanghai, or Shenzhen, thus reinforcing China's current uneven R&D landscape. Finally, although government has an impact on development of all returnee companies, the involvement of government has varied according to different sectors. Each issue will be discussed in detail in this section.

First, although the six companies studied in this chapter are from high-tech sectors with different characteristics, they all have played a key role to either transform, start-up, or greatly facilitate the development of their respective sectors. While some sectors have a high barrier of entries, such as health biotech and telecom, which involve large amounts of capital investment and long-term development cycle, others have a low barrier of entries, such as internet service. In addition, the cases from this chapter indicate that returnee companies tend to enter the sectors that have weak presence of MNCs and domestic companies in China, due to undeveloped domestic market at the time. All these companies have been the pioneers in their relative sectors, although these sectors may not be so new in Western countries. Asiainfo was the first to build China's internet infrastructure, whereas Soho and Baidu were among the pilots that provided internet services in China. UTStarcom started with a relatively backward mobile technology, but it was new and appropriate for the Chinese market at the time. For the health biotech sector, Sibiono was the first firm approved for commercialization of gene therapy in China and one of the pioneers in the world (Frew et al., 2008), whereas Sandia was among the first Chinese companies that provided contract research services to global clients.

Second, returnee companies have sought financing mainly through foreign VCs or capital market. Due to the weak VC market in China, foreign VCs as well as IPOs at technology stock markets (such as NASDAQ) have played a key role in starting up and growing star returnee companies in some of China's high-tech sectors, such as internet infrastructure and service and telecom equipment. Out of the 3200 companies listed on NASDAQ, 137 are from mainland China and 18 from Hong Kong (China Stock Digest, 2013). Similar to the experience of Taiwan and India (Saxenian, 2002, 2006), the overseas professional experience of returnees certainly has made foreign financial capital more accessible to returnee companies than their domestic counterparts. Nevertheless, the experience of Sibionio highlights that government grant and direct financial support have been particularly important for innovative health biotech companies due to the high risk and longer product development cycle. It is interesting to note that several CEOs of returnee companies have become the first generation of VCs in China after they left their ventures, such as: Ying Wu from UTStarcom, who established his own VC company; Hongwei Zhou from Yahoo China; and Lei Jun from Kingsoft. This trend further indicates the urgency to develop China's own VC sector.

Third, returnee companies have in general taken advantage of locations of Beijing, Shanghai, and Shenzhen, the most R&D-intensive city regions in China. These returnee companies have led in the development of ICT and biomedical industries, confirming the gravity effect. These city regions took diverse paths for developing innovation capabilities in these industries and have different implications for returnee companies.

Beijing leads in R&D capacity and is the largest base of China's high-tech industries as it topped other cities in revenue and number of employees in high-tech parks in 2009. Depending little on foreign capital or markets, Beijing's ICT industry is featured by a large number of small and medium-sized domestic companies in software and computer services (Zhou et al., 2010). Beijing is by far the most attractive place to settle down for returnee companies, especially in the ICT industry that mainly orients to domestic market (companies such as Sohu and Baidu), but also for internet infrastructure and telecom companies such as Asiainfo and UTStarcom due to their closeness to the regulatory government agencies.

Shanghai possesses strong manufacturing capacity of ICT and biotech industries, especially in integrated circuit (IC) and biomedical sectors, with the largest and the most complete and technologically advanced IC industrial cluster in China, producing over 70 percent of national total output and sales. Most of Shanghai's high-tech manufacturing capacity comes from large state-owned enterprises (SOEs) and MNCs, with little contribution from non-governmental firms due to an unsupportive institutional environment for them (Breznitz and Murphree, 2011). With the support of Shanghai's municipal government for the health biotech sector and its pro-MNC policy, it is ideal for returnee companies in contract services, such as Sandia. Although we did not examine the IC sector in this chapter, there are a significant number of returnee companies in the IC sector that have been quite active and which deserve further research attention.

Shenzhen developed from a small fishing village in 1978 to one of the most exciting metropolises in China within three decades. It is the third-largest high-tech industrialization base in China, featured mostly by domestic firms actively involved in indigenous innovation (Zhou et al., 2010). Accessible venture capital from Hong Kong, the completeness of the ICT industrial value chain, and active private entrepreneurship have been attributed to prosperity of domestic high-tech firms in Shenzhen (Breznitz and Murphree, 2011). The experience of Sibiono reflects the attractiveness of Shenzhen for innovative domestic companies. In fact, with the support of local government in indigenous innovation, Shenzhen has bred other well-known innovative domestic companies, such as Huawei and ZTE, in the telecom equipment sector.

Fourth, although returnee companies generally enjoy benefits provided by government policies specifically targeted for returnee companies and individuals, government involvement in different sectors has been quite different in terms of financing and regulatory support. Since the early 1990s the central government and local governments have instituted preferential policies such as tax breaks, tax-free imports of automobiles and computers and certain equipment for returnee companies, as well as subsidized housing, schooling for children of returnees, jobs for spouses, and long-term residence permits (Zweig, 2006). Local governments of Beijing, Shanghai, and Shenzhen have been particularly proactive in attracting returnees to set up high-tech companies in their respective high-tech zones. Many cities, including these three, have set up incubators for returnee companies in their economic development zones, which help returnees to deal with the large amount of paperwork involved in setting up a company, and also provide spaces, networks, management, and human resources training to returnee companies. Beijing's Zhongguancun Science Park, the earliest science park in China dubbed as China's Silicon Valley (Zhou, 2008), has been aggressively attracting returnees to set up companies in one of the 23 incubators (Chen, 2008; Dai and Liu, 2009). Shanghai and Shenzhen have put in place similar efforts to compete with Beijing. Further, Shanghai was one of the first group of cities that issued long-term residences for returnees with foreign passports. For Shenzhen, under the regulation of August 1989 (Zweig, 2006: 75):

> Returnees could come directly to the Special Economic Zone (SEZ), legally change their residence and that of their family, keep any foreign currency they earned in Shenzhen (even if they left the SEZ), buy a new house at near cost,

> establish a private business, and "enjoy precedence over ordinary people with similar conditions and qualifications in the use of scientific and technological development funds."

Nevertheless, while preferential policies look similar in various locations, government involvement in financing and regulation varied distinctively according to sectors or even subsectors. For instance, returnee companies in the internet infrastructure (for example, Asiainfo), telecom (for example, UTStarcom), and innovative products of the health biotech (for example, Sibiono) sectors have been heavily affected by government intervention through direct financing (Sibiono), regulation and policies (UTStarcom, Sibiono), or the provision of a consumer market (Asiainfo, UTStarcom). In contrast, returnee companies in the internet services (for example, Soho and Baidu) and contract services in health biotech (for example, Sandia) have had little government intervention or government financing. Although it is beyond the scope of this chapter to discuss the reasons for different degrees and types of intervention, the diversity exhibited by these six cases reminds us that we need to be careful not to generalize government intervention on returnee high-tech companies.

9.6 CONCLUSIONS AND IMPLICATIONS

This chapter has studied returnee firms using six cases: Asiainfo, Sohu, Baidu, Sibiono, Sandia Medi Tech, and UTStarcom. Overall, the paths of returnee companies differ significantly according to their sectors and the context of the Chinese market. Some developed the most forefront technologies such as gene therapy (Sibiono), whereas others provided somewhat backward (by Western standards) but nevertheless suitable mobile technologies to Chinese consumers. Some have been behind the scenes, working for the service providers by building the most-needed infrastructure to develop ICT-based knowledge economy (Asiainfo), whereas some have been upfront, directly facing the consumers by providing internet services with unique Chinese characteristics. While most oriented themselves to serve the domestic market, some entered the global innovation network by providing contract research services to global clients, using China's unique human talents and the resources of a large reserve of patient population.

This chapter particularly emphasizes the importance of accessing financing, either through overseas financing or through government grant, to establish and grow returnee companies. Further, returnee companies have located mostly in Beijing, Shanghai, and Shenzhen, the most

R&D-intensive city regions in China with leading high-tech industries. Each city region has its own development pathway and unique strength to support returnee high-tech companies. Finally, although there have been preferential government policies for returnee companies and entrepreneurs at both central and local levels, significant variation exists in government involvement in different sectors or subsectors.

NOTES

1. Although UTStarcom is still headquartered in the US, it is counted as a returnee company as the founder Ying Wu returned to China to explore business opportunity in 1992 and since then most of his company's business has been based in the Chinese market. Ying Wu has been frequently quoted as an example of overseas returnees by the Chinese media as well (see Mingren, 2014).
2. In 1995, deciding to also invest in the internet, China Telecom rented 64K international routes from Sprint and asked Sprint to build a test internet in Beijing and Shanghai. Sprint contracted this task to Asiainfo. This initiated the building of the internet of Asiainfo in China. While the contract led to small revenue, it was the first one in China and involved various aspects of the internet, including DNS analysis in China and applying for IP addresses for China, and so on. Following the initial success, Asiainfo signed the contract to build the internet in Guangdong for close to $3 million in August 1995; then they built ChinaNet (with China Telecom) for $8 million in November 1995.
3. The 863 Program, also called State High-Tech Development Plan, was initiated in 1986 as a government-funded program intended to stimulate the development of several key technologies in China, including biotechnology, space, information technology, laser technology, automation, energy, new materials, telecommunications, and marine technology.
4. Although China has regulations regarding to clinical trial, the large accessible population with increasingly Westernized diseases in China present as an attractive asset for CROs (see Lee, 2012).

REFERENCES

Alexa Internet, Inc. (2013). 'How popular is sohu.com?'. Available at: http://www.alexa.com/siteinfo/sohu.com (accessed 20 October 2014).

Hudong Baike (2013). 'Wang Xiaochuan: Idealism Entrepreneur'. Available at: http://www.baike.com/wiki (accessed 20 October 2014) (In Chinese).

Barboza, D. (2010). 'Baidu's gain from departure could be China's loss'. 13 January, *The New York Times*.

Breznitz, D. and M. Murphree (2011). *Run of the Red Queen: Government, Innovation, Globalization, and Economic Growth in China*. Yale University Press.

Cao, C. (2008). 'China's brain drain at the high end: why government policies have failed to attract first-rate academics to return'. *Asian Population Studies*, **4**(3) 331–45.

Chang, P. and Deng, Z. (1992). 'The Chinese brain drain and policy options'. *Studies in Comparative International Development*, **27**(1), 44–60.

Chang, S. (1992). 'Causes of brain drain and solutions: the Taiwan experience'. *Studies in Comparative International Development*, **27**(1), 27–43.
Chen, Y.C. (2008). 'The limits of brain circulation: Chinese returnees and technological development in Beijing'. *Pacific Affairs*, **81**(2), 195–215.
Qin, S. (2010). 'The rise and fall of UTStarcom'. *China Entrepreneurs*, 8 September. Available at: http://finance.sina.com.cn/leadership/20100908/15058624000.shtml (accessed 20 October 2014) (in Chinese).
China Stock Digest website (2013). 'China stocks listed on the NASDAQ'. Available at: http://chinastockdigest.com/other-resources/china-stock-market/china-stocks-listed-on-the-nasdaq/ (accessed on 17 May 2014) (in Chinese).
Choi, H. (1999). 'Shifting human resources in South Korean science and technology'. *Comparative Education Review*, **43**(2), 212–32.
Dai, O. and Liu, X.H. (2009). 'Returnee entrepreneurs and firm performance in Chinese high-technology industries'. *International Business Review*, **18**(4), 373–86.
DeVoretz, D. and Zweig, D. (2008). 'An overview of twenty-first-century Chinese "Brain Circulation"'. *Pacific Affairs*, **81**(2), 171–3.
Fan, P. (2011a). 'Innovation, globalization, and catch-up of latecomers: cases of Chinese telecom firms'. *Environment and Planning-Part A*, **43**(4), 830–49.
Fan, P. (2011b). 'Innovation capacity and economic development: China and India'. *Economic Change and Restructuring*, **44**(1–2), 49–73.
Filatotchev, I., Liu, X., Lu, J. and Wright, M. (2011). 'Knowledge spillovers through human mobility across national borders: evidence from Zhongguancun Science Park in China'. *Research Policy*, **40**(3), 453–62.
Fortune (2009). 'The world's 3rd fastest-growing company in 2009'. Available at: http://archive.fortune.com/magazines/fortune/fortunefastestgrowing/2009/snapshots/1.html (accessed 20 October 2014).
Fortune (2010. 'The world's 12th fastest-growing company in 2010'. Available at: http://archive.fortune.com/magazines/fortune/fortunefastestgrowing/2010/full_list/ (accessed 20 October 2014).
Frew, S.E., Sammut, S.M., Shore, A.F., Ramjist, J.K., Al-Bader, S., Rezaie, R., Daar, A.S. and Singer, P.A. (2008). 'Chinese health biotech and the billion-patient market'. *Nature Biotechnology*, **26**(1), 37–54.
Huang, F. (2008). 'The dynamics of regional advantages and disadvantages: Beijing as a center of high-tech entrepreneurship'. Dissertation at University of Southern California.
Jonkers, K. and Tijssen, R. (2008). 'Chinese researchers returning home: impacts of international mobility on research collaboration and scientific productivity'. *Scientometrics*, **77**(2), 309–33.
Kenney, M., Breznitz, D. and Murphree, M. (2013). 'Coming back home after the sun rises: returnee entrepreneurs and growth of high-tech industries'. *Research Policy*, **42**, 391–407.
Kshetri, N. (2007). 'Institutional changes affecting entrepreneurship in China'. *Journal of Developmental Entrepreneurship*, 12(4), 415–32.
Lee, J. (2012). 'Regulatory and practical considerations'. Available at: http://www.iconplc.com/icon-files/insight-newsletter/March12/conducting.html (accessed 17 May 2014.)
Li, Z., Zhang, J., Wen, K., Thorsteinsdottir, H., Quach, U., Singer, P.A. and Daar, A.S. (2004). 'Health biotechnology in China – Reawakening of a giant'. *Nature Biotechnology*, 22(Supplement), DC13–DC18.

Mingren (2014). 'Introduction of Ying Wu, founder of UTStar.com'. Available at: http://mingren.maigoo.com/1361.html (accessed 17 May 2014) (In Chinese).
Ministry of Education of China (MOE) (2013). 'Reports on overseas students'. Available at: http://news.sciencenet.cn/htmlnews/2013/2/275135.shtm (accessed 4 March 2013) (in Chinese).
People's Daily (2004). 'Chinese firm develops gene therapy injection'. Available at: http://english.peopledaily.com.cn/200403/23/eng20040323_138238.shtml (accessed 30 November 2013) (In Chinese).
Ragin, C. C. (1994). *Constructing Social Research: The Unity and Diversity of Method*. Thousand Oaks, CA: Pine Forge Press.
Reuters (2013). 'Research and markets: China Contract Research Organization (CRO) Industry Report', 2012–2013. Available at: http://www.reuters.com/article/2013/03/18/research-and-markets-idUSnBwb6bJ8da+108+BSW20130318 (accessed 20 October 2014).
Rusconi, S. (2003). 'Towards first commercialization of a gene medicine product in China'. Available at: http://www.unifr.ch/nfp37/sibiono.html (accessed 20 October 2014) (In Chinese).
Saxenian, A. (2002). 'Transnational communities and the evolution of global production networks: the cases of Taiwan, China and India'. *Industry and Innovation*, **9**(3), 183–202.
Saxenian, A. (2006). *The New Argonauts: Regional Advantage in a Global Economy*. Cambridge, MA: Harvard University Press.
Saxenian, A., Motoyama, Y. and Quan, X. (2002). *Local and Global Networks of Immigrant Professionals in Silicon Valley*. San Francisco, CA: Public Policy Institute of California.
Solimano, A. (ed.) (2008). *The International Mobility of Talent: Types, Causes, and Development Impact*. New York: Oxford University Press.
Tian, F. (2011). *Emigration of Chinese Scientists and Its Impacts on National Research Performance From a Sending Country Perspective*. PhD dissertation, Fairfax, VA: George Mason University.
Wang, H. (2010). *Chinese Returnees: Impact on China's Modernization and Globalization*. Presentation made at Brookings Institute, Washington DC, 6 April.
Yoon, B.S.L. (1992). 'Reverse brain drain in South Korea: state-led model'. *Studies in Comparative International Development*, **27**(1), 4–26.
Zhou, Y. (2008). *The Inside Story of China's High-Tech Industry: Making Silicon Valley in Beijing*. Lanham, MD: Rowman & Littlefield.
Zhou, Y., Y. Sun, Y.D. Wei, and G.C. Lin (2011). 'De-centering 'spatial fix' – patterns of territorialization and regional technological dynamism of ICT hubs in China'. *Journal of Economic Geography*, **11**, 119–50.
Zhou, Y. and Hsu, J.Y. (2011). 'Divergent engagements: roles and strategies of Taiwanese and mainland Chinese returnee entrepreneurs in the IT industry'. *Global Networks*, **11**(3), 398–419.
Zweig, D. (2006). 'Competing for talent: China's strategies to reverse the brain drain'. *International Labour Review*, **145**(1–2), 65–89.
Zweig, D. Chung, S.F. and Vanhonacker, W. (2007). 'Rewards of technology: explaining China's reverse migration'. *Journal of International Migration and Integration*, **7**(4), 449–71.

10. Evolving global innovation networks of Indian pharmaceutical companies

Peter Kedron and Sharmistha Bagchi-Sen

10.1 INTRODUCTION

This chapter addresses how Indian pharmaceutical companies are expanding their innovation spaces.[1] India has become a leading source of emerging-market outward foreign direct investment (OFDI) (Athreye and Kapur 2009; UNCTAD 2013). However, unlike many other emerging nations, Indian OFDI targets advanced economies and knowledge-intensive industries (Kumar 2008; Pradhan 2008). One of India's most active outwardly investing industries is the pharmaceutical sector. From 1961 to 2007, the pharmaceutical industry accounted for 14 percent of all Indian OFDI, 94 percent of which occurred between 2000 and 2007. Similarly, Indian pharmaceutical companies completed 87 foreign acquisitions from 2000 to 2007, representing 31 percent of all Indian foreign acquisitions (Pradhan 2007). The Indian pharmaceutical industry is now a major international contender of worldwide generic drug production, advanced pharmaceutical ingredient (API) development, and contract manufacturing. India is the world's third largest pharmaceutical producer by volume, accounting for 10 percent of global production (Department of Pharmaceuticals 2012). In 2012, India accounted for 20 percent of global generic production. Compared to advanced economies, establishing facilities, producing bulk drugs, and conducting clinical trials are between 40 and 70 percent less expensive in India (Wadhwa et al. 2008). Those cost advantages reflect firm-specific advantages of Indian pharmaceuticals with reverse-engineering capabilities and a high-skilled but low-cost labor force developed during a period of industrial regulation that indirectly fostered capacity development (Hoffman 2005; Pradhan 2008). In recent years, Indian pharmaceuticals not only expanded their sales to the US and European Union (EU), but also used their accumulated wealth to acquire patented knowhow and increase engagement in upstream innovation.

Foreign companies have forged partnerships with Indian companies to gain entry into the lucrative Indian market (Athreye and Kapur 2001; Athreye and Godley 2009). This typifies oligopolistic reaction within the pharmaceutical industry, and demonstrates the expansion of spaces of innovation and production through globalization.

This chapter examines OFDI trends of Indian pharmaceutical companies. Indian pharmaceuticals are innovative firms that generate non-traditional ownership-specific advantages in varied geographic contexts (Kedron and Bagchi-Sen 2012) from the earliest stages of their internationalization. These companies continually augment their firm-specific advantages by adapting organizational structures and strategies to shifting conditions within the global pharmaceutical industry. Responsiveness to national regulatory and market environments allow Indian pharmaceuticals to overcome latecomer disadvantages in an oligopolistic global pharmaceutical industry. Access to broadly applicable firm- and country-specific advantages (for example, reverse engineering, low cost skilled labor pool) allows Indian pharmaceuticals to simultaneously pursue multiple internationalization strategies, and to shift between strategies as conditions change. This flexibility has allowed leading Indian firms to not only geographically expand their operations, but also to enter new segments of the pharmaceutical value chain and alter global patterns of pharmaceutical innovation.

Building on a major theme of this book, this chapter addresses how Indian pharmaceutical companies are expanding their innovation spaces through a variety of strategies such as partnerships and acquisitions. In the context of this chapter, we take global innovation to mean the product and process innovations that Indian pharmaceuticals develop in dispersed global contexts as well as the organizational and strategic innovations these companies use to achieve their goals. In doing so, this chapter moves beyond existing studies that attribute the success of Indian pharmaceuticals primarily to national policy environments or production cost advantages in order to develop a discussion of the role of opportunities available in advanced nations and the ability of Indian pharmaceuticals to absorb and use these opportunities to develop their own innovation network.

This chapter is organized into the following five sections. Sections 10.2 and 10.3 position the experience of Indian pharmaceuticals within the existing literature on multinational enterprise (MNE) and foreign direct investment (FDI), as well as in the pharmaceutical value chain. Section 10.4 addresses the firm-specific advantages of Indian pharmaceuticals, suggesting they are connected to the historic context of Indian industrial development. Section 10.5 analyzes firm-level adjustments of leading Indian pharmaceutical companies as they overcome foreign-market entry barriers and establish international operations. By augmenting and exploiting firm-specific advantages developed both in India and abroad, these companies became part of the global organization of pharmaceutical innovation. A concluding Section 10.6 discusses the contributions of this chapter's synthesis and analysis to the existing literature.

10.2 OUTWARD FOREIGN DIRECT INVESTMENT AND THEORIES OF MULTINATIONAL ENTERPRISE

A key conceptual critique of FDI theory is that explanations fail to provide a dynamic perspective on the relationship between strategies and structures of multinational firms (Benito et al. 2009; Santangelo and Meyer 2011; Meyer and Thaijongrak 2013). In practice, studies commonly match firm-specific advantages to country-specific advantages, transforming strategic and organizational decisions dependent on the past and important for the future into static or cross-sectional analysis (Asmussen et al. 2009; Hennart 2009). In doing so, studies often provide limited theoretical insight into the how and why of continuous strategic and organizational change by multinationals following initial entry. Such a static conceptualization is particularly problematic for the study of knowledge-intensive industries where innovation is central to production. Development of knowledge-intensive products such as pharmaceuticals involve processes that extend over long periods of time, recombine knowledge from multiple sources, and are regulated differently in different times and places. Understanding the dynamics of multinational firms and the global innovation spaces they help create requires explanations that account for this variation.

Perhaps the most widely adopted conceptualization of FDI is Dunning's eclectic paradigm (Dunning 1980; Dunning and Lundan 2008). The paradigm provides a multidisciplinary framework that aggregates theoretical insights to explain foreign market entry. To outcompete host-country rivals, firms must possess access to firm- or home-country-specific advantages that allow them to overcome barriers to entry (Hymer 1976; Dunning and Lundan 2008). To maintain their competitive edge firms chose to protect proprietary advantages by internalizing foreign transactions through FDI. Firms also use investments to facilitate learning and acquisition from hosts (Caves 1971; Dunning 2000; Luo 2000; Luo and Tung 2007). Asset-augmenting investments facilitate knowledge acquisition and transfer (for example, patents, access to human capital) through integrated corporate networks and established 'pipelines' giving rise to geographies of innovation such as close connections across clusters of innovation (Cantwell and Janne 1999; Bathelt et al. 2004; Gomes-Casseres et al. 2006). Research employing asset-exploiting and asset-augmenting versions of the paradigm commonly focus on only a single period in time, namely, the initial entry processes. However, innovation outcomes depend on the interactions between firm- and country-specific advantages not only at initial entry, but as long as the OFDI remains viable (Ernst 2002; Hennart 2009).

Globally, the pharmaceutical industry presents numerous non-static entry barriers to multinational firms pursuing foreign market expansion. Three entry barriers are of particular importance. First, the global pharmaceutical industry is an oligopolistic industry. Multinational pharmaceuticals face significant market-based barriers to foreign entry because their profit-maximizing decisions are dependent on the decisions of incumbent rivals (Matraves 1999; Morton 2000). Second, foreign pharmaceutical sales require regulatory approval by host-country governments. Pharmaceutical regulations require foreign drugs be tested for safety and efficacy prior to their approval for sale in a domestic market. Similarly, regulations commonly provide patent protection for drugs, which allows patent holders recourse in the form of infringement suits when foreign competitors enter protected drug markets (EMA 2014; USFDA 2014a). Third, pharmaceutical production requires knowledge (for example, human capital) that is often difficult to transfer across space and organizations (Feinberg and Majumdar 2001; McKelvey et al. 2003).

Collectively, the three barriers constrain the company access to geographic clusters of innovation within the pharmaceutical industry. To overcome these limitations, emerging-market pharmaceutical firms use acquisitions and formal partnerships to expand their innovation space into established clusters. Focusing on Indian pharmaceuticals, Kedron and Bagchi-Sen (2011, 2012) identify specific regulations (for example, Abbreviated New Drug Approvals) and strategies (for example, acquisition) that facilitate foreign market entry and innovation access. The authors demonstrate that initial entry strategies provide access to knowledge and expertise used as inputs for further in-house research and development (R&D). Pradhan and Abraham's (2005) study of Indian software companies demonstrates how foreign acquisitions designed to access technology unavailable in India became the platform for subsequent export success.

Horner (2013) ties the development of the global production chains of Indian pharmaceuticals to firm-specific advantages developed under different domestic institutional regimes. Those findings echo Kumar's (2008) argument that the success of Indian multinationals depends on their development of broadly applicable low-cost production advantages and accumulation of wealth over time in a regulated but captive market. Gubbi et al. (2010) similarly tie development of those advantages to institutional change within India and a need to look outside the country for new business opportunities. Analyzing foreign acquisitions of Indian multinationals, Gubbi et al. (2010) also note a preference for advanced-economy entry, particularly the US and European markets.

Ramamurti and Singh (2009) propose five internationalization strategies for emerging-market multinationals based on target market, industry conditions, and country- and firm-specific advantages. Of these five types, three are applicable to the experience of Indian pharmaceuticals. First, local optimizers base foreign-market entry on low-cost product development primarily for sale in other emerging markets. Second, multinationals using a low-cost partner strategy leverage labor-cost advantages to position themselves as component suppliers to developed markets within selected value-chain segments. Firms employing this strategy use upstream OFDI to move up their respective industry value chains, and downstream OFDI to extend production-cost advantages. Finally, global first movers are innovators operating at international technological and organizational frontiers, with global operations functionally integrated across the entire industry value chain. Aligning their conceptual framework with other emerging work on the dynamics of MNEs, Ramamurti and Singh suggest firms are capable of shifting between generic strategies and may simultaneously employ several strategies.

10.3 THE GLOBAL PHARMACEUTICAL VALUE CHAIN

Four stages of product development characterize the pharmaceutical value chain. Figure 10.1 summarizes each of these phases and possible drug development pathways (Wadhwa et al. 2008; Kedron and Bagchi-Sen 2012). New drug development begins during the discovery phase, when companies identify and isolate potential pharmaceutical compounds. Selected drug candidates then enter clinical trials to establish the safety and efficacy. Those small production runs are increased in size during a manufacturing phase reliant on high-yield chemistry. Finally, drugs are marketed and distributed globally through channels that vary according to differences in national regulatory structures (EMA 2014; USFDA 2014a).

The progress of drugs through the phases of the pharmaceutical value chain depends on candidate novelty. Innovative new concept drugs representing a first attempt to treat a disease face an extended discovery phase

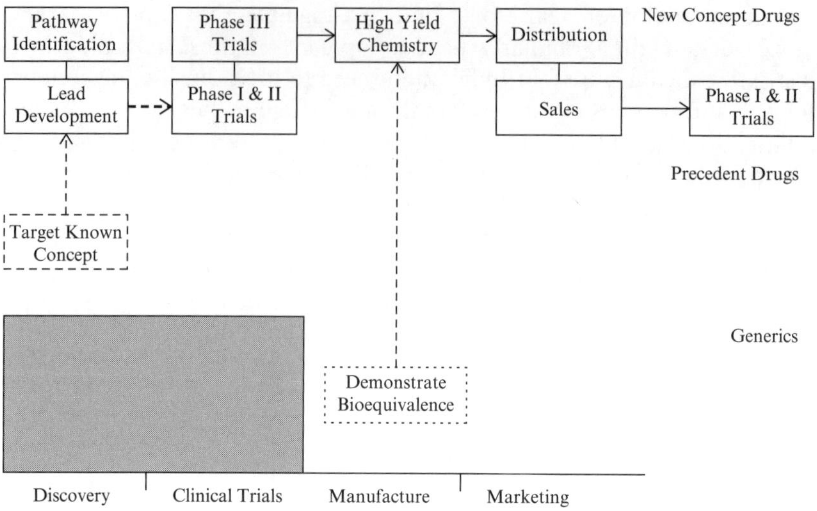

Source: Adapted from Wadhwa et al. (2008) and Kedron and Bagchi-Sen (2002)

Figure 10.1 Pharmaceutical industry value chain

as scientists test and re-test candidate drugs. In contrast, precedented drugs build on established concepts, which lessen innovation costs by narrowing the set of potential candidates. After leading candidates are identified, new concept and precedented development paths merge through the value chain. Alternatively, a firm could bypass both the discovery and clinical trials phases by developing a generic version of an off-patent drug. To successfully sell that generic drug in the world's largest pharmaceutical markets, the US and EU, the manufacturer must demonstrate bioequivalence. This involves small-scale clinical trials establishing that the generic drugs do not produce significantly different health outcomes from the original (Wadhwa et al. 2008; USFDA 2014a).

The pharmaceutical value chain is global in scope, and spatial variations in regulatory, market, and knowledge-sharing barriers require emerging-market firms to readjust their innovation strategies to local contexts (Zeller 2004; Wadhwa et al. 2008; Kedron and Bagchi-Sen 2012). The capacity of Indian pharmaceuticals to make those adjustments depends on their existing firm-specific advantages and the barriers to entry they must overcome.

10.4 EVOLUTION OF THE INDIAN PHARMACEUTICAL INDUSTRY

Hansen (2010), Mathew (2010), and Nayak (2011) segment industrial development within India into periods following passage of the 1977 Industrial Policy Statement. The authors identify a common first period characterized by shifting government policy and minimal OFDI preceding the Industrial Development Policy of 1991. Examining the Indian pharmaceutical industry, Horner (2013) notes that few Indian pharmaceuticals were directly involved in drug development or OFDI prior to 1970. The Patent Act of 1970 established a process patent regime in India that allowed drug reformulation without the threat of prosecution. Simultaneous government-led reductions in international competition through the Foreign Exchange Regulation Act of 1973 and the New Drug Policy of 1978, which increased the availability of drug targets by encouraging foreign multinationals to partner with Indian pharmaceuticals rather than maintain independent operations. In the next decade, the number of foreign pharmaceuticals operating in India decreased from 45 to 6 (Sahu 1998). Indian firms responded by entering the industry (Glenmark, DRL) or expanding operations (Ranbaxy). As Horner notes, many of India's largest pharmaceuticals were founded during this period, and developed firm-specific advantages by either supplying or learning from foreign multinationals.

Following passage of India's Industrial Development Policy in 1991, the country entered a period of industrial expansion. That policy broadly opened India's economy to trade by reducing licensing requirements and liberalizing FDI limits. Service-oriented multinationals pursuing singular investment motivations in developed markets characterized Indian OFDI during the 1990s (Hansen 2010). In 1995, Indian joined the World Trade Organization (WTO) and agreed to implement the organization's Agreement on Trade Related Aspects of Intellectual Property Rights (TRIPS) by 2005. Beginning in 2000, Indian firms started to pursue OFDI in developed and developing markets using a range of investment mechanisms. Preceded by generic export sales, selected Indian pharmaceutical firms began using greenfield investments, joint ventures, and acquisitions in OFDI. When TRIPS replaced India's process patent regime with a product patent regime in 2005, Indian pharmaceutical OFDI increased. For example, combining Ranbaxy's 2006 acquisition of Terapia with DRL's 2006 acquisition of Betapharm would account for 6 percent of India's total 2005 OFDI.

Mathew (2010) identifies the period after the implementation of TRIPS as the beginning of an innovation era within India. In particular, Mathew cites passage of the National Innovation Act of 2009, rising public sector investment in R&D, and increasing university involvement as signs of an emerging Indian national innovation system. Specific to the pharmaceutical industry, the government created institutions like the National Institute of Pharmaceutical Education and Research, to develop training and research centers in India's major drug-producing regions and fill knowledge gaps in the pharmaceutical innovation system. Indian drug producers expanded R&D expenditures to 5 percent of sales during this period, which is a large increase from an average of 2 percent from 1995 to 2005, but still below leading advanced-country incumbents that spend up to 25 percent of revenues on R&D (Department of Pharmaceuticals 2012).

Despite these advances, India's Department of Pharmaceuticals identifies a number of barriers to pharmaceutical innovation. First, overall spending within the pharmaceutical sector remains low. Total research spending by all Indian pharmaceuticals was less than research spending by Pfizer in 2012. Aggregate research spending is low in part because only 20 to 30 firms have the capacity to fund and manage the lengthy drug development process (Pharmaceutical Working Group 2012). Second, despite government efforts, the number of university–industry partnerships within the sector remains small. Third, regulatory hurdles within India hinder pharmaceutical innovation by creating delays in the drug development process.

With an innovation system with competitive advantages in development and production of bulk and generic drugs, Indian pharmaceuticals target OFDI and innovation efforts on advanced markets seeking to reduce costs through generic substitution. Sales and entry opportunities vary across nations with differing rates of generic penetration of the total market for drugs. The United States, United Kingdom, and Germany all have penetration rates of over 50 percent, while France, Italy, and Spain all have rates below 30 percent (IRCA Limited 2012). However, as numerous drugs with large markets lose patent protection a new generic market opportunity is emerging across all advanced countries. For example, Pfizer's patent for Lipitor, which expired in 2011, opened an $11 billion a year market to generic competition.

10.5 BARRIERS TO MARKET ENTRY WITHIN THE GLOBAL PHARMACEUTICAL VALUE CHAIN

A limited number of Indian pharmaceuticals have the capacity to undertake advanced-market OFDI. Approximately, 300 medium and large Indian pharmaceutical firms account for 75 percent of domestic sales and

Table 10.1 International activities of leading Indian pharmaceutical companies[1]

Company	Sales ($m)	Foreign Sales (%)	ANDA	NDA	US Patents	FDA Facilities
Ranbaxy Laboratories	1,956	76	138	20	139	7
Dr Reddy's Labs	1,580	82	162	21	131	12
Cipla	1,338	55	52	13	92	6
Sun Pharmaceuticals	1,249	51	125	4	28	8
Lupin	1,246	70	106	4	53	6
Zydus Cadila	978	71	3	0	46	13
Aurobindo Pharma	957	25	185	12	26	11
Wockhardt	819	72	171	0	46	9
Jubilant Life Sciences	750	69	23	6	20	1

Note: [1]Sales data from 2011. ANDA, NDA, patent, and facilities data as of 2013

Sources: Compiled by author from Indian Department of Pharmaceuticals (2012), USFDA (2014), USPTO (2014)

90 percent of exports (Department of Pharmaceuticals 2012). Of those firms, the top 50 exporters account for 75 percent of export sales, with the top ten accounting for 40 percent of the market (Greene 2007). Table 10.1 presents the international activities of leading Indian pharmaceuticals. In 2011, nine Indian pharmaceuticals had sales of over $750 million, of which only four had over 70 percent of revenues derived from international sales.

Lead firms target the United States and EU for export sales, with the US alone accounting for over 20 percent of sales. Access to the US market occurs primarily through the Hatch-Waxman Act's abbreviated new drug approval (ANDA) process, a regulatory barrier/facilitator, which provides a six-month period of market exclusivity for successful first-to-file generic versions of off-patent drugs. Indian pharmaceuticals have over a hundred approved ANDA in the United States, with a large portion of those filing after 2009 (USFDA 2014b).

Indian pharmaceuticals have gradually improved their positions as generic producers in international markets such as the United States and the EU. For example, Lupin is the fifth-largest generic drug firm in the United States, with 14 market-leading products. DRL has 25 drugs among the top three sellers in the US generic market and over $60 million in over-the-counter sales revenues (ICRA 2012). At the same time Indian pharmaceuticals have expanded their international activities up the value chain into R&D and innovation. Ranbaxy, DRL, and Cipla have translated generic success into new drug approvals and approved US patent applications. Balancing innovation with generic sales depends on ownership of US FDA-approved production facilities in India (to manage costs) and abroad (market entry points) that produce generics acceptable to advanced-country markets and serve as collaborative platforms. For Indian pharmaceuticals, variation in US and European entry barriers are particularly important, and lead firms use acquisitions and partnerships to overcome entry barriers and expand innovative activities.

10.5.1 Regulatory Barriers to Foreign Market Entry

Bundles of related patents applying to different portions of a drug formulation and expiring at different times protect pharmaceuticals from market competition. Foreign firms introducing competing or generic drugs are likely to face litigation from patent-holding rivals that claim infringement on some patent within the bundle. In Europe, drug approval is further complicated by the lack of a single pan-European patent system with the capacity to grant approvals that do not have to be mirrored by independent national patents. Indian pharmaceuticals must manage potential patent litigation in multiple markets, with resolution often delayed as national judiciaries await each other's rulings. To balance the innovation incentive patents provide with the need for consumer access, major pharmaceutical markets have introduced regulations offering market incentives for first-to-file generic producers seeking to replace expiring patents. In the US, the Hatch-Waxman Act creates incentives for generic entrants, but also encourages incumbents and third-party entrants to use litigation to slow generic entrants (Kedron and Bagchi-Sen 2011). For example, litigation delayed DRL's introduction of a generic version of Zofran for five years as GlaxoSmithKline, Teva, and Apotex brought patent infringement lawsuits against the firm in attempts to either defeat or defend ANDA approval.

Indian pharmaceuticals that overcome patent barriers register their production facilities with the US and EU regulatory agencies to facilitate sales and value-chain coordination. In 2014, 509 Indian pharmaceutical facilities were registered with the USFDA (USFDA 2014a). Once in a market, pharmaceuticals must also navigate regulatory variations in marketing and distribution. For example, many European governments use a tendering system to procure drugs for hospital care, and in some cases ambulatory care, and restrict distribution to pharmacies. This system privileges government contracts and procurement, placing an emphasis on marketing networks. In contrast, the US system of drug distribution restricts some pharmaceutical products to prescription-only purchasing, while allowing over-the-counter purchases of drugs with broader safety margins.

Indian pharmaceuticals develop distribution networks and overcome regulatory barriers in advanced countries by acquiring firms with existing marketing networks and drug pipelines. Table 10.2 presents selected examples of the largest acquisitions of Indian pharmaceuticals with operations in multiple advanced economies.

While acquisitions vary in value, each serves as an entry point into national markets or specific market segments. For example, acquisitions by Ranbaxy and DRL provided those firms with production facilities, approved drug pipelines, and marketing networks in Germany, France, and Italy. Ranbaxy used its 2006 acquisition of Terapia to establish a European production hub at two facilities in low-cost Romania. The company used a similar strategy to support US entry in 1995, when the firm purchased Ohm Labs. In contrast, Sun Pharmaceuticals used a 2010 acquisition of Israeli pharmaceutical Taro to augment assets in US and Canadian production. Smaller acquisitions like Wockhardt's 2003 purchase of German producer Espharma or Lupin's purchase of Hormosan Pharma target specific market segments.

10.5.2 Incumbent Response as a Barrier to Foreign Market Entry

The oligopoly structure of the pharmaceutical market incentivizes incumbent producers to use price and non-price mechanisms to control Indian pharmaceutical entry. When patent litigation fails or a generic drug is poised to enter a market, incumbent producers license their patents to Indian entrants as approved generics. Licensing allows incumbents to recuperate revenue losses and limit competition from foreign entrants to partner firms, while allowing selected Indian generic partners expedited market entry. Incumbent producers combine pharmaceutical licensing with price reductions and reformulation of challenged drugs. By reformulating licensed or off-patent drugs, incumbent producers gain patent protection for competing pharmaceuticals marketed as an improved product. This strategy mitigates incumbent market share loss and directs innovation toward incremental adjustment of existing drugs. For example, although both DRL and Ranbaxy received licensing approval to sell generic versions of the drug Imitrex, GlaxoSmithKline (GSK) released a reformulation of that drug prior to generic entry that eroded generic market share.

Table 10.2 Mergers and acquisitions of selected Indian pharmaceuticals, 2000–2014

Date	Company	Acquisition	Country	Value	Description
2003	Wockhardt	Espharma	Germany	$11m	Entry point to German market with a focus on urology and neurology
2004	Ranbaxy Laboratories	RPG Aventis	France	$70m	Acquires 52-drug generic product line and 6 percent French generic market share
2006	Ranbaxy Laboratories	Terapia SA	Romania	$324m	Purchases two production facilities with EU accreditation, used as European sales hub
2006	Ranbaxy Laboratories	Allen SPA	Italy	–	Acquires two Italian production facilities and large marketing and distribution labor force
2006	Dr Reddy's Laboratories	Betapharm	Germany	$570m	Acquires 145-product portfolio for the German market along with sales network and production facilities
2007	Wockhardt	Negma	France	$256m	Purchase of fourth-largest French pharma group made Wockhardt largest Indian pharmaceutical in Europe
2008	Dr Reddy's Laboratories	Dow Small Molecule	UK	–	Gains compound development capabilities and a platform to partner with incumbent producers
2008	Lupin	Hormosan Pharma	Germany	$10m	Serves as European sales outlet for Lupin and gained licensing experience in nervous system drugs
2008	Aurobindo Pharma	TAD Italy	Italy		Gains access to 70 product pipeline and creates entry point to the Italian market
2010	Sun Pharmaceuticals	Taro	Israel	$500m	Purchase gives Sun production facilities in the US and Canada to complement existing US facilities
2011	Zydus Cadila	Nesher Pharma	US	–	Purchases manufacturing and marketing capabilities in the US generic controlled-release drug market

Source: Compiled by author from Factiva, Lexis-Nexis, and company websites and annual reports

While a barrier to generic sales of Indian pharmaceuticals, interdependent oligopoly pricing facilitates expansion of Indian pharmaceuticals in other value-chain segments. Incumbent producers contract Indian pharmaceuticals to manufacture advanced pharmaceutical ingredients (API) and complete early stage clinical trials at FDA-registered facilities in India. This strategy minimizes the loss of firm-specific advantages through uncompensated knowledge spillovers or the reverse engineering of patented drugs, that is, incumbent pharmaceuticals use formalized partnering agreements with Indian pharmaceuticals to control knowledge transfer. In those agreements Indian pharmaceuticals leverage expertise in reverse engineering and high-yield chemistry to develop new drug formulations based on the API. Sales rights for drugs developed in these ventures are shared between incumbent and Indian pharmaceuticals and often segmented by geography. Table 10.3 outlines selected partnerships between incumbent producers and Indian pharmaceuticals. The partnerships presented were chosen to illustrate the variety of collaborative arrangements among the subset of Indian pharmaceuticals with foreign market operations in multiple advanced markets.

Partnerships between Indian and incumbent pharmaceuticals range in form from joint ventures to co-marketing and licensing agreements. For example, Ranbaxy's 2003 joint venture with GSK restricted the firm's role to the development of high-yield chemistry processes for selected drug candidates. A 2007 renewal of that agreement expanded that role to include clinical trials and candidate selection. While still tying compensation from GSK to milestones in the candidate development process, the new arrangement did expand Ranbaxy's role across the drug development segment of the value chain. Jubilant's similar agreement with Eli Lilly called for the Indian pharmaceutical's development of candidate drugs, which upon reaching a phase II milestone would be out-licensed to Eli Lilly for sale. Joint venture agreements between DRL and GSK, Sun Pharmaceuticals and Merck, and Cadila and Abbott all seek to leverage Indian production cost advantages and incumbent drug pipelines for sales in emerging markets.

Some partnerships have reversed the outward orientation of Indian investment and sales. Rather than generating drugs for sale in markets outside of India, agreements between Lupin and Eli Lilly, and Cadila and Bayer use Indian sales networks to allow incumbent producers to enter the Indian market. The 2008 purchase of Ranbaxy by Daiichi Sankyo for \$4.6 billion and the 2010 purchase of Piramal Healthcare by Abbott for \$3.7 billion gave those companies access to the Indian market and generic product pipelines (Ranbaxy 2008; Lee and Pettypiece 2010). The capabilities of Indian pharmaceuticals to produce low-cost generic drugs are

Table 10.3 Partnership activities of selected Indian pharmaceuticals, 2000–2014

Date	Company	Partner	Country	Venture	Description
2003	Ranbaxy Laboratories	GlaxoSmithKline	US	Joint Venture	Ranbaxy develops manufacturing processes for candidates, role expanded in 2007 to include to drug selection
2008	Jubilant Life Sciences	Eli Lilly	India	Joint Venture	Jubilant to develop molecule through phase II clinical trials, which are then out-licensed to Eli Lilly for marketing and sale
2009	Dr Reddy's Laboratories	GlaxoSmithKline	Multiple	Joint Venture	Venture where DRL manufactures GlaxoSmithKline products for sale in emerging markets
2010	Zydus Cadila	Abbott Laboratories	Multiple	Licensing	Cadila licenses branded generics to Abbott for sales in emerging-market economies
2011	Sun Pharmaceuticals	Merck	Multiple	Joint Venture	Venture to market and manufacture generic versions of Merck products in emerging markets
2011	Wockhardt	Sheffield Bioscience	UK	Co-Market	Sheffield agrees to sell Wockhardt's insulin products in a strategic marketing and development alliance
2011	Zydus Cadila	Bayer	India	Joint Venture	Marketing and sales agreement that allows Bayer to sell drugs into the Indian market
2011	Lupin	Eli Lilly	India	Co-Market	Agreement to bring Eli Lilly's insulin products into the Indian market through Lupin's sales network
2013	Sun Pharmaceuticals	Intrexon	US	Joint Venture	Development agreement to produce gene-based therapies for ocular diseases

Source: Compiled by author from Factiva, Lexis-Nexis, and company websites and annual reports

attractive assets as best-selling drugs lose patent protection. For example, Daiichi Sanyo's acquisition of Ranbaxy included the company's six-month exclusivity for US sales of generic Lipitor, the world's best-selling drug. Similarly, Abbott's purchase of Piramal Healthcare provided the company the rights to 350 brands and generics and a 7 percent share in the Indian pharmaceutical market (Pharmaceutical Working Group 2012).

10.6 CONCLUSIONS AND IMPLICATIONS

This chapter shows the emergence and impact of Indian pharmaceutical companies on innovation, production, and investment within the global pharmaceutical industry. Indian pharmaceutical investment accumulated wealth from selling generics and with this wealth, they now target developed countries to acquire knowledge, patents, and so on, to augment their in-house innovation. Indian companies have undertaken multiple forms of investment guided by multiple strategies tied to firm-specific advantages developed under varied regulatory environments. Indian pharmaceuticals do not follow a single foreign market entry strategy or pathway. Rather Indian pharmaceuticals use flexible OFDI arrangements to both exploit and augment existing firm-specific advantages.

Expanding on a major theme of this book, this chapter explores how Asian firms are creating global innovation spaces by crossing geographic boundaries. The innovative activities of Indian pharmaceuticals depend on their ability to engage in global partnerships, execute international acquisitions, and maintain locally developed firm-specific advantages. To develop and convert innovations into business success, Indian pharmaceuticals must overcome spatial, regulatory, and market barriers that inhibit their participation in global innovation networks and their expansion into new segments of their industry's value chain. Flexibility is critical to overcome entry barriers tied to market structure, regulatory environments, and limits to knowledge transfer. Moreover, entry barriers are themselves dynamic as incumbents adjust their strategies (for example, incremental adjustment of existing drugs) to use barriers in different combinations and inhibit Indian entry. Acquisitions provide a platform for both generic sales and innovation partnerships. In the United States, Indian firms now hold not only generic drug approvals, but also new drug approvals and patents. In Europe, Indian pharmaceuticals operate sales and distribution networks in leading markets, and have approved drug pipelines for future sales.

Although infrequently discussed in individual studies, the research literature on Indian pharmaceuticals collectively illustrates that the innovative activities of these firms are built on innovation systems that are international in scale. While studies commonly emphasize the production cost and reverse-engineering advantages of Indian pharmaceuticals developed within a nationally specific system of innovation, they commonly underplay how leveraging those advantages allow these firms to position themselves within the global pharmaceutical value chain using international alliances centered on drug development. The balance that Indian pharmaceutical strike between regional, national, and global innovation networks supports the notion that Asian innovation spaces must be understood beyond the regional or national scale, and must be thought of as more than centers of incremental low-cost innovation.

Looking forward, both the global operations of Indian pharmaceuticals and their position within the industry's value chain appear uncertain. Production advantages, innovation success, and expansion into advanced markets have made Indian pharmaceuticals attractive acquisition targets for incumbent producers. Ongoing growth of the Indian pharmaceutical market, the capacity to develop generic drugs at a time when advanced nations are lowering drug costs through generic substitution, and rights to generic sales to off-patent blockbuster drugs have motivated acquisition of leading Indian pharmaceuticals like Ranbaxy and Piramal. Identifying these opportunities remaining Indian pharmaceuticals are now also acquiring other Indian pharmaceuticals, continuing the evolution of the country's influence in the industry's global value chain. However, competition from incumbents remains (for example, incremental adjustments to off-patent drugs) and will continue to pose challenges to market expansion in the United States and the EU.

NOTE

1. The authors would like to acknowledge the support of our respective universities. Moreover, Professor Bagchi-Sen would like to acknowledge the support of the companies and foundations who support the Visiting Professor's Program at the School of Business, Economics, and Law, University of Gothenburg. See the Acknowledgments section at the beginning of this book for more details.

REFERENCES

Asmussen, C., G. Benito and B. Petersen (2009), 'Organizing foreign market activities: from entry mode choice to configuration decisions', *International Business Review*, **18** (2), 145–55.

Athreye, S. and A. Godley (2009), 'Internationalization and technological

leapfrogging in the pharmaceutical industry', *Industrial and Corporate Change*, **18** (2), 295–323.
Athreye, S. and S. Kapur (2001), 'Private foreign investment in India: pain or panacea?', *The World Economy*, **24** (3), 399–424.
Athreye, S. and S. Kapur (2009), 'Introduction: the internationalization of Chinese and Indian firms – trends, motivations and strategy', *Industrial and Corporate Change*, **18** (2), 295–323.
Bathelt, H., A. Malmberg and P. Maskell (2004), 'Clusters and knowledge: local buzz, global pipelines and the process of knowledge creation', *Progress in Human Geography*, **28** (1), 31–56.
Benito, G., B. Petersen and L. Welch (2009a), 'Towards more realistic conceptualizations of foreign operation modes', *Journal of International Business Studies*, **40**, 1455–70.
Cantwell, J. and O. Janne (1999), 'Technological globalization and innovation centres: the role of corporate technological leadership and locational hierarchy', *Research Policy* **28** (2–3), 119–44.
Caves, R. (1971), 'International corporations: the industrial economics of foreign investment', *Econometrica*, **38** (149), 1–27.
Department of Pharmaceuticals (2012), *Annual Report 2011–12*, New Delhi: Ministry of Chemicals and Fertilizers, Government of India.
Dunning, J. (1980), 'Toward an eclectic theory of international production: some empirical tests', *Journal of International Business Studies*, **11** (1), 9–31.
Dunning, J. (2000), 'The eclectic paradigm as an envelope for economic and business theories of MNE activity', *International Business Review*, **9** (2), 163–90.
Dunning, J. and S. Lundan (2008), *Multinational Enterprises and the Global Economy*, Northampton, MA, USA and Cheltenham, UK: Edward Elgar.
Ernst, D. (2002), 'Global production networks and the changing geography of innovation systems: Implications for developing countries', *Economic Innovation New Technology*, **11** (6), 497–523.
European Medicines Agency (EMA) (2014), 'Human medicines: regulatory information'. Available at: http://www.ema.europa.eu/ema/index.jsp?curl=pages/regulation/landing/human_medicines_regulatory.jsp&mid= (accessed 20 October 2014).
Feinberg, S. and S. Majumdar (2001), 'Technology spillovers from foreign direct investment in the Indian pharmaceutical industry', *Journal of International Business Studies*, **32** (3), 421–37.
Gomes-Casseres, B., Hagedoorn, J. and A. Jaffe (2006), 'Do alliances promote knowledge flows', *Journal of Financial Economics*, **80** (1), 5–33.
Greene, W. (2007), 'The emergence of India's pharmaceutical industry and implications for the US generic drug market', Office of Economics Working Papers, US International Trade Commission.
Gubbi, S., Aulakh, P., Ray, S., Sarker, M. and R. Chittoor (2010), 'Do international acquisitions by emerging-economy firms create shareholder value? The case of Indian firms', *Journal of International Business Studies*, **39**, 1220–35.
Hansen, M. (2010), 'In search of the "Indianness" of Indian multinational enterprises: is there anything special about the Indian path to outward foreign direct investment?', in Sauvant, K., Pradhan, J., Chatterjee, A. and B. Harley (eds), *The Rise of Indian Multinationals: Perspectives on Indian Outward Foreign Direct Investment*, London: Palgrave Macmillan, pp. 25–56.

Hennart, J. (2009), 'Down with MNE-centric theories! Market entry and expansion as the bundling of MNE local assets', *Journal of International Business Studies*, **40**, 1432–54.

Hoffman, J. (2005), 'Generics growth in the US and the EU: the role of India', *Journal of Generic Medicines*, **3**, 7–19.

Horner, R. (2013), 'Strategic decoupling, recoupling and global production networks: India's pharmaceutical industry', *Journal of Economic Geography*, **14** (6), 1117–40.

Hymer, S. (1976), *The International Operations of National Firms: A Study of Direct Investment*. Cambridge, MA: MIT Press.

IRCA Limited (2012), 'Indian pharmaceutical sector – industry update'. Available at: http://icra.in/Files/ticker/Indian%20Pharmaceutical%20Sector.pdf (accessed 15 November 2013).

Kedron, P. and S. Bagchi-Sen (2011), 'US market entry processes of emerging multinationals: a case of Indian pharmaceuticals', *Applied Geography*, **31**, 721–30.

Kedron, P. and S. Bagchi-Sen (2012), 'Foreign direct investment in Europe by multinational pharmaceutical companies from India', *Journal of Economic Geography*, **12**, 809–39.

Kumar, N. (2008), 'Internationalization of Indian enterprises: patterns, strategies, ownership advantages, and implications', *Asian Economic Policy Review*, **3**, 242–61.

Lee, L. and S. Pettypiece (2010), 'Abbot's Piramal deal makes it India's biggest drug maker'. Available at: http://www.bloomberg.com/news/2010-05-21/abbott-laboratories-to-buy-piramal-healthcare-solutions-for-2-12-billion.html (accessed 20 June 2014).

Luo, Y. (2000), 'Dynamic capabilities in international expansion', *Journal of World Business*, **35** (4), 355–78.

Luo, Y. and R. Tung (2007), 'International expansion of emerging market enterprises: a springboard perspective', *Journal of International Business Studies*, **38**, 481–98.

Mathew, G. (2010), *India's Innovation Blueprint: How the World's Largest Democracy is Becoming an Innovation Superpower*. Oxford: Chandos Publishing.

Matraves, C. (1999), 'Market structure, R&D and advertising in the pharmaceutical industry', *The Journal of Industrial Economics*, **47** (2), 169–94.

McKelvey, M., H. Alm and M. Riccaboni (2003), 'Does co-location matter for formal knowledge collaboration in the Swedish biotechnology-pharmaceutical sector?', *Research Policy*, **32** (2), 483–501.

Meyer, K. and O. Thaijongrak (2013), 'The dynamics of emerging-economy MNEs: how the internationalization process can guide future research', *Asia Pacific Journal of Management*, **30**, 1125–53.

Morton, F. (2000), 'Barriers to entry, brand advertising, and generic entry in the US pharmaceutical industry', *International Journal of Industrial Organization*, **18** (7), 1085–104.

Nayak, A. (2011), *Indian Multinationals: The Dynamics of Explosive Growth in a Developing Country Context*. London: Palgrave Macmillan.

Pradhan, J. (2007), 'Trends and patterns of overseas acquisitions by Indian multinationals', Institute for Studies in Industrial Development Working Paper Series No. 2007/10.

Pradhan, J. (2008), 'The evolution of Indian outward foreign direct invest-

ment: changing trends and patterns', *International Journal of Technology and Globalization*, **4** (1), 70–86.
Pradhan, J. and V. Abraham (2005), 'Overseas mergers and acquisitions by Indian enterprises: patterns and motivations', *Indian Journal of Economics*, **85**, 365–86.
Pharmaceutical Working Group (2012), 'Working groups/steering committees for the twelfth five-year plan (2012–2017)', New Delhi: Planning Commission, Government of India. Available at: http://planningcommission.gov.in/aboutus/committee/index.php?about=12strindx.htm (accessed 20 November 2013).
Ramamurti, R. and J. Singh (2009), *Emerging Multinationals in Emerging Markets*. New York: Cambridge University Press.
Ranbaxy (2008), 'Ranbaxy and Daiichi Sankyo successfully complete landmark deal'. Available at: http://www.ranbaxy.com/ranbaxy-and-daiichi-sankyo-successfully-complete-landmark-deal/ (accessed 20 June 2014).
Sahu, S. (1998), *Technology, Transfer, Dependence, and Self-Reliant Development in the Third World: The Pharmaceutical and Machine Tool Industries in India*. Westport, CT: Praeger Press.
Santangelo, G. and K. Meyer (2011), 'Extending the internationalization process model: increases and decreases of MNE commitment in emerging economies', *Journal of International Business Studies*, **42**, 894–909.
United Nations Conference on Trade and Development (UNCTAD) (2013), 'UNCTAD Stat'. Available at: http://unctadstat.unctad.org/wds/ReportFolders/reportFolders.aspx?sCS_ChosenLang=en (last accessed 16 July 2013).
US Food and Drug Administration (USFDA) (2014a), 'Development and approval process (drugs)'. Available at: http://www.fda.gov/Drugs/DevelopmentApprovalProcess/default.htm (accessed 20 March 2014).
USFDA (2014b), 'Drug approval reports'. Available at: http://www.accessdata.fda.gov/scripts/cder/drugsatfda/index.cfm?fuseaction=Reports.ReportsMenu (accessed 10 January 2014).
US Patent and Trademark Office (USPTO) (2014), 'Patent full-text and image database'. Available at: http://www.uspto.gov/patft//help/images.htm (accessed 20 October 2014).
Wadhwa, V., B. Rissing, G. Gereffi, J. Trumpbour and P. Engardio (2008), *The Globalization of Innovation: Can India and China Cure the Global Pharmaceutical Market?* Kansas City, MO: The Kauffman Foundation.
Zeller, C. (2004), 'North Atlantic innovation relations of Swiss pharmaceuticals and the proxies with regional biotech areas', *Economic Geography*, **80** (1), 83–111.

11. The emergence of Indonesian multinational enterprises

Teuku Arckyansyah Meraxa and Sharmistha Bagchi-Sen

11.1 INTRODUCTION

This chapter examines innovation spaces developing in Indonesia, through multinational enterprises (MNEs).[1] Hence, the purpose of this chapter is to examine the characteristics of a selected group of multinationals from Indonesia to understand how they have leveraged country-specific advantages to draw attention from the international business community to Southeast Asia. In recent years, the focus of research on emerging markets in Asia has been China and India. Indonesia, with a population size of over 240 million, ranks fourth in the world with less than 15 percent involved in agriculture, a gross domestic profit (GDP) growth rate of about 5.8 percent and a country where trade imbalance is not an issue. The potential size of the market is large with about 43 percent of the population below 25 years of age. Indonesia receives more foreign direct investment (FDI) than its outward FDI (OFDI). Although most of its major trading partners are in Asia, slightly over 6 percent of both exports and imports, respectively, is with the United States (CIA, 2013). The goal of this chapter is to study key characteristics of a selected group of Indonesian multinationals with a specific goal of identifying their innovative strategy pertaining to their emerging-market multinational status.

In 2012, the Boston Consulting Group (BCG) published a report entitled *The Companies Piloting a Soaring Region* (Chin et al, 2012). In this report, BCG identified 50 companies from Southeast Asia, coined as the "Southeast Asia Challengers." The companies, originated in Indonesia, Vietnam, Thailand, Malaysia, Singapore and the Philippines; these companies are presumed to be fast-growing and expanding rapidly within the region or globally. Boston Consulting Group developed its own method of analysis, which classifies the companies as "Challengers" (Chin et al., 2012, p. 9). The major criteria to become the Asian Challengers are: (1)

"They are expanding twice as fast as their 'global peers' – multinationals in similar industries headquartered in developed markets – and are more profitable"; and (2) these companies typically have completed major international acquisition deals in the past few years, to increase their share in the overseas market.

11.2 BACKGROUND: THE GROWTH OF EMERGING-MARKET MULTINATIONAL ENTERPRISES

Research in international business on latecomers has now spanned more than two decades. The emergence of global networks shows how theoretical developments based on Western MNEs can be used to explain the geographic patterns. At the same time, authors have asked if we need new theoretical arguments to explain the emergence of latecomers in global business or new multinationals. The coinage of the term "BRIC," for Brazil, India, Russia, China brought further attention to specific cases from China and India – these two countries have a large population and with regulatory changes, they have emerged as major contenders of the Triad (United States, the European Union [EU] and Japan) in terms of trade and direct investment. China's rise is explained through its experimentation with modern capitalism and the entry to World Trade Organization (WTO) to boost trade (O'Neill, 2001); India's ascendance in international business is analyzed using its historical roots in colonial trade (Chattopadhyay et al., 2012; Ramamurti, 2009 and 2012).

Since the 1990s, technological advancements, political shifts, and trade negotiations have been unprecedented. Following the collapse of communism, the conclusion of the Uruguay Round Trade deal, and the realization of the WTO, globalization took off. The deregulation and privatization of telecommunications in many countries, radical changes in computing technology and the rise of the Internet dramatically altered cross-border business strategies, including the formation of global production and innovation networks. These developments fueled the outsourcing and off-shoring trends of multinationals, resulting in complex network development to accommodate various types of organizational structures for supply chain development and distribution networks with the end result of lowered entry barriers for emerging-market multinationals in the highly regulated countries of Western Europe and the United States (Evans and Wurster, 2000; see also Chapter 10 in this book). Meanwhile, capital markets also became more open and integrated than ever before, making it easier for emerging-market multinationals to raise foreign equity capital and debt, or to list their shares on foreign stock exchanges (Farrell et al., 2005). Easier access to capital through the global integration of financial centers such as London, Singapore, Tokyo and New York allows companies to rapidly expand their presence and convert into a global player (Sassen, 2006). Multinationals from emerging markets also benefit from the emergence of a global labor market – they can hire senior managers with international experience (Michaels et al., 2001) and establish offices around the world to create new geographies of knowledge and technology-intensive industries. See Chapters 2, 3, 14 and 15 in this book regarding patents and also software and information technology.

In their home countries, emerging-market multinationals have historically worked with poor infrastructure (for example, limited number of airports), limited institutional support and low income levels. These economic, institutional and infrastructural barriers forced emerging-market multinationals to become innovative (for example, reaching the bottom of the pyramid) to overcome weaknesses in their home countries and adapt their operations and innovations for similar economies (Khanna and Palepu, 2010; Ramamurti, 2012). Ramamurti (2012) contextualizes the emerging-market multinationals from India and China within the well-established eclectic paradigm developed by Dunning (1981; see also Dunning and Lundan, 2008). The eclectic paradigm includes offering a discussion of competitive advantages that allow firms to become multinationals.

Emerging-market multinationals typically have two types of competitive advantages. First, country-specific advantage (CSA) is a type of advantage that is offered by a particular country, which may help strengthen operations. Examples of CSAs include cheap labor and abundant natural resources, which can also be accessed by foreign firms, and are therefore not exclusively for local or domestic use (Rugman, 2009; Ramamurti, 2012). Secondly, firm-specific advantages (FSAs) include several capabilities that undergird firms' international competitiveness (Guillén and García-Canal, 2009): (1) insight into customer needs; (2) ultra-low cost production; (3) frugal innovation; (4) operational excellence in adverse environments; (5) privileged access to resources and markets; and (6) traditional first-mover advantages (Ramamurti, 2009).

Table 11.1 shows how Ramamurti has classified emerging-market multinationals and discussed these advantages altogether for each category (see Table 11.1 for definitions): (1) Local optimizer takes advantage of cheap labor at home; (2) Low-cost partner takes on contractual work from foreign multinationals and provides entry of foreign multinationals to the home market; (3) Global consolidators are those with accumulated wealth can acquire foreign operations or undertake new investments to

Table 11.1 Foundations of internationalization by emerging-market multinational enterprises

Generic internationalization strategy	Home country-based competitive advantage	Reasons for internationalization	Examples	Possible entrants from Indonesia
1. Local optimizer	• Ability to optimize products for home market by making them ultra-cheap and easy to use • Consumer insight on how to trade off price and product features	• To introduce products optimized for the home market into emerging markets	HiSense, Mahindra & Mahindra, Tata Motors, Shoprite, Marcopolo	Mayora, Wings, ABC Group, GarudaFood, Indofood, Djarum
2. Low-cost partner	• Ability to leverage home country's talent to engage in labor-cost arbitrage despite adverse business environment • Project management skills	• To diversify the low-cost location from which customer in high-cost countries can be served • To move up the value curve by getting closer to key customers in developed countries	Wipro, Infosys, TCS, GVK Bio, WEG, Sabo	Astra Otoparts (Astra International)

Table 11.1 (continued)

Generic internationalization strategy	Home country-based competitive advantage	Reasons for internationalization	Examples	Possible entrants from Indonesia
3. Global consolidator	• Production and project execution excellence in capital-intensive process industries • Late-mover advantages in scale, organizational processes, and technology • Strong position in large and growing home market, resulting in strong cash flows	• To build scale and lower cost • To acquire complementary technologies and capabilities • To improve (turn around) the performance of acquisitions abroad	Tata Steel, Hindalco, South African Breweries, Lenovo, Wanxiang, Cemex, Haier	Aneka Kimia Raya
4. Natural-resource vertical integrator	• Privileged access to natural resources • Privileged access to large home markets	• To enable forward integration to secure markets • To enable backward integration to meet growing domestic demand for natural	Gazprom, Lukoil, Norilsk, Vale, Anglogold, PTT; Petrobras, Oil & Natural Gas Commission, Indian Oil, China National Offshore Oil	Adaro, Bumi Resources, Bayan Resources, Bakrie Oil & Gas, Medco Energy

		resource-based products	Corporation, Chinalco	
5. Global first mover	• Large home demand that helps local firms(s) launch in an emerging industry • Sometimes aided by home country leapfrogging to latest technology (for example, wireless telephony or renewable energy) • Sometimes significant state support	• To acquire key missing technologies and capabilities necessary to compete in new emerging industries • To acquire global customers, scale and reach	Embraer, Huawei, ZTE, Suzlon Energy, Dr. Reddy's Pharmaceuticals	Not available yet

Source: Based on Ramamurti (2012: 246) and adapted by authors to Indonesia

consolidate markets which may have been export markets; (4) Natural resources vertical integrator are involved in integration of backward links or supply chains into production and distribution; and (5) Global first mover are leaders in certain niche markets.

The BCG classification of Southeast Asian multinationals matches the strategies discussed by Ramamurti for each of his categories, and an overview can be found in Table 11.1

According to BCG, the Southeast Asian Challengers[2] have unique growth strategy, which is labeled by BCG as: (1) Natural resources play – companies controlling large natural resources extraction zone, and may have developed sophisticated supply chain system; (2) Regional rollout – companies that create regional leadership in the market, through understanding consumer knowledge and operations within them; (3) Masters of emerging markets – companies that actively pursue expansions in emerging markets, by utilizing the practice, business models and consumer perspective gained in their home markets; and (4) Global niches – companies that dominate the worldwide niche industry. Examples of Indonesian multinationals may not fit every group (for example, global niches category) or depending on their strategy for market expansion beyond Indonesian borders, they may fit more than one group (see Table 11.1). One major reason is that Indonesia is a latecomer catching up with China, India and others in the region and will show some level of unique characteristics given its roots in resource-based trading or commodity trading (for example, oil and gas, electrical appliances, plywood, textiles, rubber). This study uses three cases which illustrate characteristics of innovative strategies used by Indonesia's emerging-market multinationals – the discussion is contextualized within the literature on emerging-market multinationals, which has its roots in Dunning (1981).

11.3 ECONOMIC DEVELOPMENT IN INDONESIA

Being the largest economy in Southeast Asia, Indonesia has been considered as a potential new entrant for BRIC (Chetan et al., 2009; Farzad, 2010; Vaswani, 2013). Economic development has been strong over the past decade since the Asian economic crisis. Indonesia has maintained a 5–6 percent GDP growth, which exceeds the rate of its regional peers. In addition, Indonesia has managed to withstand the global economic crisis of 2008 due to its robust domestic consumption pattern. Indonesia's growth has been fueled by the rise of the middle class, an increase in domestic consumption, and labor productivity improvements (Oberman et al., 2012). Rising sectors are consumer goods, manufacturing and financial services.

Meanwhile, Indonesia also expects to have greater growth in industrial zones and cities (Rastogi et al., 2012). Indonesia has stabilized after the Asian financial crisis; its remarkable recovery and political environment of democracy makes it a key player in ASEAN (Association of Southeast Asian Nations) and APEC (Asia-Pacific Economic Cooperation) (Srinivas, 2010; Barton, 2011). Its 2013 GDP was about $868 billion, the highest in the region (World Bank, 2013).

Similar to India, Indonesia emerged from the colonial era. The nation had been colonized by the Dutch since the seventeenth century. Indonesia's economic output was very important for the Dutch kingdom – the archipelago was a primary exporter of nutmeg, pepper, coffee and spices to Europe. The Dutch controlled major ports such as Makassar, Batavia (the former name of Jakarta) and other ports in Sumatra (Dick et al., 2002). The Dutch boosted Indonesia's export and trade networks in the eighteenth century. Indonesia gained independence in 1945, and the Dutch presence lasted until 1949. Remnants of the Dutch rule are port development, trade networks, agricultural production and transportation infrastructure (Dick et al., 2002).

After a power struggle between Indonesia's first President, Sukarno, and Suharto, who came next, Indonesia entered the New Order Era in 1966. Indonesia began its industrialization era following the 1967–73 oil boom and deregulation in the 1980s (Hill, 1997). Target oil revenue to develop state-owned enterprises (SOEs), targeted industrial upgrade, food processing, footloose labor-intensive industries, heavy processing, metal goods, oil and gas became the key to rapid industrialization. By mid-1997, Indonesia's trade balance had a positive outlook with $900 billion surplus, low inflation and $20 billion foreign exchange reserves (Pepinsky, 2008). However, the currency contagion effect, which had started with the collapse of the Thai baht, and after it was floated against the US dollar, caused Indonesia's rupiah to decline. By 1998, Indonesia's economy had collapsed. Most Indonesian banks borrowed US dollars, which caused the financial system to collapse as banks closed down. Crony capitalism, where close ties existed between the private sector and the government under Suharto's ruling through patronage, had also contributed to this financial collapse because loans were given to cronies without guarantees or extensive reviews (Pepinsky, 2008).

Indonesia has stabilized over the past decade and has opened up opportunities for businesses to flourish despite internal issues with disputes regarding land. The International Monetary Fund (IMF) has praised Indonesia for its stability in maintaining democracy and robust economic growth for a decade after the 1997 Asian economic crisis (Rumbaugh, 2012).

Scholars have done limited research on Indonesia's business growth. Past studies have focused on crony capitalism, political economy, Suharto's regime and Indonesian Chinese and military-owned companies (Robison, 1986; Dick et al., 2002). However, information about, and analysis of, Indonesia's corporations in the modern era is still unexplored. Data are scarce and fragmented, with majority of information sources being Indonesia's business media, such as *Bisnis Indonesia*, equity research publications or other forms of news. One of the most extensive scholarly coverages on Indonesian companies was an article written by Lindblad in 2007, which provided extensive discussion on major Indonesian conglomerates by offering historical, business and political perspectives. According to Lindblad (2007), Indonesia's top 200 conglomerates contributed to about 3 percent of the country's GDP in the year 1991. The top 40 largest private companies in Indonesia generated $20.6 billion of revenues in 1988 (Lindblad, 2007 p. 72). In 2011, the landscape has changed as there have been many new major companies which are now active in Indonesia. Fortune Magazine Indonesia published the 2011 list of the nation's top 100 companies based on revenue size (cited in Partawidjaja, 2012). In 2011, the top 40 private Indonesian companies made an estimated US$97.5 billion in revenues (IDR 1336 trillion, at currency conversion rate at 31 December, 2011 US$1 = IDR8968.61). When compared to Lindlad's 1988 estimation, over the period of 23 years, the increase is quite significant.

New names and players such as Adaro, Medco Energy and Bank Mega have also emerged. In Indonesia, SOEs also play major roles as they control several of the major industries, while enjoying the benefits of strong government relations such as financing and policies (Delteil et al., 2012). Some of the famous SOEs include Pertamina (oil and gas), Aneka Tambang (mining), Pusat Listrik Negara (electricity and utilities), Bank Mandiri, Bank Negara Indonesia and Bank Rakyat Indonesia (banking and financing), and Garuda Indonesia (Indonesia's national airline carrier). The total number of SOEs in Indonesia is 140 and the combined revenue was Rp 968 trillion (18 percent of the GDP) in 2009 (Abubakar, 2010).

However, despite the presence of large corporations in Indonesia, none emerged to become major multinationals such as those present in Singapore, Malaysia or Thailand (Carney and Dieleman, 2011). Two hypotheses have been developed to explain this: (1) *hidden dragons* – international activities done by Indonesian corporations which are not officially revealed, perhaps due to taxation purposes; and (2) *missing dragons* – simply because Indonesian firms do not have connections overseas and they are not willing to expand to become a multinational corporation (Carney and Dieleman, 2011).

Indonesia is a unique market: SOEs control most of the energy and welfare industries.

Domestic business control and MNEs control most of trading houses and services, which are less regulated. Several private companies and conglomerates compete in an oligopolistic market (Delteil et al., 2011). In the context of technological innovation, Indonesia has strong dependence on international influences on the development of technology-oriented industries. The use and growth of science and technology mainly depend on purchases from overseas. Machineries, computer systems and capital goods are mainly imported from countries such as Germany, the United States, Korea, Japan and China. Indonesia has thus far missed the opportunity to develop indigenous technological advances and human capital (Lundvall et al., 2006).

Being an emerging market, Indonesia's lower-class population segment is huge. The demography is attractive for business because based on its population size of over 240 million, the county's consumer spending is increasing. Indonesia is a unique location for innovations that would fit the market. Most Indonesian consumers are not yet able to purchase products that are bought in developed markets (Govindarajan and Ramamurti, 2009). Thus, some of the ways firms benefit is through the development of products that fit local needs and incorporate local input (Li, 2012). Firms operating in Indonesia have developed their competitive advantages by being able to offer low-cost imitations, or benefit from accelerated learning such as technology licensing for original equipment manufacturing (OEM) of parts. Some firms operating in emerging markets can innovate further by adapting their business in similar developing markets (Chin et al., 2013), which is a path some Indonesian multinationals have chosen.

11.4 DATA AND METHODOLOGY

This chapter utilizes the case study method to illustrate the evolution of Indonesian multinationals. A case-study approach is suitable to international business if consistent and quantitative data are not available (Tsang, 2012). The argument for using such an approach is based on a discussion of replication logic (Eisenhardt, 1989), which means each case serves as a distinct experiment that stands on its own as an analytic unit. Multiple cases are discrete experiments that serve as replications, contrasts and extensions to the emerging theory (Yin, 1994). Case study is also popular in theory building (Eisendhardt and Graebner, 2007, p. 27). Ramamurti and Singh (2012) introduced cases of Indian companies to illustrate their taxonomies. Williamson and Zheng (2012) used a case-study approach to present their findings on Chinese multinationals. In the current study of Indonesian multinationals, "Cases are selected because they are particularly suitable for illuminating and extending relationships, and logic among constructs" (Eisendhardt and Graebner, 2007, p. 27). The constructs used are based on Dunning's eclectic paradigm: country- and firm-specific advantages. These constructs are the basis of Ramamurti's taxonomy of emerging-market multinationals. These constructs can also be seen as strategies, often innovative.

Three companies from Indonesia are selected to extend Ramamurti's category to examine the Indonesian context of multinationals (see Table 11.1): (1) Adaro Energy – a mining and energy company which has recently gained focus as one of Indonesia's top companies; (2) Mayora – consumer goods, which was also highlighted by BCG as a Southeast Asia Challengers; and (3) Astra – a conglomerate with strong orientation in automotive, which has often described as the pinnacle of Indonesian corporation (Robison, 1986; Linblad, 2007). Adaro Energy fits the natural resources vertical integrator model, Mayora fits the emerging-market model and Astra fits multiple models at the same time. Meanwhile, there has not been any potential candidate that would fit the global niche category given that Indonesia does not have an equivalent of Brazil's Embraer, India's Tata or Taiwan's Asus.

11.5 ANALYSIS OF INDONESIAN MULTINATIONAL ENTERPRISES

11.5.1 Adaro Energy

Adaro Energy is an Indonesian coal mining and energy company. It is known as one of the largest mining companies in Indonesia, competing with other giants such as Bumi Resources, Berau and Indika Energy (GBR, 2012). According to Reuters, Adaro is currently the second-largest mining company in the nation (Latul and Jensen, 2013). Adaro has coal reserves of 1.1 billion tons as of end of year 2012, and is growing tentatively through acquisitions. Adaro began its operations in the 1990s, after a consortium of Indonesian and Australian investors purchased 80 percent of a mining concession that had belonged to Spain's Enamdisa in Kalimantan on the island of Borneo. Enamdisa had initially purchased the mining rights from the Indonesian government as mining districts in Kalimantan were being tendered for investors in the 1970s.[3] Adaro has attempted to internalize its production supply chain from since it started its mining operations. Its first development was to build its own 80 km private coal haul road from its mining pit, west of the Barito River. This was an investment that cost a substantial amount of money, and was unfortunately rejected by banks. Shareholders shelled out $20 million of personal funds to fund the development.

Adaro grew steadily starting in its early years with 1 million tons of coal annually, up to 24 million tons by 2004. In 2005, Adaro and its major investors conducted a deal with multiple investment banks and private equities which among them are Citigroup and Goldman Sachs, and Indonesia's Saratoga Capital and Farallon Capital to raise a $973 million debt-funded leveraged buyout with an entailing equity investment of $50 million in order to execute major acquisitions of assets to establish a vertical integration (Miller, 2012; Adaro, 2011, 2012).

Using the raised funds Adaro restructured its debts, working capital conditions and financed further expansions. Adaro acquired a 36 percent stake in Coaltrade Services International Pte Ltd in Singapore to run its sales and marketing operations; 85 percent of Saptaindra Sejati (SIS), a mining contractor to handle its mining and hauling operations, thus increasing control over its core business;[4] as well as a 51.2 percent stake at Sarana Daya Mandiri, a channel dredging operator which helps to maintain barges shipment over the Barito river in South Kalimantan. In that year, Adaro's assets grew 8000 percent, and they immediately translated into higher net income. While in the following 2006, Adaro's production grew 28 percent (Adaro, 2012).

In 2009, Adaro secured another round of financing through syndicated loans, valued at $500 million. The loan was used to finance the company's capital expenditure. Adaro purchased a Singaporean-based shipping and logistics company, Orchard Maritime Logistics Pte Ltd, in which its assets, consisting of barges and shipping vessels, were integrated into a shipping company that belonged to Adaro, Maritim Barito Perkasa (MBP) (Data Consult, 2010).

By 2010, Adaro had completed its acquisitions of a "pit-to-port" vertical integration. One other part of Adaro's supply chain component is the Indonesia Bulk Terminal, which is located in Pulau Laut, South Kalimantan. The port has a capacity of 12 million tons, and is able to accommodate ships with a capacity of 80000 deadweight tons. The terminal typically loads up to 35000 tons of coal per day. In the terminal, Adaro had also formed a joint venture with the Royal Dutch Shell in building a $40 million fuel storage facility. The project was completed in 2010. Since then another round of strategic programs has been undertaken to expand its downstream business. Adaro is focused on a completion of pit-to-power vertical integration, that is, the company wants to integrate its supply chain operations from coal-mining operations all the way to building coal-fueled power plants. Thus, Adaro had also begun a series of power plant constructions across Indonesia.

Adaro acquired a $122 million loan from the International Finance Corporation in 2010 for capital expenditure and further expansions. A portion of the loan was used to build a power plant in Kalimantan. Additionally, Adaro has also initiated international joint ventures with Japanese and Korean companies. Recently, Adaro, J-Power (Japan) and Itochu (Japan) put in a stake in building Bhimasena Power, a 2 × 1000-megawatt coal-powered electric plant in central Java. Bhimasena Power Plant is set to get a power purchase agreement with Pusat Listrik Negara (PLN), Indonesia's state-owned electric utilities company. Meanwhile, Adaro also entered a venture agreement with Korea East-West Power (subsidiary of the public-owned Korea Electric Power Corp.) in 2012, to build a 200-megawatt coal-fueled power plant in South Kalimantan (Yonhap, 2012).

From 2011 until 2013, Adaro had another round of acquisitions to expand its mining assets and improve efficiency. Adaro spent over $720 million to acquire mining concessions in various areas in Kalimantan and South Sumatra. Part of the purchases also includes a 25 percent joint venture with BHP Billiton, under the IndoMet Coal Projects in Kalimantan. In 2011 Adaro acquired 35 percent of Servo Meda Sejahtera, a coal-hauling road and port operator, in order to improve its supply chain operations.

Adaro accommodates a large international sales network, which consists of over 50 customers in 14 countries. In mid-year 2013, Adaro's major international buyers came from India (22 percent), Japan (12 percent) and Hong Kong (9 percent), whereas 20 percent of their customers are from Indonesia (Adaro, 2013). Adaro typically targets known power-generation utilities companies such as Tata Power for its Mumbai power station in India and Castle Peak Power in Hong Kong. In Indonesia, Adaro's major buyer is PLN, the nation's state-owned utilities company. In 2013, *Fortune Magazine Indonesia* placed Adaro as one of the country's most admired companies. The company's strong revenue generation, increased production capacity and positive growth outlook, allowed Adaro to acquire lower cost of capital.

In terms of FSA, Adaro has privileged access to its natural resources. It has the mining license to Indonesia's second-largest mining site in the country, which the company has obtained since 1982 under the Indonesian government's Coal Cooperation Mining Agreement for mining rights. It is one of the most difficult licenses to achieve in Indonesia. Meanwhile, the media boasts that Adaro's success can also be significantly based on the managerial talents of the company directors Sandiaga Uno and Edwin Soeryadjaya of the private equity firm

Saratoga Capital. During Saratoga's involvement in expanding Adaro, their financial and managerial talent helped Adaro in acquiring capital by deciding to pursue the leveraged buyout and initial public offerings to initiate a major corporate restructuring program. This eventually led to a massive approach of improving Adaro's cost structure, and the improved production has been attributed by the media to the key people of Saratoga (Miller, 2012). Soeryadjaya also happens to be the son of William Soeryadjaya, the founder of the Indonesian conglomerate Astra International. Such ties also helped Adaro's FSAs. For example, in 2010, Adaro secured a $95 million leasing facility from Astra for heavy mining machineries, as Astra is the country's dealer for mining equipment Komatsu (Adaro, 2010).

At the moment, Adaro also benefits from significant CSAs. Indonesia's major infrastructure development plan and direction has been targeted towards domestic electric generation. Indonesia's recent economic development plan has also been created through its national agenda of furthering development across the archipelago. Access to electricity is one of the key infrastructure development hindrances, so the Indonesian government has been attempting to enlist Adaro to start to supply their products domestically.

11.5.2 Mayora Group

Mayora Group is an Indonesia-based global food producer. The company produces snacks, energy drinks, biscuits and cookies. Its brands are major household brands in Indonesia and many parts of the world. These brands include Kopiko (coffee candy), Roma (biscuits), Energen Sereal (breakfast cereal drink), Beng-Beng (chocolate bar), Torabika (instant coffee) and many more. Mayora originally started as a family-owned homemade biscuit producer in 1977. The company began its major production when it established its first production facility in the Tangerang industrial district on the outskirts of Jakarta. In 1990, Mayora went public, and has now become one of the fastest-growing food companies in Indonesia. Mayora maintains an annual net income increase of 30 percent. In 2011, it booked $810 million dollar revenue.

Mayora is an example of a local optimizer in which "firm specific advantages are derived from optimizing products and production processes for the distinctive conditions of the home market, that is, serving low income countries with underdeveloped hard (for example roads, ports etc.) and soft (for example, human capital, institutions etc.) infrastructure" (Khanna and Palepu, 2005). Mayora's product is highly suited to the local taste and preferences of the lower- and middle-income classes in Indonesia, which represents the majority of the domestic mass market. The company's existence in the Indonesian market for 30 years allows it to fully understand the customer insights and demography. Mayora Sales Director Theo Gazali stated in 2007 that one of its approaches to understand the market is to observe how consumers behave in between their activities and think about how they insert their products within those activities, a method he called "24 hours consumer model". As a result Mayora managed to identify where they should sell their products through pairings, for example, they would sell Torabika instant coffee through newspaper vendors as they see that this is a matching pair among Indonesian consumers (Markplus, 2007).

When it comes to rolling out new products, Mayora has significant focus on its R&D innovations, production and speed to market. Mayora is known to be capable of launching new products or brands in just a few months, where it typically takes years for a large multinational corporation to do so. To ensure the effectiveness of this production and operational excellence, Mayora owns all of its production facility to maintain full control. Mayora has nine plants in Indonesia, of which seven are located in Tangerang and Bekasi (Greater Jakarta Area) and two in Medan, Sumatra (*Indonesia Finance Today*, 2013). Mayora also has production plants in Thailand and recently, Nigeria and India. Through control over their production facilities, Mayora also has great control over their costs. Mayora boasts that its biscuit plant in Tangerang is the largest in Southeast Asia. Mayora has strong ties with its distribution and retail networks to ensure speed to market and strong execution. Currently, Mayora has major distribution networks in Indonesia which amount to around 150,000 outlets through Inbisco Niagatama, Mayora's distribution arm (Agustian, 2013).

Mayora has a strong presence in the international market. Mayora targets countries that have similar emerging-market conditions, as it wants to be able to replicate its successes from Indonesia through the same ways of producing, marketing and executing product sales. Formal distribution and sales offices are set up in China, India, Philippines, Thailand, Malaysia, Singapore, Vietnam and Nigeria, while Mayora also maintains export and marketing supports to other developing and also developed countries around the world. Mayora's products are available in 80 markets around the world. Mayora's total sales in the international market represent 25 percent of its total company revenue. Of this, 5 percent is from India and China (Irene and Chopra, 2010; Chin et al. 2012). The BCG praised Mayora for being able to adapt well into emerging-market conditions. The company is capable of removing bureaucracy, estimating cost structures and pricing, tailoring its products to local tastes, and finding market niches. It also deploys both traditional and modern trade channels (Chin et al., 2012).

In regards to CSA, Mayora benefits from the Indonesian market as a developing country with a population that is about a third of India's. Indonesia's mass market has a demography that exhibits growing consumption levels each year. Indonesia's middle class is also expected to double in a decade (Oberman et al., 2012; Rastogi et al., 2013). Since the target market is highly sensitive towards price, Mayora has been able to suppress price changes over time, keeping prices constant. This way, Mayora maintains its consumer base and can challenge competitors by being a leader in cost structure.

11.5.3 Astra International

Astra International is an Indonesian conglomerate corporation, which operates in multiple industries. Astra had its humble beginnings as a distribution center for building materials in 1957 founded by the Soeryadjaya family. It initially operated as a trading company. The company went public in 1990 (Astra International, 2011). Currently, it holds about $30 billion market capitalization. In 2013, Astra posted $15.8 billion (Rp 193.9 trillion) in sales, the largest figure ever posted by a firm in Indonesia. Astra is now majority owned by Jardine Cycle & Carriage, Singapore, a subsidiary of the Hong Kong conglomerate firm, Jardine Matheson. The rapid expansion of Astra began in the 1970s with the formation of joint ventures with Toyota, Honda and Komatsu, in automobile, motorcycle and tractor businesses (Robison, 1986; Lindblad, 2007). The company now has many subsidiaries involved in various business segments. Publicly listed Astra currently runs six business segments: automotive; financial services; heavy equipment and mining; palm oil; toll road; and document solution (Lubis, 2013).

Automotive is Astra's core business. In 2013, the company earned about $802 million (Rp 9.8 trillion) net income from its automotive business. Astra has multiple joint ventures and strategic alliances with foreign companies. The Astra automotive group owns subsidiaries that partner with foreign car manufacturers under joint-venture agreements (Toyota, Isuzu and Honda) and distribution licenses (Daihatsu, UD Trucks, Peugeot and BMW). Meanwhile, Astra Otoparts (AOP) is a subsidiary that manufactures and supplies car parts and spare parts for its foreign partners. More than ten foreign automotive companies have partnerships with AOP. It is the only group that also sells to international markets. Around the world, AOP serves 49 countries. It also has representation offices in Singapore, Dubai and Australia. In 2013, AOP planned future joint ventures with Italy's Pirelli, to build a motorcycle tire plant in Indonesia (Pirelli, 2012).

Under its financial services business, Astra owns: Asuransi Astra Buana (AAB) which offers health and automobile insurance; and Permata Bank, Indonesia's eighth-largest bank, which Astra shares its ownership with the British-based Standard Chartered Bank. Astra also owns various motor vehicle financing companies, which helps their customers acquire financing for Astra's personal vehicle, motorcycle and heavy machineries financing. This business model creates good synergy for Astra's business. Under heavy machineries, Astra owns United Tractors (UT), as the exclusive distributor for Komatsu in Indonesia. United Tractors also has partnerships with UD Trucks, Scania and Bomag. Astra's heavy machineries products cover a wide variety of industry, which includes mining, plantation, construction and forestry. Recently, Astra has expanded into the mining business. In the period 2008–11, Astra purchased eight mining companies, located in six sites. Astra also owns a mining contractor firm to run the operations, called Pamapersada Nusantara (Pama).

As a conglomerate, Astra is able to pursue multiple emerging-market multinational enterprise strategies at once. Astra Automotive group utilizes the "low-cost partner" strategy (Ramamurti, 2009). Astra Otoparts partners with more than 20 Japanese subsidiaries in manufacturing car parts to be sold domestically in Indonesia, as well as globally. The partnership has led Astra to become the biggest automotive-component manufacturer in Indonesia, with a total employment of over 36000 employees. Meanwhile, Astra's finance division could well qualify as a local optimizer (Ramamurti, 2009), as it offers a specialized financing scheme for motorcycle buyers in Indonesia. It targets low-income people who are able to afford motorcycles. Astra uses indigenous Indonesian managers (Lindblad, 2007).

In its mining operation, Astra is becoming a natural resources vertical integrator. From an established relationship with mining machineries firm such as Komatsu, UD Trucks and a few others, Astra benefits from the flexibility in channeling mining contractor and machineries services to any operations within Indonesia. Astra have now expanded upstream by starting their own mining concessions. Astra seems to be aiming to fully integrate its mining operations, particularly with the increasing coal demand from Indonesia (Habir and Rajendran, 2007).

Astra's CSA varies depending on the line of business. As a low-cost partner, AOP benefited from Indonesia's sizeable cheap workforce. Meanwhile, low-income customers, supported by the lack of public transport in Indonesia, allow Astra's motor financing company, Federal International Finance, to be highly profitable in extending motorcycle credit to customers. Astra's mining division, on the other hand, benefits from available access to coal supplies in Indonesia.

The variation of multinational strategies employed by Astra provides the company to acquire different FSAs. From being in the business for over 30 years, AOP have developed production and operational experience. A trusted partner for foreign car manufacturers, AOP has even attracted more foreign companies to start joint ventures in Indonesia. As aforementioned, Pirelli announced their plan to build a tire facility in Indonesia together with Astra Otoparts in 2012 (Pirelli, 2012). The combined investment is valued at $120 million, to serve the Southeast Asia motorcycle tire market. Astra's mining division benefited from having privileged access to Indonesian coalmines. Meanwhile, Astra truly takes advantage from its local knowledge and consumer-demand conditions in Indonesia in making their motorcycle credit facility one of the most profitable businesses for the group.

11.6 CONCLUSIONS AND IMPLICATIONS

This chapter has explored the emerging innovation space related to Indonesia. Three cases were selected to discuss different types of MNEs, which are emerging in Indonesia with possibilities of becoming more than regional players. Table 11.2 provides a summary of how they have leveraged their country- and firm-specific advantages.

Astra, Mayora and Adaro's internationalization paths began with activities in other countries within the same region, such as Adaro's acquisition of Orchard Maritime Logistics in Singapore or Astra's regional sales network in Southeast Asia. Carney and Dieleman (2011) argue that Indonesia is still lacking multinationals when compared to the nations surrounding it. India, China, Thailand and Singapore all have global players that have entered multiple overseas markets. Perhaps in the future, the direction of Indonesia's firms will follow similar strategies.

Indonesia has shown resilient economy growth within the past decade. Its presence in the global economy is increasing rapidly. Nonetheless, challenges are still faced by Indonesia as an emerging economy. Institutional problems persist such as a lack of infrastructure, regulations and human capital (such as R&D). These issues need to be addressed in the future, as they could well prevent the growth of other potential multinationals in the nation. Future direction of this research will consider how local knowledge is translated into strategic advantage for expansion of trade and direct investment within Asia and elsewhere, especially the United States.

Table 11.2 Country-specific and firm-specific advantages of Indonesian companies

Company	Type	Country-specific advantage	Firm-specific advantage
Adaro	Natural-resources vertical integrator	(i) Vast natural resources deposit in Indonesia (ii) Strong demand in home market for energy, due to rising economy (increase in power plants)	(i) Privilege access to mining rights (ii) Ability to control cost and production overheads (iii) Managerial talents, which helped Adaro's acquisition and financial strategy to be very successful
Mayora	Local optimizer	(i) Indonesia's rising consumer market, due to increase in income (ii) Underdeveloped hard and soft infrastructure, which allows Mayora to learn in adverse market condition	(i) Products suited to local market – enhanced knowledge in consumer taste (ii) Production and operational excellence, Mayora fully controls its manufacturing and sales (iii) Adversity advantage from being a major player for over 30 years – Mayora has earned consumers' trusts as well as strong brand recognition
Astra	Low-cost partner (emerging to natural-resources vertical integrator and local optimizer as well)	(i) Indonesia's cheap wages, which strengthen Astra's low-cost partner position (ii) Rising consumer market, which increases purchases of products (iii) Vast natural resources (specifically for Astra's mining operations)	(i) Operational and production excellence, supported by skills and expertise transferred from JVs with foreign companies (ii) Adversity advantage, being a one of the major players in the automotive and car parts industry, Astra has established a strong name in the local and global market

Source: Compiled by authors, based upon case studies from this chapter

NOTES

1. The authors would like to acknowledge the support of the State University of New York at Buffalo. Moreover, Professor Bagchi-Sen would like to acknowledge the support of the companies and foundations who support the Visiting Professor's Program at the School of Business, Economics, and Law, University of Gothenburg. See the Acknowledgments section of this book for more details.
2. The BCG lists 12 Indonesian challengers: ABC Group, Adaro Energy, Aneka Kimia Raya, Astra Otoparts, Bayan Resources, Djarum, GarudaFood, Golden Agri-Resources, Indofood, Kalbe Farma, Mayora and Wings Air.
3. The global oil crisis of the 1970s changed Indonesia's government policy on energy, which was initially limited to gas and oil, to include coal for domestic use. The Department of Mining divided the East and South Kalimantan into several coal districts, and invited investors to purchase mining rights.
4. As of 2010, SIS also has mining contracts other than Adaro in South and East Kalimantan.

REFERENCES

Abubakar, M. (2010), *Towards World Class Corporations*. Ministry of State-Owned Enterprises: Indonesia. Available at: http://www.euromoneyconferences.com/downloads/asia/2010/indonesia10/soe.pdf (accessed 20 October 2014).

Adaro Energy (2008), *Annual Report 2008*. Available at: http://www.adaro.com/investing/financial-reports/ (accessed 20 October 2014).

Adaro Energy (2011), *Annual Report 2011*.

Adaro Energy (2012), *Annual Report 2012*. Available at: http://www.adaro.com/investing/financial-reports/ (accessed 20 October 2014).

Agustian, W. (2009), 'Rating Mayora Indah Turun Jadi Underperform', *Okezone*. Accessed 30 January 2014 from http://economy.okezone.com/read/2009/10/06/278/262922/rating-mayora-indah-turun-jadi-underperform/large.

Astra International (2011), *Annual Report 2011*. Available at: http://www.astra.co.id/index.php/investor_info/annual_report (accessed 20 October 2014).

Astra International (2012), *Annual Report 2012*. Available at: http://www.astra.co.id/index.php/investor_info/annual_report (accessed 20 October 2014).

Barton, D. (2011), 'Indonesia in the New World', *Strategic Review*, August–October.

Carney, M. and M. Dieleman (2011), 'Indonesia's missing multinationals: business groups and outward direct investment', *Bulletin of Indonesia Economic Studies*, **47** (1), 105–26.

Central Intelligence Agency (CIA) (2013), *The World Factbook*. Available at: https://www.cia.gov/library/publications/the-world-factbook/ (accessed 20 October 2014).

Chattopadhyay, A., R. Batra and A. Ozsomer (2012), *The New Emerging Market Multinationals: Four Strategies for Disrupting Markets and Building Brands*. New York: McGraw-Hill Professional.

Chin, V., A. Koch, M. Meyer, E. Tan and B. Waltermann (2012), *The Companies Piloting a Soaring Region*, Boston Consulting Group. Available at: http://www.bcg.com/expertise_impact/capabilities/globalization/publicationdetails.aspx?id=tcm:12-100136&mid=tcm:12-100220 (accessed 20 October 2014).

Chetan, A., D. Tan and S. Singh (2009), 'Indonesia economics: adding another "I" to the BRIC story', 22 June, Morgan Stanley Research.

Data Consult (2010), 'Company profile: Adaro Group', *Indonesian Commercial Newsletter 2010*, pp. 27–33.

Delteil, B, K. Gan, and K. Chandra (2011), *Emerging Market Entry Candidates Indonesia*. Jakarta: Accenture.

Dick, H.W., V.J.H. Houben, J.T. Linblad and K.W. Thee (2002), *The Emergence of a National Economy: An Economic History of Indonesia, 1800–2000*. Honolulu, HI: University of Hawaii Press.

Delteil, B., K. Gan and K. Chandra (2011), *Emerging Market Entry Candidates – Indonesia*. Jakarta: Accenture.

Dunning, J.H. (1981), *The Eclectic Theory of the MNC*. London: Allen &Unwin.

Dunning, J. and S. Lundan (2008), *Multinational Enterprises and the Global Economy*. Northampton, MA, USA and Cheltenham, UK: Edward Elgar.

Eisenhardt, K.M. (1989, 'Building theories from case study research', *Academy of Management Review*, **14** (4), 532–50.

Eisenhardt, K.M. and M.E. Graebner (2007), 'Theory building from cases: opportunities and challenges', *Academy of Management Journal*, **50** (1), 25–32.

Evans, P. and S. Wurster (2000), *Blown to Bits: How the New Economy of Information Transforms Strategy – Indonesia*. Boston, MA: Harvard Business School Press.

Farrell, D., A.M. Key and T. Shavers (2005), 'Mapping the global capital markets', McKinsey Global Institute.

Farzad, R. (2010), 'The BRIC debate: drop Russia, add Indonesia?', *Bloomberg BusinessWeek*, 18 November.

Gazali, T. (2007), 'Mayora: Menciptakan Saluran Distribusi Yang Kreatif', *Markplus Conference*, MarkPlus Insight.

Global Business Report (2012), 'Mining in Indonesia', *Engineering and Mining Journal.* Accessed at http://www.panasiacorp.com.au/_content/documents/657.pdf.

Govindarajan, V and R. Ramamurti (2011), 'Reverse innovation, emerging markets and global strategy', *Global Strategy Journal*, **1** (3–4), 191–205.

Guillén, M.F. and E. García-Canal (2009), 'The American model of the multinational firm and the "new" multinationals from emerging economies', *The Academy of Management Perspectives*, **23** (2), 23–35.

Habir, A.D. and K. Rajendran (2007), 'The changing face of human resource management in Indonesia'. In C. Rowley and S. Abdul-Rahman (eds), *The Changing Face of Management in South East Asia*, Abingdon: Routledge, pp. 30–58.

Hill, H. (1997), *Indonesia's Industrial Transformation*. Singapore: Institute of SEA Studies.

Irene, J. and S. Chopra (2012), *Mayora Indah: Confectioner's Resolve*. CLSA Asia-Pacific Markets.

Kedron, P. and S. Bagchi-Sen (2011), 'US market entry processes of emerging multinationals: a case of Indian pharmaceuticals', *Applied Geography*, **31** (2), 721–30.

Khanna, T. and K. Palepu (2005), *Spotting Institutional Voids in Emerging Markets*, note no. 9–106–014. Boston, MA: Harvard Business School Publishing.

Khanna, T. and K.G. Palepu (2010), *Winning in Emerging Markets: A Road Map for Strategy and Execution*. Boston, MA: Harvard Business School Publishing.

Latul, J. and F. Jensen (2013), 'Burned by China, Indonesia coal miners turn to power stations', Reuters, 17 September. Available at: http://www.reuters.com/article/2013/09/16/indonesia-coal-idUSL4N0GM28820130916 (accessed 20 October 2014).

Li, P.P. (2012), 'Disruptive innovation by emerging multinational latecomers: the mechanism underlying the trajectories of catching-up and leapfrogging', paper presented at the 3rd Copenhagen Conference on Emerging Multinationals: Outward Investment from Emerging Economies, Copenhagen, Denmark, 25–26 October.

Lindblad, J.T. (2007), 'Conglomerates in Indonesia: the road to power and beyond'. In A.E. Fernandez Jilberto and B. Hogenboom (eds), *Big Business and Economic Development: Conglomerates and Economic Groups in Developing Countries and Transition Economics Under Globalisation*, London and New York: Routledge, pp. 65–85.

Lubis, A.M. (2013), 'Astra to spend Rp 60t on vehicle financing', *Jakarta Post*. Available at: http://www.thejakartapost.com/news/2013/11/29/astra-spend-rp-60t-vehicle-financing.html (accessed 18 January 2014).

Lundvall, B., P. Intarak and J. Vang (2006), *Asia's Innovation Systems in Transition*. Northampton, MA, USA and Cheltenham, UK: Edward Elgar.

Michaels, E., H. Handfield-Jones and B. Axelrod (2001), *The War for Talent*, Boston, MA: Harvard Business School Press.

Miller, M. (2012), 'Saratoga's Sandiaga Uno has bounced back from the Asian crisis', *The Deal*. Available at http:www.thedeal.com/magazine/ID/045165/2012/. . ./sandiaga-uno-has-bounced-back-from-the-asian-crisis.php (accessed 11 November).

Oberman, R., R. Dobbs, A. Budiman, F. Thomson and M. Rosse (2012), 'The archipelago economy: unleashing Indonesia's potential', McKinsey Global Institute. Available at: http://www.mckinsey.com/insights/asia-pacific/the_archipelago_economy (accessed 20 October 2014).

O'Neill, J. (2001). 'Building better global economic BRICs'. Goldman Sachs Global Economics Paper No: 66.

Partawidjaja, D. (2012). 'Inilah 100 perusahaan terbaik Fortune Indonesia', *Kontan*. Accessed 18 January 2014 from http://industri.kontan.co.id/news/inilah-100-perusahaan-terbaik-fortune-indonesia.

Pepinsky, T. (2008), 'Institutions, economic recovery, and macroeconomic vulnerability in Indonesia and Malaysia'. In A. MacIntyre, T.J. Pempel and J. Ravenhill (eds), *Crisis as Catalyst: Asia's Dynamic Political Economy*, Ithaca, NY: Cornell University Press, pp. 231–50.

Pirelli (2012), 'Pirelli and PT Astra Otoparts TBK to establish a joint venture to build moto tire factory in Indonesia'. Available at: http://www.pirelli.com/corporate/en/channel/sustainability/2012/04/24/pirelli-and-pt-astra-otoparts-tbk-establish-a-joint-venture-60-pirelli-40-pt-astra-otoparts-to-build-new-moto-tyre-factory-in-indonesia (accessed 20 October 2014).

Ramamurti, R. (2009), 'What have we learned about emerging market MNEs?' In R. Ramamurti and J.V. Singh (eds), *Emerging Multinationals in Emerging Markets*, Cambridge: Cambridge University Press, pp. 399–426.

Ramamurti, R. (2012), 'Competing with emerging-market multinationals', *Business Horizons*, **55**, 241–9.

Ramamurti, R. and J.V. Singh (2009), *Emerging Multinationals in Emerging Markets*. Cambridge: Cambridge University Press.

Rastogi, V., E. Tamboto, D. Tong and T. Sinburimsit (2013), *Asia's Next Big*

Opportunity – Indonesia's Rising Middle-Class and Affluent Consumers. The Boston Consulting Group.
Robison, R. (1986), *Indonesia: The Rise of Capital*, London: Allen & Unwin.
Rumbaugh, M.T. (ed.) (2012), *Indonesia: Sustaining Growth During Global Volatility*. Washington, DC: International Monetary Fund.
Rugman, A.M. (2009), 'Theoretical aspects of MNEs from emerging economies', in R. Ramamurti and J.V. Singh (eds), *Emerging Multinationals in Emerging Markets*, Cambridge: Cambridge University Press, pp. 200–25.
Sassen, S. (2001), *The Global City New York, London, Tokyo*, Princeton, NJ: Princeton University Press.
Srinivas, P. (2013), 'Indonesia's financial sector: a half-full glass', *Strategic Review*, January–March.
Vaswani, K. (2013), Should Indonesia be the I in BRICS? Available at: http://www.bbc.co.uk/news/business-21921593 (accessed 20 October 2014).
Williamson, P., and M. Zeng (2009), 'Chinese multinationals: emerging through new global gateways', in R. Ramamurti and J.V. Singh (eds), Emerging Multinationals in Emerging Markets, Cambridge University Press, pp. 81–109.
World Bank (2013). *World Development Indicators*. Accessed at http://data.worldbank.org
Yin, R.K. (1994), *Case Study Research: Design and Methods*, 2nd edn.Thousand Oaks, CA: Sage.
Yonhap News (2012), 'South Korean firm to build Indonesian thermal power plant'. *Yonhap News Agency*. Accessed 1 February 2014 at http://english.yonhapnews.co.kr/business/2012/01/19/42/0501000000AEN20120119003400320F.HTML

12. Tapping into Western technologies by Chinese multinationals: Geely's purchase of Volvo Cars and Huawei's hiring of Ericsson employees in Sweden

Ethan Gifford, Marcus Holgersson, Maureen McKelvey and Sharmistha Bagchi-Sen

12.1 INTRODUCTION

This chapter focuses on how firms can access, and tap into, relevant external knowledge for innovation, examining how this happens within two industries that are highly dependent upon scientific and technical knowledge for innovations within their products.[1] One aspect of innovation spaces in Asia is how Western companies are moving into Asia. This chapter addresses another aspect, how Asian companies are searching for, and accessing, technological opportunities and market opportunities in the West. The underlying mechanism studied is how and why Asian companies establish subsidiaries in Europe, with a focus on Sweden. This chapter provides two case studies, which provide insights and understanding about two different mechanisms by which firms from emerging markets can use their establishment, networks, and resources (including human resources) as vital sources of especially technological and scientific knowledge. Specifically, these are mechanisms whereby Chinese companies can use the geographical and knowledge capabilities in Sweden.

The two sectors studied are automobile and telecommunication. The chapter presents two case studies of Chinese firms moving into Sweden, including Geely's purchase of Volvo Car Corporation (VCC) in the automotive industry and Huawei's hiring of Ericsson's previous employees in the telecommunication industry.[2] Chinese companies are increasingly moving abroad to access both advanced technologies and new markets.

Both sectors are characterized by significant spending in research and development (R&D) in order to remain competitive in terms of technological developments in fast-moving environments. The challenge for this type of industry – and hence for these firms – is to expand the market enough in order to reap scale economies from significant investments into R&D. The company needs to sell enough units to recuperate that investment and also generate more cash for future investments into R&D. Both sectors have faced challenges from financial crises (for example, the Asian crisis, information technology (IT)-related financial downturn in 2000, and the 2008 financial crisis), shortened product cycles based on technological advancements, and related developments in technical standards in telecommunication.

The studied industries are part of the Chinese 12th Five-Year Plan for national strategic emerging industries (APCO Worldwide, 2010). Chapter 2 of this Five Year Plan addresses Chinese innovation policy. The automobile industry, VCC, and Geely are part of the "new energy auto industry." Ericsson, Huawei, and the telecommunication industry are part of the "new generation information technology industry."

At the national level, we can ask why Sweden is of interest to Chinese firms in these strategic industries. Our view is that Sweden can be seen as a particularly interesting national context for tapping into technological and industrial knowledge, given the high educational achievements, context of innovativeness, and its long-term international orientation. Sweden is a small country of approximately nine million inhabitants at the far north of the European Union (EU). The country combines market capitalism with a social welfare system with attributes such as free tertiary and university schooling and public health care for the entire population. The development of the Swedish welfare model has been closely tied to expanding access to education among its population. Sweden implemented widespread education in the late 1880s, and a major expansion of university education has more recently occurred, starting in the 1970s and rapidly expanding from the 1990s.

In international comparisons, Sweden often ends up at the top of rankings of innovativeness on the national level, such as in The Innovation Union Scoreboard[3] (European Commission, 2013). The scoreboard has for the last several years identified Sweden as the most innovative country in Europe, labeling the country as one of few 'innovation leaders.' Sweden is highly R&D intensive, with approximately 2.5 to 3.7 percent of gross domestic profit (GDP) spent on it, which consists primarily of industrial R&D. Sweden's large multinational firms are highly export oriented, and highly participative within the broader global economy. Many of these firms exist in the export-led industrial sectors like telecommunications (Ericsson), engineering (SKF, ABB), transportation (AB Volvo and VCC), and pharmaceuticals (AstraZeneca). Since the mid-1980s, many successful Swedish multinationals were bought and merged into industrial groups headquartered elsewhere (such as VCC into Ford and recently, Geely) and merged with foreign competitors (such as Astra and Zeneca forming AstraZeneca).

The first case in this chapter describes Geely's purchase of VCC and challenges that arose with the new ownership. Volvo Car Corporation (VCC)[4] is specialized in car production, with a reputation for quality and safety. As recounted in detail below, VCC has had difficulties in financing R&D and selling enough cars over several decades, leading to different ownerships before Geely. Geely has a wide range of vehicles, which are sold in China, but has traditionally been associated with a reputation of producing lower-quality automobiles.

The second case describes Huawei's hiring of former Ericsson employees. Ericsson is an international corporation within telecommunications, and although it was initially highly successful in mobile phones (handsets) with early standards, the firm has retreated to telecommunication equipment. This industry has had frequent shake-outs of the players, due to the rapid pace of change in customer preferences (which went from high-price gadgets to consumer products) and in technology (with changes in standards as well as in overlaps with other technologies). Within a very short time, Huawei has become a major competitor in telecommunications including sub-sectors of telephones and network equipment. Huawei characterizes itself as a collective of professionals who employ a rotating chief executive officer (CEO) system under the decision rule of the board of directors, rather than a private company (Huawei, 2011).

A key challenge for Sweden – and for small, export-led countries more generally – is how to maintain their base for R&D expenditures, either through nationally owned companies or through attracting international R&D labs. For an international firm, Sweden then provides access to advanced technology and skilled engineers – both of which are mostly concentrated in Swedish multinationals. Therefore, accessing advanced technology by tapping into the knowledge of a Swedish firm can be considered as valuable for a Chinese firm – or for any firm from an emerging market. Sweden also educates many skilled engineers, who later work and obtain significant industrial and international experience.

12.2 ACCESSING NEW KNOWLEDGE BY TAPPING INTO WESTERN TECHNOLOGIES

Our starting point is the observation that tapping into Western industries and technologies can be vital for increasing the competitiveness of firms from emerging markets like China. Learning from and imitating innovating firms and countries are important means for catching up (Granstrand and Holgersson, 2014). The extensive literature on catch-up suggests different pathways for companies from emerging markets to enhance their capabilities, and these pathways involve learning processes (Mathews, 2006a; Malerba and Nelson, 2012). Earlier literature assumes that these firms from emerging markets initially work with mature technologies, and only at a late stage do they move from imitation to innovation.

From an innovation management perspective (Dodgson et al., 2013), firms realize that they increasingly need to learn about, and develop new knowledge and technology, that can be used in product and process innovations.[5] Innovations are widely recognized as crucial for competitiveness, for both firms and economies. Decades ago, researchers would often assume that advanced technology came from the developed countries whereas the developing countries imitated or followed. Today it is not that simple. For instance, China has become one of the world leaders in terms of innovation and R&D activities and has experienced rapid growth concerning the proportion of economic activity composed of knowledge and technology-intensive industries (National Science Board, 2014).

The question of why many Chinese firms are highly interested in tapping into technology and industries from innovative economies like Sweden relates back to the possibility to access technological opportunities through global network relationships.

Accessing knowledge resources is a type of foreign direct investment (FDI). This literature stresses that FDI is an important key element in the emergence and restructuring of spatial and organizational patterns of economic activity and received attention in economic geography, international business and other disciplines (Buckley, 2002; Peng, 2004). Commonly, this takes the form of joint equity ventures, acquisitions, mergers, or establishing or expanding operations in a host country. In recent years, the literature has examined emerging-market multinational enterprises (EMMNEs). One motivation for foreign market entry by EMMNEs commonly identified is to use or learn from the location-specific advantages of foreign markets to augment or develop ownership-specific advantages (such as skills and proprietary technologies) that the EMMNEs lack (Dunning, 1995; Luo, 2000; Rugman and Li, 2007; Narula and Dunning, 2010). Augmenting resources and acquiring strategic assets, including access to knowledge and talent, can serve as a springboard for foreign market entry and further expansion through established channels of marketing and distribution (Mathews, 2006b; Luo and Tung, 2007).

China is quite active in these processes. China and Chinese companies have been frontrunners in FDI by emerging economies. Empirical evidence from DeBeule and Duanmu (2012) suggests that these firms tend to target large markets and that their main motivation for FDI is geographical expansion. According to an econometric study of Chinese investment patterns by Buckley et al. (2008, p. 136), "general market seeking motives underpin much of Chinese investment behavior". However, Chinese EMMNEs target high-tech manufacturing companies to acquire strategic assets in form of patents and knowledge, although it is not clear whether these acquisitions impact resource augmentation if there is a lack of absorptive capacity (DeBeule and Duanmu, 2012). According to Nicolas (2014), Chinese direct investments to the EU have increased from $150 million in 2004 to $7.6 billion in 2011. There have been a higher number of greenfield investments, albeit lower investments, and a smaller number of more significant mergers and acquisitions (M&As).

This chapter focuses mainly on the latter aspect, how Chinese firms tap into Western industries to access new technologies to improve in-house innovativeness. Work by Granstrand et al. (1993, p. 417) states that "the basic aim of the traditional forms of foreign R&D was to enhance the value of existing parent-company technology. Today, a new pattern can be increasingly be discerned . . . many R&D units are also charged with the creation and renewal of core technological capabilities". The case studies examined here are about tapping into existing capabilities by other firms, and thereby the acquisition of these resources is indeed intended to create and renew core technological capabilities of the Chinese firms. Two main mechanisms are identified by Granstrand et al. (1993), namely to make foreign acquisitions and to tap into foreign scientific infrastructures.

These two mechanisms for how a foreign MNE could tap into technology correspond to the two case studies here, respectively, Geely with VCC and Huawei with Ericsson's former employees. In terms of methodology, these cases are based upon a combination of interviews (Volvo/Geely) and archival material written primarily in Swedish (Volvo/Geely and Ericsson/Huawei).

12.3 GEELY'S ACQUISITION OF VOLVO CAR CORPORATION

On August 2, 2010 the Chinese firm Zhejiang Geely Holding Group Co., Ltd (Geely) acquired the Swedish automobile manufacturer Volvo Car Corporation (VCC) from the US automotive giant Ford. This acquisition was important for Geely in its ambition to build up its brand, technology base, and engineering capabilities, which makes it an interesting example of Chinese firms acquiring Western technologies and R&D capabilities, in this case by means of corporate transactions. Geely's principal products are automobiles, motorcycles, engines, and transmissions. The company as of 2014 sells passenger cars under five brand names: Emgrand, Englon, Geely, Gleagle, and Volvo. Just as any acquisition, Geely's acquisition of VCC was to a large extent shaped by the history of VCC. Therefore, this section starts with briefly describing the history of VCC, before looking into the specific deal in which VCC was sold from Ford to Geely.

12.3.1 History of Volvo Cars (VCC)

Volvo's heritage traces back to 1915, when the Volvo trademark (Volvo is Latin for "I roll") was registered by the Swedish ball-and-roller-bearing manufacturer SKF. The trademark first came into use when SKF initiated automobile production under the brand Volvo; the first series production of cars was initiated in 1927, and the production of the first car – a Volvo ÖV4 – was finalized on 14 April 1927. In 1928 Volvo diversified into trucks, a business that became an immediate success. A few years later, in 1935, Volvo was divested from SKF and listed on the Swedish stock exchange as an independent company. In the following decades Volvo diversified further, typically by means of acquisitions, into buses, construction equipment, marine engines, and aircraft engines.

By the end of the twentieth century, the management of the Volvo group decided to focus solely on commercial vehicles rather than automobiles. Therefore, what was once the core of Volvo, the passenger car business, became subject for divestment in the late 1990s, and was sold to Ford in 1999 for $6.5 billion. The Volvo Group at that time retained the other businesses. Leif Johansson, who was president and CEO of the Volvo Group at the time, explained why Ford became the selected acquirer: 'We went through a list of different options, and believe me we have tried many different options, and we have arrived at the conclusion that Ford is the best ally, the best partner and the best owner of the Volvo car business' (CNNMoney, 1999). The key idea was that Ford would integrate VCC in its Premier Automotive Group (PAG), together with the other premium Ford brands: Aston Martin, Jaguar, and Land Rover. Ford's president and CEO at the time, Jacques Nasser, especially acknowledged the value of VCC's brand, engineering, and R&D in the purchase (CNNMoney, 1999):

> What we are really buying here is generations of hard work and dedication and brand building and ingenuity that has been put together over decades and decades . . . We are buying the strength of the brand, the reputation of the brand, and we are buying a team that is incredibly best in breed in terms of its worldwide capacity and research and development . . . Volvo is a premium automotive brand with unique appeal that represents a good opportunity to profitably extend our lineup.

Key resources that Ford acquired in the deal were thus the brand, the engineering resources (especially personnel), and the technological resources of VCC.

Hence, the Volvo trademark was clearly an important part of the purchase. However, since the brand Volvo was used both by Volvo's car business and by Volvo Group's remaining businesses (for example, trucks and buses), the ownership of the Volvo trademarks could not simply be transferred to Ford in the deal. Instead, the holding company Volvo Trademark Holding AB was set up, co-owned by Volvo Group and VCC/Ford. The Volvo trademarks were transferred to this holding company, and the trademarks were then (and still are) licensed to Volvo Group's various businesses and VCC, respectively.

The technological resources and the related intellectual property rights (IPRs) included in the deal were owned centrally within the Volvo group before the purchase. Therefore, these assets had to be reviewed before the deal to decide how to separate the portfolio of technologies and IPRs, that is, in order to deal with the so-called intellectual property (IP) disassembly problem (Granstrand and Holgersson, 2013). Typically, patents that were mainly important to passenger cars were transferred to Ford while all other patents, that is, patents mainly related to other businesses, were kept within the Volvo Group and then licensed to Ford if they were also used in passenger cars. The acquired technological IP was placed within one of Ford's subsidiaries, Ford Global Technologies (FGT), where ownership of the technological IP of all Ford's businesses was also collected. This integration of technologies within Ford was noticeable also in the VCC products within a few years, and in 2010 almost all Volvo cars were built on Ford platforms. The C1 platform was used for VCC's smaller cars and the EUCD platform was used for VCC's larger cars (ibid.).

12.3.2 Acquisition by Geely Holding Group and Assessing Technological Assets

In 2010, after only one decade within Ford, VCC was sold again, now to Geely, for $1.5 billion, a price $5 billion below what Ford paid for VCC roughly ten years earlier. An important background to this deal was the global recession following the financial crisis in 2007–2008. This led to a sharp decline in the demand for cars. This crisis hit the large American automotive groups hard, and both Ford and General Motors started to divest some of their brands and parts of their businesses to cut costs and raise cash (Ford had actually initiated the divestment of Aston Martin before the financial crisis). Ford initiated the divestment process of VCC in 2008–2009, and Geely was announced the preferred buyer in October 2009. Geely already had its own Chinese automotive business, Geely Auto.[6]

The founder and chairman of Geely, Li Shufu, argued that the acquisition would give VCC access to Geely's low-cost production and market coverage in China, while Geely Auto would gain access to Volvo's high-quality brand and frontline technologies (*China Daily*, 2009; Fangfang, 2010). The new Chinese owners and the European management team emphasized Volvo's core values at the time of the purchase. Li Shufu stated that "this famous Swedish brand will remain true to its core values of safety, quality, environmental care and modern Scandinavian design". Stefan Jacoby, who became the CEO of VCC in connection to Geely's acquisition, stated that "our employees, suppliers, dealers – and above all our customers – can be confident Volvo will preserve its special status as industry leader in vehicle safety and innovation" (Arnott, 2010).

At the time of the purchase, VCC was not profitable, and Li Shufu expected that up to $900 million would be spent in the turnaround process, aiming to reach profitability for VCC. He stated that the main issue was the large R&D costs, considering the low volumes of Volvo cars produced and sold (McDonald, 2010). The yearly sales were roughly 400 000 in the early 2000s, but only 300 000 in 2009 due to the crisis. These volumes were too small to get feasible returns from the necessary R&D investments. The automotive industry involves a high level of knowledge within many complex technologies in various areas. This, in turn, requires that the firms make large R&D efforts and partner with other firms in the industry. The industrial situation can be seen as a typical situation of economies of scale, whereby large R&D investments require large sales to reduce total average costs per unit. A potential benefit from the purchase was therefore the possibility to eventually use the same architectures and technologies in both Volvo and Geely cars.

However, Ford was restrictive in terms of how its technologies could be used after the purchase. As described above, the Volvo cars were built on Ford platforms at the time of Geely's acquisition, and Ford did not want a new major competitor on the Chinese market, especially not one that used Ford's technologies. Therefore, extensive efforts were taken in order to structure the deal to safeguard Ford's competitive position, while still allowing VCC's business to continue. In a large, complex, and technology-based business like this, such contracting becomes complex, and Li Shufu even expressed that the IPR contracting and negotiations (that is, the processes of disassembling the IP) in the deal were more complicated than the price (*China Daily*, 2009).

The IP issues were dealt with by categorizing the relevant IP and technologies, and matching the categorization with suitable provisions, such as ownership transfers or licenses.[7] First, technologies that had been developed by VCC before being acquired by Ford were to be transferred back to and follow VCC into the purchase. Second, technologies that had been developed after VCC was acquired by Ford were kept within Ford, while VCC received different types of licenses for the technologies that were used by its businesses at the time of the purchase. Third, technologies that had been developed independently by VCC during the time with Ford ownership were to be transferred to VCC, while being licensed back to Ford. Finally, the Volvo trademarks were kept within the joint venture Volvo Trademark Holding AB (co-owned with the Volvo Group). Since VCC's 50 percent share of that subsidiary followed the firm in the deal with Geely, the trademark issues were fairly easily handled.[8]

A timeline of the ownership changes in VCC since the 1990s is shown in Figure 12.1.

Although there was initially a clear separation between VCC and Geely Auto, largely due to the contracts with Ford, the firms eventually started to integrate. A couple of examples of how VCC's R&D also started to become utilized in Geely Auto can be given as indications of this new trend. First, VCC's head of design, Peter Horbury, was transferred from VCC to Geely in 2011, being succeeded at VCC by Thomas Ingenlath. Second, and probably more important, China-Euro Vehicle Technology (CEVT), a joint engineering and development center, was set up in VCC's hometown Gothenburg in 2013, aiming at developing the next-generation modular architectures for both VCC and Geely Auto (for B- and C-segment cars) and at transferring knowledge from VCC to Geely Auto (Pröckl, 2013). At the same time, Geely's chairman, Li Shufu, became more involved in the management of VCC, and the cultural integration was reportedly not entirely smooth.

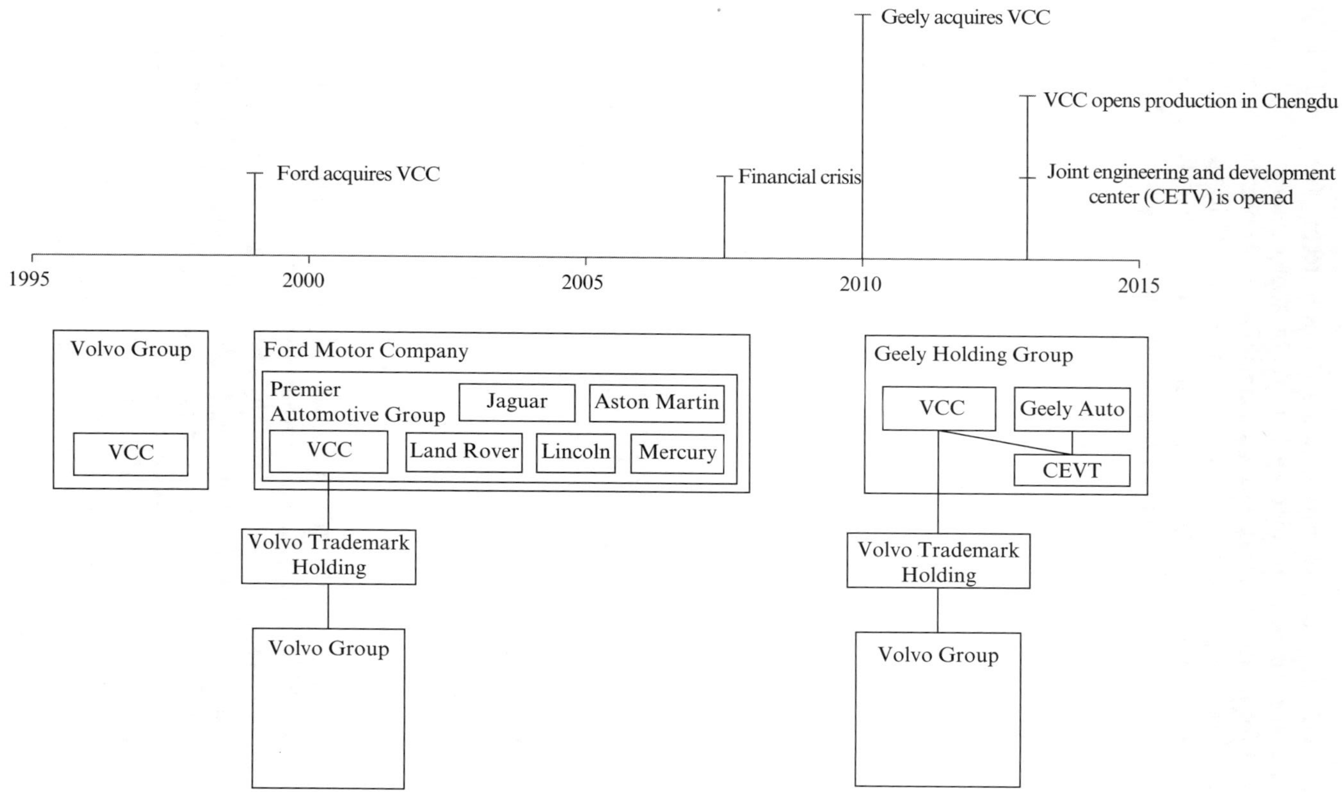

Figure 12.1 A timeline of ownership changes in VCC since the 1990s

Some issues related to different perceptions of future design and the "core values" of Volvo also became more evident in the media. As Shirouzu (2013) puts it: "Safe, solid and understated in a Swedish way, or an aspiring luxury brand eager to pander to a Chinese taste for excess and 'bling': that's the debate causing ripples at the top of Volvo Car Corp., which has been under Chinese ownership since 2010."

While the VCC management wanted to stick with VCC's traditional core values, including Scandinavian designs, Li Shufu emphasized the importance of adapting the designs to the Chinese market to be able to reach a goal of 800000 sold cars in 2020. Li Shufu commented on this discussion (Matson, 2014): "It is necessary to secure that we have a clear and conscious process for the product strategy [. . .] In that process people will have opinions, but only by having open and honest discussions is it possible to reach a wise decision on products and development" [translation by authors].

As of 2014 the VCC management has according to Li Shufu reached a conclusion on how to move forward, and the disagreements have been put aside (Matson, 2014). In line with this progress was the opening of a new VCC production facility on 5 June 2013 in Chengdu, China, initially used for producing a limousine version of the Volvo S60 (S60L). This in turn goes in line with the joint strategy for the future, one where Geely Auto focuses on the mass market while VCC focuses on the premium segment, while at the same time utilizing a joint technology base. Production in China is intended to facilitate the exploitation of the market opportunities associated with the large, and growing, market for high-end automobiles.

12.4 HUAWEI HIRING OF HUMAN CAPITAL IN RELATION TO ERICSSON

This section provides a case study of the second mechanism, "tapping into a foreign scientific infrastructure," as a way for a Chinese firm to acquire talent in Sweden. This mechanism could be seen as very broad, such as interaction with universities or engaging in open innovation. We have a more narrow focus, namely, the hiring by Huawei of talented and educated engineers in Sweden, who previously worked for a competitor, Ericsson, within the same industry. Short histories of both Ericsson and Huawei are given. Figure 12.2 presents the main timeline for Ericsson (top) and Huawei (bottom) for the period 2001 to 2013 in Sweden. The details are provided in the following sections.

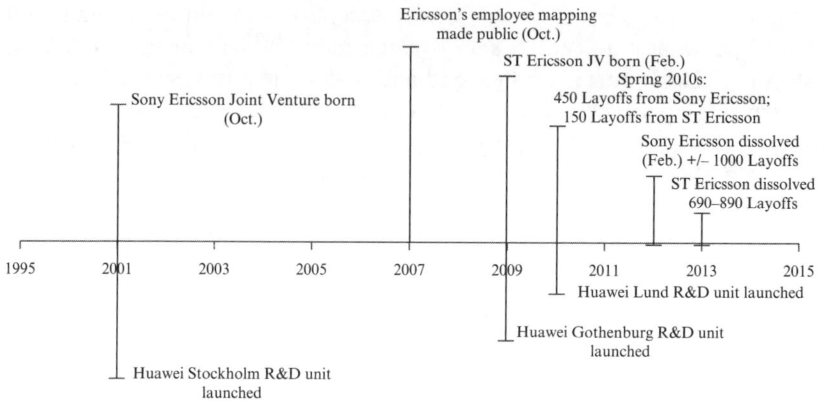

Figure 12.2 A timeline of Ericsson and Huawei in Sweden

12.4.1 The History of Ericsson

Ericsson was founded in 1876, by Lars Magnus Ericsson.[9] Using technologies developed by Bell, but not patented in Sweden, it quickly moved into telephones, electromechanical switches and military radio equipment. Ericsson also had two jointly owned subsidiaries through the years that are relevant here as well. The first was SRA (Swedish Radio AB); originally with ASEA, AGA and later Marconi for land mobile radio equipment and naval communications, which Ericsson later purchased and renamed Ericsson Radio Systems. The second was Ellemtel, a manufacturer of computer-controlled telecommunication switches (owned together with the public company Televerket).

While always engaged in telecommunication as the primary focus, the company has shifted foci during recent decades, reflecting its responses to opportunities opening globally through new technologies and markets. In the 1970s, Ericsson had two divisions, where the largest one was involved in switching (especially for public telecommunications operators) and the smaller one was involved in business radio communication. The business radio communication division included production of military radio equipment and consumer electronics (radio and television). At that time, about 2 percent of sales were in mainly military communications in products related to what later became mobile telephony. The 1980s saw the global deregulation of the telecommunication industry and following that, the rapid development of different types of telecommunication operators, including ones for mobile telephony. By the late 1990s, Ericsson had shifted to 70 percent of sales in mobile telephony, and it had three business areas: (1) mobile phone and terminals, for consumer products; (2) infocom systems, providing operators with networks and services; and (3) mobile systems, producing mobile communication systems. Turnover also grew dramatically up through the late 1990s.

Like the automobile industry, firms in telecommunication have invested much into R&D in order to keep up with changing industry standards and developments. In the early 1990s, three separate mobile telecommunication standards were being developed. Ericsson, through Lars Ramqvist, who became president and CEO of Ericsson in 1990, invested heavily in R&D, especially in Sweden. Ramqvist had previously been head of Ericsson Radio System and had seen the explosive growth of mobile communications, at about 40 percent annual per year (Skidé, 1994, p. 13). Ramqvist's strategy was to pursue R&D and to go for all the international markets and hence develop equipment for the three major standards: 'I had to explain to shareholders that I was proposing to increase R&D costs very considerably, possibly by as much as 50 percent a year over two years,' Ramqvist says. 'We would be spending 15 times as much on R&D as on dividends at a time when recession had hit the industry and we were faced with falling profits' (*International Management* 1994, p. 27).

However, by the end of that decade, Ericsson faced falling sales globally, due to a variety of factors. These include the Asian crisis (late 1990s), over-investment in some markets followed by restricted access to capital for operators and vendors, lagging sales in the US, as well as increasing competition from especially Asian competitors like Samsung and LG. The net income swung dramatically in some years. In SEK millions, Ericsson went from a profit of SEK21 018 million in 2000 to a loss of SEK21 264 million in 2001. It was profitable again but at a low level in 2003. Some of the losses depended upon the switches and networks that it installed for operators and vendors.

During this period, the phones (handsets) became an increasingly difficult market for Ericsson, given that the company was attuned to producing technically sophisticated phones with high prices rather than the low-cost and consumer-friendly design of new competitors. Ericsson tried to respond, by developing phones in the new Sony Ericsson joint venture, which started in October 2001. In 2002, Ericsson reduced its R&D budget by approximately US$1.1 billion out of approximately US$7.8 billion on R&D, and it reduced employees (globally) by approximately 17 000 individuals, including in R&D (Ahlbom, 2002). More than half of the 80 global research and development labs were closed, a key motive of which was to create more effective innovation processes.

In February 2012, Ericsson exited the phone segment all together, when it divested their 50/50 mobile handset joint venture called Sony Ericsson to the Sony Corporation. "We have been in the handset business since it started so this is an emotional day for us", Hans Vestberg, chief executive of Ericsson, told the *Financial Times*. "On the other hand it's very logical what is happening right now, how we evolve this partnership . . . The whole handset industry has evolved to smartphones, and we see the market being so many more devices, not only mobile phones, using the network. So the importance of having the handset business is less" (Palmer and McCarthy, 2011). In 2012, Ericsson budget for R&D was approximately US$6 billion, again reduced by 5–10 percent from the previous year (Ahlbom, 2012). Global services continued to grow as a business area during this period.

12.4.2 Huawei and Telecommunication

Founded in 1987 in Shenzhen, China, Huawei is currently among the forerunners in telecommunications equipment manufacturing, having beat out Ericsson in terms of size in 2012 and thus securing its position as the industry's largest actor (*The Economist*, 2012). At the close of 2010, Huawei's international employment was self-reported as 110000 individuals, 46 percent of whose job entailed some degree of engagement in R&D or product development (Boutellier, 2008; Huawei, 2010). Huawei became a large domestic player in the Chinese telecommunications market roughly five years after its inception, and continued to gain mainstream market share at home, eventually winning its first international contract in 1997 (Bloomberg News, 2010). Since then, it has become active in markets spanning every region of the globe. By 2000, Huawei's presence on the global telecommunications market was recognized globally (Boutellier et al., 2008), and major competitors such as Ericsson, Cisco, Qualcomm, and Alcatel began to view the company as an up-and-coming threat due to their sizeable market shares. Huawei went from competitive advantages in selling to and sharing design with the influx of domestic newcomers in the Chinese telecommunication industry to becoming the leading overall vendor in the industry (ibid.).

Though Huawei's international expansion has been felt all over the globe, its investment in Sweden is of particular interest in this chapter. One of the main challenges facing the company has been the availability of talent in terms of science, R&D, and engineering (Boutellier et al., 2008). In Sweden, the supply of this type of talent is in relative abundance, due to the location of telecommunications giant Ericsson (and its difficulties) as well as the supply of trained and specialized engineers from the country's universities. Since Huawei's first establishing of an international R&D facility in Stockholm in 2001, Sweden has been an attractive destination for the company. Gordon Luo, Huawei's Nordic chief officer, remarked

(Ahlbom, 2013; quotation translated from Swedish by authors) that Sweden and the Swedish government "make decisions based on facts, not based on protectionism," this combined with the liberal trade politics has created what he feels is a favorable investment climate for Huawei to invest in Sweden.

Despite these statements showcasing Huawei's favorable view of directly investing in R&D in Sweden, the road has not been without hindrance, or controversy, due to competition among the firms. Global competition between Sweden's telecommunications giant Ericsson (along with its joint ventures, Sony Ericsson and ST-Ericsson) and Huawei has intensified in recent years, not least in terms of service contracts in target markets, both geographical and technological. Additionally, much of Huawei's strength in the Swedish market could, we argue, stem from its hiring of former Ericsson personnel who have experience in key technological fields related to core business areas that Huawei is already in or is entering into.

Since its entrance into Sweden in 2001, Huawei has been acquiring a wealth of R&D knowledge through the hiring of employees and managers who have built up competence in key areas while working for Ericsson or one of Ericsson's joint ventures. Many previous Ericsson employees have found themselves lacking in gainful employment during cutbacks, which was something that Huawei may have been quick in offering. Other employees may have been offered more favorable conditions than those they previously enjoyed at Ericsson.

This ebb and flow process of competition deserves further research, and the next section will map the activities of Ericsson, and the hiring and expansion of Huawei, in Sweden's three main IT/telecommunication clusters, namely Kista (a district of the Stockholm municipality), Lund (near Malmö), and Gothenburg. These are the three main metropolitan areas of Sweden, and centers for Ericsson R&D. The next section will focus on the establishment of Huawei's subsidiaries in these locations and how some of its business areas match those of Ericsson on a region-by-region basis.

12.4.3 Huawei's Strategic Regional Co-location to R&D Knowledge in Sweden

Three regions are discussed in turn – Kista, Gothenburg, and Lund – in relation to Huawei's strategic co-location to R&D knowledge in Sweden.

Lying on the edge of the Stockholm municipality, Kista is one of Sweden's largest industrial districts for telecom and related technologies. It has been and remains an important area of operations for Ericsson, as well as Stockholm centrally. The Kista operations began to grow exponentially during the IT boom of the late 1990s/early 2000s. The core competence of this area of Ericsson's business lay in wireless communications and information technologies.

Huawei first established its own Kista-based R&D facility in 2001. As of 2010, this business unit focuses on several key technologies, including mobile system design, algorithm/IRF design, and chipset design (Huawei, 2010). In March 2007, Urban Fagerstedt, an Ericsson employee of 28 years, went to work at Huawei. He split his time working as manager of the R&D office in Kista, in the Stockholm area, and as a management consultant for Huawei's management board in China. He left Ericsson in winter 2004, when the Swedish telecom giant reorganized large portions of its R&D activity. He had spent his last four years at Ericsson as the head of development of radio systems, an area that then stood for 65 percent of Ericsson's sales (Ahlbom, 2007). He now supplies key knowledge and skills within his areas of expertise to Huawei.

In addition, Jack Järkvik was pointed out by Fagerstedt as a person holding key engineering competence who now takes part in training the Chinese portion of Huawei's R&D staff how to effectively organize and run R&D projects within research and development. Järkvik, during his time as an Ericsson consultant, helped develop the concept of "system anatomy," which became a widespread analytical tool in R&D projects both in Ericsson and other major competitors. While expressing Huawei's interest in pursuing human resource opportunities further in Sweden concerning R&D, Fagerstedt hinted that he knew where the key personnel were located in the Ericsson corporate hierarchy that are best within Huawei's areas of interest, though he would never contact them and encourage them to come over to the Chinese contender. "It's not my style, and would be very immoral. However, should they send me their résumé, that's something else altogether," he told *Ny Teknik* (from Ahlbom, 2007; translation from Swedish by the authors).

In the year that followed, Ericsson did not appear (from observed press correspondence) to favor the growing interest Huawei was taking in its current and former employees. There was an ensuing war of words in the press and of publicly stated intentions concerning the movement of key industry knowledge via personnel from Ericsson to Huawei. In late 2007 it was made public via an Ericsson employee that the Swedish company had conducted a mapping procedure regarding former and current employees who had gone to work for Huawei. The mapped information included subjects such as family, friends, economic standing, and free-time activities. This information then found its way into reports compiled within the company (Aftonbladet, 2007; Fröberg, 2007).

Henry Sténson, Ericsson's director of communications during the event, was candid about how Ericsson viewed the competitive development between the two companies in Sweden, calling it a "surgically precise"

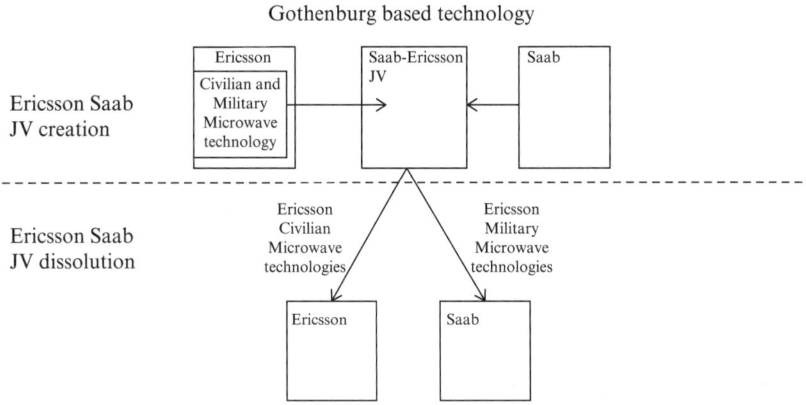

Figure 12.3 Microwave technologies in the Gothenburg region

acquisition of key personnel. "Many of our employees are angry about how Huawei is behaving. They lure people with salaries up to 50 percent more than what we are able to pay. Their acquisitions are a threat to jobs at Ericsson." As noted above, Fagerstedt was one of the mapped former Ericsson employees who had gone over to Huawei. He maintains that he received neither higher wages nor that Huawei was actively attempting to poach key personnel from Ericsson (ibid.).

Since establishing the group's first international R&D facility in the Kista, Huawei has been aggressively starting new R&D facilities in Sweden as well as globally. The next major step for Huawei in Sweden occurred on the country's west coast, in Gothenburg, in the fall of 2009. Here, Huawei's decision to co-locate its own activities with Ericsson competence areas and personnel is noteworthy and can be more specifically related to the technological competencies in the region. Figure 12.3 provides an overview of the key knowledge base in the Gothenburg region in microwave technologies, between Ericsson and Saab.

Gothenburg has a concentration of microwave technology business, as visualized in Figure 12.3. In the Gothenburg metropolitan area, Ericsson's regional offices had traditionally worked with microwave technology relating largely to military and civilian communication, as well as with wireless technologies. Concentrated on this area, a specific joint venture of note occurred between Ericsson and the Saab group. Saab Ericsson Space was founded in 1992, combining Saab's computer technology and mechanical systems with Ericsson's expertise in microwave equipment. In 2006, after a reorganization of the Ericsson's Gothenburg-based activities, the Ericsson-owned military communications unit was mainly sold to the Saab group, thus ending the joint venture, while Ericsson retained ownership of its civilian communications technology, of which microwave and base station technology played a key role (RUAG Space Sweden, 2001; Ericsson, 2006).

In 2009, Huawei opened its own facility in Gothenburg and its research orientation there was directed towards three types of technology: general development of base station technology, development of microwave products, and core network technology (Ahlbom, 2009). There is a close match to Ericsson's (and the Saab microwave division's) competences in the area.

Moreover, in terms of leadership, Huawei recruited Mats Andersson as site and system manager of radio base systems in Gothenburg. Andersson was formerly systems engineer, senior research engineer, project manager, department manager, and general manager at Ericsson from 1986–2006. Additionally he was an antenna engineer from 1982–83. Andersson led Saab Ericsson Space's antenna department from 1995–98.

The southern city of Lund, near to Malmö, has also been a key location for both Huawei and Ericsson. Figure 12.4 visualizes Ericsson units in the development of chip and mobile platform technologies in Lund. Lund has been the site of two joint ventures, between Sony and Ericsson and between Ericsson and ST Microelectronics, and both were later dissolved.

Sony and Ericsson created their joint venture for phones, Sony Ericsson, in 2001. At that time, Ericsson did not include the transfer of handset core technology (chipset technology), or its core R&D in this area, in the arrangement. This division of Ericsson would later become a separate subsidiary of Ericsson, namely, Ericsson Mobile Platforms (EMP).

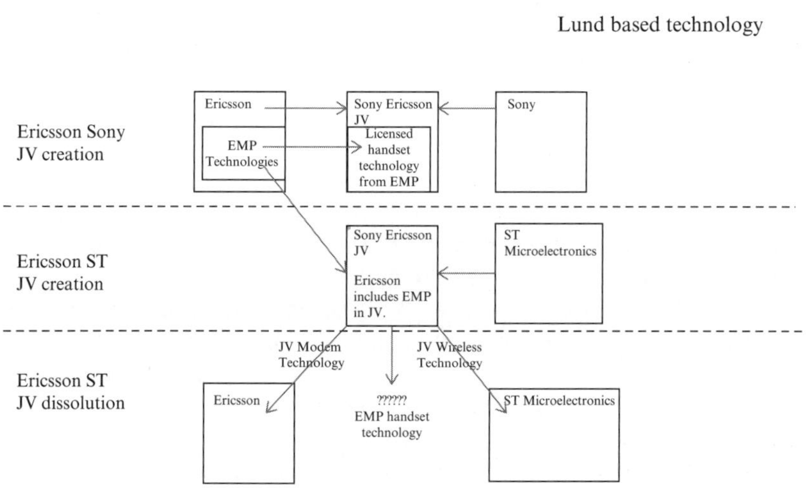

Figure 12.4 Chip and mobile platform technologies in the Lund region

This company then licensed its technology to Sony Ericsson along with various other actors in the mobile telecommunications industry worldwide. Ericsson's main non-developmental contributions to the joint venture were then general excellence in mobile telecommunications, business infrastructure knowledge, and operator relations (Sigurdson, 2004). Later, EMP became part of Ericsson's joint venture with the Swiss-based ST Microelectronics, becoming ST-Ericsson. This company maintained its operations in Lund, Sweden, in the area of chipset technology. When this joint venture was dissolved, the area of modem technology was reappropriated into Ericsson, while ST Microelectronics reassumed portions of the joint venture's wireless business.

Both joint ventures in Lund were affected by layoffs, and both have since been dissolved. During 2010, the two joint ventures were forced to make substantial personnel cuts, about 450 from Sony Ericsson and 150 from ST-Ericsson (Ahlbom, 2010). In 2011, Sony Ericsson became a full subsidiary of Sony Mobile, with Ericsson selling its shares to Sony. Although Ericsson sold all of its assets in the joint venture, Sony Mobile and Ericsson set up a cross-licensing agreement concerning wireless connectivity for a variety of platforms (Ericsson, 2012). Sony Mobile announced in late 2012 that it would relocate its head of operations from Lund to Tokyo, resulting in around 1000 employees in Lund being laid off. In the fall of 2013, the dissolution of ST-Ericsson left roughly 400–600 employees affected, with an estimated layoff of 290 employees in the Lund area (Kalin, 2014). Between 2010 and 2013, both joint ventures reduced their head count by a total of between 2100 and 2300 employees in engineering and consulting-related fields.

Thus, an abundance of skilled and experienced engineers with competence in chipset technology, platform design, and operational and infrastructural knowledge who had worked (and presumably most also resided) in the Lund area were available in the form of ST-Ericsson and Sony Mobile employees. Huawei established its own Lund R&D facility in early 2010, around the time of the first wave of layoffs by Ericsson's joint ventures. Huawei's Lund center is involved in "the development of mobile phones; technology that is closer to platform development than to applications" (ibid.).[10] It is expressed via Huawei's (2012) Europe fact sheet that the Lund center is involved in "terminal chipset design."

12.5 CONCLUSIONS AND IMPLICATIONS

This chapter has analyzed how firms access and develop knowledge and opportunities across innovation spaces. The cases are Chinese firms

(Geely and Huawei), which have undertaken investments in Sweden that have a specific focus on human capital and knowledge, specifically to access managerial, business, market, scientific, and technical knowledge.

Opportunities and challenges are discussed as well. These come to the forefront with Geely's acquisition of VCC, which is motivated by a need to tap into talent in technological and industrial fields in Sweden, a country (and its MNEs) known for high technology and high-quality products and processes. Geely purchased VCC in 2010. Approximately 20000 people work within VCC, and Geely accessed individuals with engineering and production experience, as well as patents, brands, other IPRs, and production facilities. Thus, a key opportunity is the possibility to acquire such assets, but there was also the need to invest significant amounts of capital by Geely. Financial capital is a generic challenge for the automobile industry, which needs a continuous flow of cash to undertake R&D and innovation. In this case, the Chinese entrepreneur who is chairman of the holding company can access significant amounts of capital.

Another challenge is the need to carefully disentangle the intellectual property and platforms between VCC and their previous owner, Ford, which did not want a new competitor in China using Ford technology. Ford previously owned VCC, and a major aspect of the purchase and later integration was how to address complex property rights (IPR, brand, platforms) sharing among the three companies. The final challenge encompasses the problems related to differences in perception about the tastes and demands of future and current customers, which in this case was formulated as safety, reliability, and Scandinavian design (i.e. elegant but minimalistic) contrasted with design that appeals to Asian high-end consumers. The reports in the media suggest that a compromise has been met, but differences will continue to be debated in the future.

Similarly, opportunities and challenges are visible in the Huawei case as well, when it acquired talent as well as technological and industrial knowledge in Sweden from Ericsson. Usually, the mechanism of accessing a foreign scientific infrastructure could be thought of as setting up R&D centers, and engaging in university–industry collaboration or other forms of open innovation. Here, we have added a new type of mechanism, when the company established co-located R&D centers with Ericsson centers to hire former employees of its rival. Huawei set up its first Swedish R&D lab in 2001, about the time that Ericsson started having severe economic problems, downsizing R&D and moving phones (handsets) to a joint venture with Sony. The main opportunity for Huawei has been to strategically hire key employees from a leading competitor, especially during times of intense global competition. Huawei has acquired many key employees and managers from Ericsson in almost every strategic area of importance to a value chain such as supply chain and logistics, to R&D and innovation project management, to development of radio systems and mobile technology.

One challenge from a company perspective is that Huawei needed to spread the R&D sites around Sweden, in order to co-locate the R&D sites with Ericsson's sites and its joint ventures in Stockholm/Kista, Gothenburg, and Lund. This is to benefit from the potential new employees and previous Ericsson employees' preference of locations. It would be much more difficult in the initial step and at this scale to hire when this necessitates a move to China and a complete change of lifestyle for the employee and its family. Once hired, many employees have moved to China for extended durations, as also happens in the VCC case. A related challenge has been to create a culture within Huawei's Swedish organization that attracts Swedish personnel. Given that Huawei expanded into areas – both geographically and technologically – that mirrored the investments made previously by Ericsson, Huawei could more easily hire people with technological and industrial experience, in the core technologies needed for telecommunication.

The Chinese firms, while keeping their headquarters in Asia, have used, respectively, foreign acquisition and hiring former employees, as mechanisms to link the innovation space of Sweden with that of China. Thus, these cases provide understanding of how Chinese companies access technology, human capital, and markets from Sweden and thereby expand the resources and opportunities available to companies located in innovation spaces in Asia.

NOTES

1. We would like to acknowledge and thank the Sten A. Olsson Foundation for Research and Culture and the Volvo Group for supporting this research. This chapter has been written in context of the research program Radical Innovations for the Enhancement of the Swedish Economy, running at the Institute of Innovation and Entrepreneurship, with Professor Maureen McKelvey as principal investigator. We would also like to acknowledge and thank the Broman Foundation, for supporting the research of Holgersson through a postdoc stipendium.
2. The case studies are here provided as illustrations, and they are based on archival and social media material for both cases, as well as interviews for the Volvo case.
3. Previously known as the European Innovation Scoreboard.
4. The brand "Volvo" is shared with AB Volvo, which makes trucks, heavy equipment, and so on.
5. Innovation used to be synonymous with R&D-intensive and high-tech products sold on a market, but the modern conceptualization is more complex and including many other types of innovations, services, bundles of services and products and innovations for consumers with low wages, known as "base of pyramid" (Dodgson et al., 2013).
6. According to its website in 2014: "Geely Automobile Holdings Limited is an automobile

manufacturer, focusing on development, manufacturing and sales of passenger vehicles. The company's shares are listed on the main board of Hong Kong Stock Exchange; the controlling shareholder of the company is Zhejiang Geely Holding Group Company Limited, a private company which is wholly-owned by Mr Li Shu Fu, the company's chairman, and his associates. With its headquarters established in Hangzhou and nine manufacturing plants in Linhai, Ningbo, Luqiao, Shanghai, Lanzhou, Xiangtan, Jinan, Chengdu and Cixi in China, the company has a total annual production capacity of 625 000 units of vehicle per year by end of 2012."

7. Note that most IP and technologies had been collected within FGT during VCC's time within Ford, as described above.
8. See Granstrand and Holgersson (2013) for a more thorough description of the management of the IP disassembly problem in this case.
9. This company overview is based upon McKelvey et al. (1998).
10. In Swedish: "grundutveckling av mobiltelefoner, teknik som är närmare plattformar än applikationer" (ibid.; author's English translation; quotation from Urban Fagerstedt).

REFERENCES

Aftonbladet (2007), 'Spionerar på avhopparna', 18 October. Available at: http://www.aftonbladet.se/nyheter/article11286294.ab (accessed 11 March 2014).

Ahlbom, H. (2002), 'Ericsson hårdbantar sin FoU', *Ny Teknik*, 2 May. Available at: http://www.nyteknik.se/nyheter/it_telekom/allmant/article223943.ece (accessed 1 June 2014).

Ahlbom, H. (2007), 'Ericssonavhopparen – Så ska Huawei bli störst', *Ny Teknik*, 3 October. Available at: http://www.nyteknik.se/nyheter/it_telekom/allmant/article257343.ece (accessed 15 March 2014).

Ahlbom, H. (2009), 'Huawei söker 30-tal ingenjörer i Göteborg', *Ny Teknik*, 19 October. Available at: http://www.nyteknik.se/nyheter/it_telekom/allmant/article264893.ece (accessed 15 March 2014).

Ahlbom, H. (2010), 'Huawei öppnar FoU i Lund', *Ny Teknik*, 13 April. Available at: http://www.nyteknik.se/nyheter/it_telekom/allmant/article268910.ece (accessed 14 March 2014).

Ahlbom, H. (2012), 'Ericsson minskar sin FoU', *Ny Teknik*, 25 January. Available at: http://www.nyteknik.se/nyheter/it_telekom/allmant/article3392805.ece (accessed 1 June 2014).

Ahlbom, H (2013), 'Sverige har ett bra affärsklimat', *Ny Teknik*, 14 February. Available at: http://www.nyteknik.se/nyheter/it_telekom/mobiltele/article3636552.ece (accessed 15 March 2014).

APCO Worldwide (2010), 'China's 12th five-year plan: how it actually works and what's in store for the next five years', 10 December. Available at: http://www.export.gov.il/UploadFiles/03_2012/Chinas12thFive-YearPlan.pdf (accessed 1 June 2014).

Arnott, S. (2010), 'China's Geely buys Volvo for $1.5bn', *The Independent*. Available at: http://www.independent.co.uk/news/business/news/chinas-geely-buys-volvo-for-15bn-2041772.html (accessed 24 March 2014).

Bloomberg News (2010), 'Huawei Technologies profit rises 30%, led by higher international sales', 17 April. Available at: http://www.bloomberg.com/news/2011-04-17/huawei-technologies-profit-rises-30-led-by-higher-international-sales.html (accessed 11 March 2014).

Boutellier, R., Gassman, O. and Zedtwitz, M. (2008), *Managing Global Innovation: Uncovering the Secrets of Future Competitiveness*, 3rd edn, Berlin: Springer.

Buckley, P.J. (2002), 'Is the international business research agenda running out of steam?', *Journal of International Business Studies*, **33**, 365–73.

Buckley, P.J., Clegg, J., Cross, A., Voss, H., Rhodes, M. and Zheng, P. (2008), 'Explaining China's outward FDI: an institutional perspective', in K. Sauvant (ed.), *The Rise of Transnational Corporations from Emerging Markets: Threat or Opportunity?*, Northampton, MA, USA and Cheltenham, UK: Edward Elgar.

China Daily (2009), 'Volvo deal to help Geely's going-global ambition', 25 December. Available at: http://www.chinadaily.com.cn/business/2009-12/25/content_9231856.htm (accessed 24 March 2014).

CNNMoney (1999), 'Ford buys Volvo car arm', 27 January. Available at: http://money.cnn.com/1999/01/27/deals/ford/ (accessed 23 March 2014).

DeBeule, F., and Duanmu, J.L. (2012), 'Locational determinants of internationalization: a firm-level analysis of Chinese and Indian acquisitions', *European Management Journal*, **30** (3), 264–77.

Dodgson, M., Gann, D.M., and Phillips, N. (2013), *The Oxford Handbook of Innovation Management*, New York: Oxford University Press.

Dunning, J.H. (1995), 'Reappraising the eclectic paradigm in an age of alliance capitalism', *Journal of International Business Studies*, **26**, 461–92.

The Economist (2012), 'Who's afraid of Huawei?', 3 August.

Ericsson (2006), 'Ericsson agrees to sell its defense business to Saab', press release, 12 June. Available at: http://www.ericsson.com/news/1054978 (accessed 1 April 2014).

Ericsson (2012), 'Ericsson annual report, 2011'. Available at: http://www.ericsson.com/res/investors/docs/2011/ericsson_ar_2011_en.pdf (accessed 9 March 2014).

European Commission (2013), *Innovation Union Scoreboard 2013*, prepared by Hollanders, H., and Es-Sadki, N.. Available at: http://ec.europa.eu/enterprise/policies/innovation/files/ius-2013_en.pdf (accessed 15 May 2014).

Fangfang, L. (2010), 'Geely buys Volvo in biggest overseas foray', *China Daily*, 29 March. Available at: http://www.chinadaily.com.cn/china/2010-03/29/content_9653260.htm (accessed 24 March 2014).

Fröberg, J. (2007), 'Ericsson-avhoppare synas i detalj', *Svenska Dagbladet*, 19 October, p. 7.

Granstrand, O., Håkanson, L. and Sjölander, S. (1993), 'Internationalization of R&D – a survey of some recent research', *Research Policy*, **22**, 413–30.

Granstrand, O. and Holgersson, M. (2013), 'Managing the intellectual property disassembly problem', *California Management Review*, **55** (4), 184–210.

Granstrand, O. and Holgersson, M. (2014), 'Multinational technology and intellectual property management – Is there global convergence and/or specialisation?' *International Journal of Technology Management*, **64** (2/3/4), 117–47.

Huawei (2010), '2010 corporate social responsibility report'. Available at: http://www.huawei.com/en/ucmf/groups/public/documents/attachments/hw_093033.pdf (accessed 15 March 2014).

Huawei (2011), 'Letter from the CEO'. Available at: http://www.huawei.com/ucmf/groups/public/documents/attachments/hw_126991.pdf (accessed 20 October 2014).

Huawei (2012), 'Huawei in Europe factsheet'. Available at: http://pr.huawei.com/ilink/en/download/HW_278575 (accessed 6 March 2014).

International Management (1994), 'Sweden's mobile leader', May, pp. 24–25, 27–29.
Kalin, K.S. (2014), 'Hit gick de telekom-anställda', *Sydsvenskan*, 3 January. Available at: http://www.sydsvenskan.se/ekonomi/hit-gick-de-telekom-anstallda/ (accessed 30 March 2014).
Luo, Y. (2000), 'Dynamic capabilities in international expansion', *Journal of World Business*, **35**, 355–78.
Luo, Y. and Tung, R.L. (2007), 'International expansion of emerging market enterprises: a springboard perspective', *Journal of International Business Studies*, **38**, 481–98.
Malerba, F. and Nelson R. (2012), *Economic Development as a Learning Process: Variation Across Sectoral Systems*, Northampton, MA, USA and Cheltenham, UK: Edward Elgar.
Mathews, J.A. (2006a), 'Dragon multinationals: new players in 21st century globalization', *Asia Pacific Journal of Management*, **23** (1), 5–27.
Mathews, J.A. (2006b), 'Catch-up strategies and the latecomer effect in industrial development', *New Political Economy*, **11** (3), 313–35.
Matson, H. (2014), 'Volvo PV tar sikte på miljonen', *Dagens Industri*, 1 April, 6–7.
McDonald, J. (2010), 'Geely-Volvo deal completed, Chinese buy Ford's struggling brand', *The Huffington Post*, 2 August. Available at: http://www.huffingtonpost.com/2010/08/02/geelyvolvo-deal-completed_n_666996.html (accessed 25 March 2014).
McKelvey, M., Texier, F. and Alm, H. (1998). 'The dynamics of high tech industry: Swedish firms developing mobile telecommunication systems', in Edquist, C. (ed.), *Innovation Systems and European Integration (ISE) Report of an EU Project*. CD-ROM. Linköping, Sweden: Department of Technology and Social Change.
Narula, R. and Dunning, J.H. (2010), 'Multinational enterprises, development and globalization: some clarifications and a research agenda', *Oxford Development Studies*, **38** (3), 263–87.
National Science Board (2014), 'Science and engineering indicators 2014'. Available at: http://www.nsf.gov/statistics/seind14/ (accessed 15 May 2014).
Nicolas, F. (2014), 'China's direct investment in the European Union: challenges and policy responses', *China Economic Journal*, **7** (1), 103–25.
Palmer, M. and McCarthy, C. (2011), 'Sony buys out Ericcsson from mobile deal', *Financial Times*, 27 October. Available at: http://www.ft.com/intl/cms/s/d7c78210-0068-11e1-ba33-00144feabdc0 (accessed 1 June 2014).
Peng, M.W. (2004), 'Identifying the big question in international business research', *Journal of International Business Studies*, **35**, 99–108.
Pröckl, E. (2013), 'Moduler är nya modellen för Volvo' *Ny Teknik*, 6 March. Available at: http://www.nyteknik.se/nyheter/fordon_motor/bilar/article3651016.ece (accessed 1 June 2014).
RUAG Space Sweden (2001), 'History'. Available at: http://www.ruag.com/space/Space_Home/Space_Division/History (accessed 16 May 2014).
Rugman, A.M. and Li, J. (2007), 'Will China's multinationals succeed globally or regionally?', *European Management Journal*, **25** (5), 333–43.
Shirouzu, N. (2013), 'Volvo execs, owner Geely clash over brand direction', *Automotive News Europe*, 10 September. Available at: http://www.autonews.com/article/20130910/OEM/130919981/volvo-execs-owner-geely-clash-over-brand-direction# (accessed 31 May 2014).

Sigurdson, J. (2004), 'The Sony Ericsson endeavor part 1', working paper no 190, Stockholm School of Economics: European Institute of Japanese Studies.

Skidé, U. (1994), 'Jan Uddenfeldt i spetsen för mobiltelefonin', *Elektroniktidningen*, **1**, 13–14.

13. Foreign direct investment in R&D and the base of the pyramid: is a new space of innovation emerging in India?

Sharmistha Bagchi-Sen, Torsten Schunder and Evangelos Bourelos

13.1 INTRODUCTION

This chapter addresses the role of new customers in developing innovation spaces, focusing upon base-of-the-pyramid (BOP) customers in India and the response of domestic and foreign enterprises and institutions to this group.[1] Foreign direct investment (FDI) in research and development (R&D) from US and European companies in emerging economies are drawing an increasing amount of attention in the literature. Historically, multinational corporation's (MNC) R&D operations with a focus on innovation were located in their home country. However, in recent years, open innovation has pushed MNCs to have global innovation networks. China and India with their large markets and success stories of their homegrown emerging multinationals in certain industries, especially information technology (IT) and pharmaceuticals, have become major destinations of FDI in R&D.

Recent literature suggests that emerging markets may offer opportunities to innovate for new types of customers using non-traditional collaborators (Doh and Teegen 2002; Lambell et al. 2008; Yaziji and Doh 2009) with implications for organizational innovation (cited in Aoyama 2012; Parthasarathy et al. 2014) and sustainable livelihoods. China and India, as well as other developing economies, are characterized by a large share of poor or low-income population with limited capacities to become producers or consumers in the globalized marketplace. This group is referred to as the BOP, the bottom or base of the pyramid. It is assumed that, while the individual purchasing power is low, the aggregate purchasing power of the population at the BOP is considerably high. The BOP, brought to attention by Prahalad (2004) and others (for example, London and Hart 2004) consists worldwide of "about 3 billion people, representing 70 percent of the developing world population with a net GDP of $12.5 trillion (PPP)." This size is considerable even if the individual customer has a low purchasing power. Prahalad proposed that serving the BOP will ultimately help alleviate poverty by providing them with consumption choices and creating more entrepreneurial opportunities. The idea of the fortune at the bottom of the pyramid has not remained unchallenged regarding (market) size and definition. For example, Prahalad himself used several definitions to delineate the group: earning of $2 per day or $2000 per year (Karnani 2006). It is also argued that the earnings of the BOP are spent, according to Maslow's pyramid of needs (Fletcher 2005), on essential goods, which would imply that a large share of their income is earmarked limiting the potential market size that was assumed by Prahalad (Karnani 2006). Furthermore, it is argued that seeing the BOP as consumers only is rather exploitative and it is necessary to incorporate the BOP as partners with ideas and inputs in the broader production and innovation system (Karnani 2006, 2007).

With increasing saturation of established markets and limits to growth, transforming the accessibility of the BOP to goods and services is seen as an important strategy for MNCs, who rarely target this particular demographic group (Fletcher 2005). Apart from adding to revenue growth, this strategy could allow the development of competitive advantages within the MNC based on the innovations necessary to serve the BOP. To this end, concepts such as pro-poor innovation and inclusive innovation are offered; the concept of *Jugaad* or improvisation for survival or frugal innovation is also brought to the forefront to demonstrate the range of possibilities for MNCs. In the absence of an organized way of identifying this group, companies struggle to maneuver local networks. Established business and R&D models of MNCs already operating in India require substantive adjustment to new networking strategies, such as co-operation with non-traditional partners (for example, non-governmental organizations (NGOs)) and the creation of local capacity (Vermeulen 2008) for launching a venture with possibilities of success. One big concern is to discourage exploitation – a proposition is to incorporate this group as producers so that they are not just a user group without possibilities of input that could improve their own situation.

This chapter explores the BOP literature alongside our understanding of FDI in R&D in emerging markets; the examples or cases are based on India. A main focus here is that innovations suitable for BOP require new modes of interactions among various constituents: the MNC, the community, NGOs, potential users, and so on. Modes of cooperation/collaboration that are emerging enable the MNCs to draw knowledge from the relevant (potential) customers and generate innovation such that the overall competitive advantage of the MNC, which may not be cost advantage, and the benefit to the target market (which is more than just a cheap product or service) are not compromised. What remains to be researched is if new organizations of innovation are occurring beyond our current understanding of formal and informal ways to augment and offer product and services for the BOP (for example, the strata within the BOP designated as poor). Also, what collaborative networks and processes of innovation are transforming non-consumers to consumers and promoting sustainability? The end outcome of the above observations seems to be a rescaling of goods and services useful for the BOP but the rescaling is not similar to reverse innovation, Jugaad and other observed concepts (see Aoyama 2014; Kolk et al. 2012). Below, this change in thinking is discussed by contextualizing the processes alongside FDI in R&D to show organizational, perceptional or cognitive, and operational challenges to the new approach.

13.2 FOREIGN DIRECT INVESTMENT IN R&D

Until recent studies acknowledged business innovation as a global phenomenon (Dodgson et al. 2013), the literature on international business discussed R&D operations under the topic of FDI. Motivations to engage in FDI are typically asset exploitation and asset augmentation (Dunning 2001; Dunning and Lundan 2009). In the context of R&D, ownership advantages are derived from protecting the knowledge base of a company. However, FDI in R&D capabilities are commonly undertaken to adapt a product and its production processes for the host market (Gassmann and Von Zedtwitz 1999). Only in recent years, host countries have started to be seen as a significant source of innovation. This is often a reflection of the emergence of homegrown multinationals with innovative capabilities in countries such as those from China and India (Buckley 2002; Peng 2004). Multinationals undertake FDI if ownership- and location-specific advantages allow benefits to exceed costs (for example, the liability of foreignness) (Dunning 2001; Hennart 2010; Oetzel and Doh 2009). Advantages can be seen as the abundant availability of comparatively cheap skilled labor in the host country or more recently, access to cheap but skilled labor. In the case of FDI in R&D, the national innovation system of the host country and its ability to foster and support science and technology advancements are important considerations to pursue asset augmentation in terms of patenting, co-product development and so on (Dunning 2001; Dunning and Lundan 2009; Serapio and Hayashi 2004).

The main focus of research on FDI has been investments originating in developed countries targeting other developed countries or developing countries but more recently FDI from emerging-market MNCs is receiving growing attention (Benito et al. 2009; Kedron and Bagchi-Sen 2012; Luo and Tung 2007). According to Serapio and Hayashi (2004), the early literature on FDI in R&D focused on the United States, Western Europe and Japan, where the majority of R&D facilities have been concentrated after the 1980s (Gassmann and Von Zedtwitz 1999). Internationalization of R&D began in the 1970s but firms started investing widely in foreign R&D during the late 1980s. In the late 1990s, R&D facilities started to emerge outside these established locations, mostly in South Asia (Cantwell 1995). Gassmann and Von Zedtwitz (1999) claim that these facilities emerged as a byproduct of international mergers and acquisitions. In the 2000s, China and India started to become very attractive locations for dedicated R&D centers of high-tech MNCs (Hall 2010; Mrinalini et al. 2013; see also Chapter 16 in this book).

Early studies related internationalization of R&D to the product life cycle in which the relocation of production in the later stages of the product life cycle is followed by R&D to support foreign subsidiaries in the adaptation of technologies (as mentioned above), but the increasing speed of product development requires simultaneous innovation, which is much more than straightforward adaptation (Reddy 2011; Serapio and Hayashi 2004). Later studies acknowledged the importance of supply-side factors such as knowledge and R&D personnel in the host country. This realization led to the differentiation of international R&D into home-base-exploiting and home-base-augmenting R&D. The former describes the exploitation of company knowledge in a host country (adaptation, ownership advantages) and the latter describes accessing knowledge/potential in the host country to enhance the overall competitiveness of a company (resource augmenting) (Dunning and Lundan 2009; Kuemmerle 1997; Serapio and Hayashi 2004).

The organizational structure that promotes FDI in R&D is industry specific (Reddy 2005). The literature generally distinguishes between centralized and decentralized modes of organization of R&D governance (Gassmann and Von Zedtwitz 1999). A centralized governance structure favors strong R&D facilities within the home country – this provides an established localized control and decision-making system, intellectual property protection, economies of scale in R&D, and a high level of specialization. Disadvantages are that a company could remain insensitive to demands of foreign markets by excluding user feedback.

Decentralized FDI in R&D can have varying degrees of freedom (for example, hub and spoke) or can form integrated research networks, and can emphasize the local context of their subsidiary location as well as provide means to incorporate new sources of knowledge (Gassmann and Von Zedtwitz 1999). The transaction costs of coordination among multiple R&D facilities may impose limits on a dispersed pattern of R&D internationalization (Gassmann and Von Zedtwitz 1999) and favor a relatively centralized R&D structure. Also the issue of intellectual property (IP) protection will determine the extent to which a company undertakes R&D in a foreign country.

The literature identifies several types of foreign R&D facilities of MNCs (Reddy 2005). Industry-specific and firm-specific conditions determine the type of the facility. *Technology transfer units* are established to transfer technology to foreign subsidiaries and adapt local production to company standards and they are likely to be co-located with production facilities. These facilities would represent a traditional approach of adaptation and home-base exploitation. *Global technology units* develop products and processes for the global market. *Corporate technology units* are responsible for the long-term development of the knowledge base of a company and perform basic research. Both types of technology units require access to skilled labor and specific knowledge and are likely to be located near relevant research facilities or research clusters. Considering their critical nature for company success, establishing such centers is seen as a major commitment. *Regional technology units* develop products for a group of markets with similar preferences whereas *indigenous technology units* develop products for local markets based on available local technologies.

Observed patterns of FDI in R&D are based on formalized and institutionalized R&D. But innovative potential can be found outside this formalized system. For example, the informal sector of emerging-market economies is considered a source of knowledge in the BOP concept. Given that BOP is not a bounded concept in terms of its location, size and other characteristics, it can be challenging for multinationals to develop organizational structure and strategies (for example, a network of institutions linking the BOP to the MNC's R&D division) suitable for harnessing the knowledge from the BOP for innovation that will benefit both the MNC and the end users.

13.3 FDI IN R&D AND THE BOP IN INDIA

Serious currency and balance-of-payment problems led post-colonial India to an economic crisis, which resulted in an International Monetary

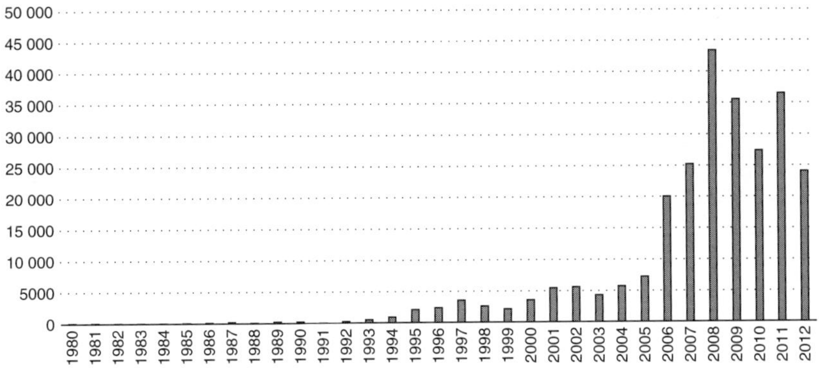

Source: Compiled by authors from World Bank Open Data (n.d.)

Figure 13.1 FDI net inflows in million US$ in India

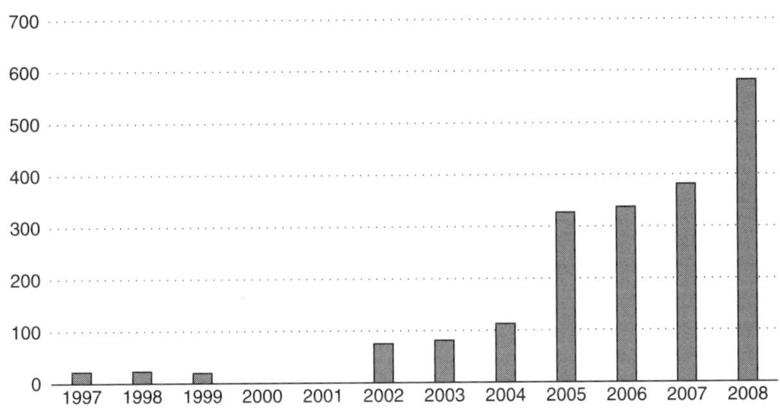

Note: no data for 2000 and 2001

Source: Based on Basant and Mani (2012: 15)

Figure 13.2 FDI in R&D of affiliates of US MNCs in India in million US$

Fund (IMF) bailout and economic liberalization and privatization in the 1990s. As can be seen in Figure 13.1, the amount of net inflow of FDI in India was marginal during the 1980s but FDI inflows rose quickly after the reforms reaching $5 billion in the early 2000s (Basant and Mani 2012; Mrinalini et al. 2013) (see also Figure 13.2). After a peak in R&D

Table 13.1 FDI in R&D in Indian industrial sectors in billion US$

Sector	2003	2004	2005	2006	2007	2008	2009	Total
Aerospace	0.02			3.25	0.02	0.23	0.1	3.62
Auto industry	0.08	0.12		0.47	0.66	0.59	0.76	2.68
Pharma and biotechnology	0.04	0.12	0.45	1.11	0.31	0.43	0.35	2.81
Machinery and equipment	0.11	0.15	0.13	0.18	0.2	0.1		0.87
Chemicals	0.03		0.01	0.02	0.22	0.02	0.06	0.36
Electronic components	0.33		0.05	0.03	0.03	0.15	0.15	0.74
Engines and turbines	0.02			0.02	0.02	0.01	0.03	0.1
Services	0.009	0.37	0.02	0.009	0.09	0.03	0.23	0.76
Metals and minerals	0.03	0.08		0.13	0.02	0.01	0.07	0.34
Software and IT	1.43	2.21	2.12	4.96	1.7	1.74	0.4	14.56
Others	0.09	0.13	0.04	0.15	0.03	0.2	1.39	2.03
Total	2.19	3.18	2.86	10.33	3.3	3.51	3.54	28.91

Source: Table created by authors based on data from Mrinalini et al. (2013: 771)

investment in 2006, the sum declined to reach in $3.5 billion in 2009 as seen in Table 13.1.

India provides several advantages to a foreign multinational company investing in R&D (Bound and Thornton 2012) and has been successful in attracting R&D in aerospace, biotech, IT and pharmaceuticals. Between 2003 and 2009, the software and IT sectors attracted over $14.56 billion of foreign R&D investments, which is more than 50 percent of the whole FDI in R&D in India during this period (Table 13.1; Mrinalini et al. 2013). This inflow represents typical FDI in R&D, which follows exploitation of internal knowledge or access to strategic locational assets (for example, skilled labor). Since these investments benefit specialized users (for example, other industries), linkages with the BOP are limited (Tiwari 2009). The BOP is not a source of labor and is bypassed as a source of innovative ideas that can benefit the MNC. However, information and communication technology (ICT) applications are critical for the BOP (Parthasarathy et al. 2014) as illustrated in the health-care cases where firms found "a sweet spot that permits a meeting of the social goals of catering to the underserved and making the provision of the product or service a sustainable proposition." In one case, working around the infrastructure (for example, lack of a steady flow of electricity led to the development of a baby warmer) paid off. In another case, low human capital (for example, low levels of literacy and training) and skeptical users (for example, needed proof of concept to be convinced – images of cure) were managed through mobile technologies to promote ophthalmic solutions.

In the world of FDI in developing countries where purchasing power is low, companies are not competing with other companies just for market share but against non-consumption (Agnihotri 2012). To motivate the BOP to purchase a product and/or take on debt to purchase, a product has to have characteristics to significantly enhance production capacities, income potential, or the ability to fulfill basic needs of the BOP. The product has to be paired with affordable pricing, low maintenance cost and high reliability providing a long-term benefit. Trust is an equally important factor affecting the decision to buy a product – this often entails the involvement of legitimate intermediaries who can convince the BOP to undertake a decision to spend scarce resources. Therefore, accessibility to finance (for example, microfinance) enabling buyers to obtain credit is critical (Agnihotri 2012). Another important factor is the need to improvise with design such that a product is suitable and reliable. Table 13.2 provides a further overview of the recent literature on BOP – these studies provide discussions of cases. Some points are listed below.

Case studies make it clear that maximization of corporate social responsibility returns did not enter the innovation or diffusion strategies of the firms (Ramani and Mukherjee 2014). The major difference between the success and failure of innovation and social entrepreneurship models at the BOP has been the perceived value of the BOP (Pervez et al. 2013). SKS Microfinance is a for-profit organization similar to the Grameen Bank. Giving out loans with a for-profit strategy and no social perspective, the organization suffered severely in the wake of suicides of customers unable to pay back their loans; customers of a province collectively refused to pay back their loans and politicians accused the organization of squeezing the poor (Pervez et al. 2013).

Agnihotri (2013) shows that about one out of eight cases is a success. Multinationals have used distinct practices parallel to their existing capabilities showing a separation of formal R&D from innovating for the BOP (Vermeulen et al. 2008). For example, the affordable ChotuKool is a cooling device "developed in close interaction with the targeted customer groups." It is able to run on batteries, making it suitable to areas with erratic power, it is lightweight and portable allowing it to be stowed away easily in small dwellings and it addresses the basic need of preventing milk and leftovers from spoiling over a short time period (Tiwari and Herstatt 2013b).

Characteristics of cost-effective frugal innovation are: value proposition,

Table 13.2 Overview of BOP literature and case studies

Paper title	Author	Methodology	Research question(s)	Cases	Conclusions
Serving the poor: Multisided mobile service platforms, openness, competition, collaboration and the struggle for leadership	Karippacheril, Nikayin, de Reuver and Bouwma (2013)	Interview, ICT	MSP strategies for serving the poor.	Airtel in India Roshan in Afghanistan Zain (now Airtel) in Africa MTN in Rwanda Vodafone/Safari in Kenya	Applying the framework of multisided platforms to mobile ecosystems, viable services for the BOP require a cooperation framework that involves all three providers in the ecosystem: operators, devices and SPs
Can breakthrough innovations serve the poor (BOP) and create reputational (CSR) value? Indian case studies	Ramani and Mukherjee (2014)	Theoretical construct, case studies	Can breakthrough innovations serve the poor (BOP) and create reputational (CSR) value?	Monsanto and Bt cotton in India Cipla and the invisible market for ARVs	Case studies make it clear that maximization of CSR returns did not enter the calculation of either of the firms in their innovation creation or diffusion strategy
International marketing at the bottom of the pyramid	Fletcher (2005)	Questionnaire / survey	How do different classes receive marketing in BOP markets?		Separate strategies needed for marketing in BOP
Innovation and social entrepreneurship	Pervez, Maritz and de Waal (2013)	Secondary data/ case studies	Which are the key success criteria for innovations	Nestle Pakistan Tata Nano	The major difference between the success

at the bottom of the pyramid – a conceptual framework			by enterprises targeting the BOP?	Grameen Bank, SKS Microfinance,	and failure of innovation and social entrepreneurship models at the BOP has been foundto be the perceived value by the BOP market.
Partnerships in BOP business	Hietapuro (2011)	Case studies	What kinds of partnerships companies form in BOP business?	20 BOP Ventures	Importance of partnerships in BOP business
Revisiting the debate over the bottom of the pyramid market	Agnihotri (2012)	Case studies	Summary of Prahalad's prepositions	Casas Bahia, Aravind hospital, Cemex, microfinance, HLL, e-Choupal, Annapurna Salt, Voxivia	Only one out of eight case studies a success
Building dynamic capabilities for the base of the pyramid	Vermeulen, Bertisen and Geurts (2008)	Secondary data	What different organizational practices do MNCs use to build collaborative BOP capabilities?	Case: Patrimonio Hoy, Cases: Annapurna Salt, HLL Soap, Case: NutriStar, Case: Energy Yoghurt, Case: Hariyali Kisaan Bazaar	Successful MNCs have used a variety of distinct practices in parallel to build their capabilities
Frugal innovations for the 'unserved' customer: An assessment of India's attractiveness as a lead market for cost-effective products	Tiwari and Herstatt (2012a)	Case studies	Assessing India's potential as a lead market for cost-effective frugal innovations	Tata Ace, Tata Swach, Vortex Gramateller, Godrej ChotuKool	Pattern of characteristics for which India has emerged as a hotbed for innovations: value proposition, volume opportunities, robustness, fault resistance, cost of ownership

Table 13.2 (continued)

Paper title	Author	Methodology	Research question(s)	Cases	Conclusions
Reviewing a decade of research on the "base/bottom of the pyramid" (BOP) concept	Kolk, Rivera-Santos and Rufín (2012)	Literature review, 104 articles	What has become of the concept over the decade following its first systematic exposition in 1999?		The BOP concept evolved dramatically following Prahalad's original call to MNCs
Encouraging entrepreneurship: Microfinance, knowledge support, and the costs of operating in institutional voids	Chakrabarty and Bass (2013)	Financial and survey data, hypothesis testing	FDI moderates the influence of an MFI's knowledge support to encourage entrepreneurship on the MFI's costs of operating at the BOP Loan defaults moderate the influence of an MFI's knowledge support to encourage entrepreneurship on the MFI's costs of operating at the BOP		The influence of an MFI's knowledge support to encourage entrepreneurship on the MFI's costs of operating at the BoP is strengthened when FDI is high and loan defaults are low
Does FDI reduce poverty? Case studies from India	Tiwari (2009)	Case studies, Regions	FDI inflows and the impact on poverty reduction	Tamil Nadu, Karnataka	Tamil Nadu show high rural multi-dimensional poverty indicators as well as high urban income poverty; Karnataka presents strong grounds for

					contesting the linkages between high growth and poverty reduction
The entry of multinational companies to the base of the pyramid: a network perspective	Tasavori (2013)	Case studies, Interviews	The relationship between MNCs and NGOs and explain how it can facilitate MNCs' entry into the BOP market	Insurance Co., Alpha Bank, FMCG Co., Home appliance	By developing their relationships with NGOs, MNCs can build trust, demonstrate their commitment and strengthen their legitimacy in the eyes of the NGOs and consequently the BOP market

Note: ARVs – antiretroviral drugs; BOP – bottom of the pyramid; CSR – corporate social responsibility; HLL – Hindustan Lever Ltd.; FDI – foreign direct investment; FMCG Co – fast-moving consumer goods companies; ICT – information and communications technology; MFI – micro finance institution; MNCs – multinational corporations; MSP – mobile service platform; MTN – a global communications company; SKS – a microfinance company; SPs - service providers

volume opportunities, robustness, fault resistance and cost of ownership (Tiwari and Herstatt 2012a). The Tata Swach water purifier is a result of several years of corporate R&D. Developed for predominantly poor people in areas without secure access to running water and electricity the device is, at $18, considerably cheaper than its competitors but has also low running costs ($0.002/l) and became a success with previous consumers (Tiwari and Herstatt 2012b).

By developing their relationships with NGOs, MNCs can build trust, demonstrate their commitment and strengthen their legitimacy in the eyes of the NGO and the BOP (Tasavori 2013). Applying the framework of multi-sided platforms to mobile ecosystems, viable services for the BOP require cooperative frameworks of all three providers in the ecosystem: operators, devices and service platforms (Karippacheril et al. 2013). For example, Nokia Lifetools provides together with several partners (such as Reuters, Syngenta) commercial services offering agricultural information and educational but also entertainment content to enhance access to information in remote areas (Hietapuro 2011). Additionally, Nokia makes its services accessible for the BOP by providing microfinance services and other financial services (for example, charging phones, money transfer) without requiring bank accounts. The services are exclusively bound to Nokia devices. Other companies offer services platform independently (Reuters Marketlife) or are operator/provider based (Airtel India) (Karippacheril et al. 2013).

While products and services may meet the needs of the BOP, Prahalad states that the BOP is also brand conscious. Therefore, the status of a brand and peer pressure may be necessary for market success; however, Agnihotri (2012) found that price sensitivity is often more important than brand reputation. Pervez et al. (2013) highlight the importance of engaging the BOP groups and considering them as part of the production (and innovation) process in order to succeed.

13.4 INNOVATION AND INVOLVEMENT WITHIN THE BOP

The number of MNCs actually involved is marginal and their success is often moderate (Karnani 2006) (see Table 13.2 for examples). Servicing the BOP as well as incorporating them in the knowledge creation or production process requires overcoming various challenges. Aside from providing a product with a high value proposition as mentioned before, functioning markets do not always exist due to various factors, such as limited availability of information or the local absence of competition or dependence on a limited number of vendors. Given the limitations in distribution systems and communication channels, marketing and distribution of an otherwise useful product can be challenging. Considering the potential small profit range of BOP products, establishing a distribution system or applying traditional marketing methods might make a product unprofitable.

Additionally, the purchasing power of a particular group may be too low to acquire even a very beneficial product. In such cases, alternative business models which support different pathways of innovation are required, which involve, for example, the cooperation with NGOs, other related institutions (for example, microfinance) or the implementation of pay-per-use solutions (Karnani 2006, 2007). The following section presents some of ways MNCs have worked with the BOP – these different ways are not mutually exclusive.

13.4.1 Types of Strategies Used to Engage the BOP

Several innovation strategies resulting in marketable products to the BOP have been identified in the literature, and this section introduces some of these concepts and few examples. A first concept is "pro-poor innovation." These originate mostly within traditional R&D facilities without embedding the poor directly in the development process. The focus of these innovations lies often in cost reduction through the removal of features or making changes in packaging (Karnani 2006). However, there have been questions about the viability of this concept due to the lack of involvement of the poor. Also the cognitive approach of producing something cheap may lead to non-acceptance within the intended target group; consumers with limited resources do not necessarily want to buy the cheapest product, they are selective buyers of good-quality items. A type of pro-poor innovation has been smaller packages of established products (for example, shampoo) making them more affordable but the success of this approach is rather questionable for the companies given that larger packages offer lower per unit cost of production and the cost savings can be then transferred to the buyer (Agnihotri 2012; Fletcher 2005; Karnani 2006).

A second concept is "grassroots innovations". These, on the other hand, are created by the poor themselves mostly out of necessity and are creative ways to solve issues at hand. An example is the Jaipur Foot invented in the context of unaffordability of artificial limbs, which affected the market for artificial limbs because of pricing ($45) and better functionality (Bound and Thornton 2012). This type of innovation is mostly reactive and facilitates the use of available resources to overcome obstacles. Generally, these innovations happen on a micro level and adaptation on larger scales can be difficult especially for service innovations which are solutions to local problems (OECD 2013; Seyfang and Smith 2007). These innovations emerge within the informal sector and outside of traditional firms or formal R&D divisions. To adapt the requirements of a product for a specific local context, it is beneficial for multinationals to work with local (micro) entrepreneurs who may have innovative ideas about product or process development, such as technology use (OECD 2013). On the other hand, involving MNCs gives entrepreneurs the ability to extend their markets. However, the process of linking the local entrepreneur with the MNC is not simple – this is where intermediaries facilitate connections (Utz and Dahlman 2007).

Other concepts are "inclusive or frugal innovation." Inclusive or frugal innovation is thought of as a wider approach than providing cheap goods, and the idea is to provide high performance and high-quality goods, at affordable prices, which actually meet the BOP demand. These innovations are not only based on technical change but include also the adaptation of new business models, which are able to overcome constraints in marketing, financing and distribution. The goal is to allow new forms of cooperation and interaction between the public and private sectors as well as non-traditional partners, such as the poor themselves and NGOs. Also the inclusion of the poor within the production process or as the originator of the innovation can be part of inclusive innovation; there is a strong focus on the interaction between R&D and the poor to create, by strong user–producer interaction, products meeting the actual demand (Guth 2005; Lundvall 2009; OECD 2013; Utz and Dahlman 2007). As a concept, inclusive innovation is able to incorporate pro-poor and grassroots innovation as well as other labels like Jugaad, which is widely used in the Indian context of frugal innovation, and is based on improvisation due to resource constraints and lack of affordable solutions (Tiwari and Herstatt 2012a). Ideally, these innovations are affordable and provide a higher use value (George et al. 2012).

In a global context, frugal innovation could provide companies with a competitive edge by increasing resource efficiency and/or product quality. General Electric (GE) and its Indian subsidiary Wipro GE Healthcare, for example, developed the baby-warmer "Lullaby" to enhance infant health care in close cooperation with the BOP without reliance on existing designs. They created an affordable system of high quality, which was sold in 2013 in 62 countries encouraging GE to invest in further cooperation with the BOP (Sehgal 2010; Tiwari and Herstatt 2012a). Another example of inclusive innovation is the Tata Nano (Dhume 2011). However, this example shows the difficulty of promoting a product as complex as an automobile despite the cost savings. While international standards have been applied in the development of the Nano, marketing the car as the world's cheapest automobile gave it a bad start; then, a series of unrelated accidents resulted in questions about the quality of the Nano. Tata was also not able to align its distribution, financing and service arrangements for its intended market (Dhume 2011; Pervez et al. 2013; Tiwari and Herstatt 2012a).

Due to the direct involvement of the ultimate consumers, a main criticism about the BOP concept is whether targeting the poor or low income might be rather exploitative. Unilever marketed a skin-whitening cream for women in India with the proposition that lighter skin would make it easier to escape poverty and marry – the message was unacceptable leading to a ban of the product (Karnani 2007). Therefore, a variety of regulations to protect the BOP innovator, producer, and user will be necessary as more and more MNC–BOP collaborations ensue (Utz and Dahlman 2007). In general, governments could play a crucial role in facilitating research for the benefit of the poor. The government has the ability to shape an inclusive innovation system by supporting not only formal R&D but also informal forms of involvement of the kind outlined above. Examples of support may include the provision of grants for prototypes as well as access to test facilities or consultancy services. Establishing seed capital and support for local entrepreneurs and creating intermediary institutions managing contact between local entrepreneurs and MNCs will be an important step to facilitate MNC–BOP connections. At the same time, reduction of entry barriers for MNCs may have beneficial effects for both the BOP and the MNC (Utz and Dahlman 2007). The core value of these concepts is asserted by Polak and Warwick (2013) in what they call zero-based design, which is motivated by "the ruthless pursuit of affordability" – that is, design or innovate from scratch with a clear understanding of the needs of the poor (as cited in Parthasarathy et al. 2014).

13.4.2 Modes of Involvement

Crucial for the creation of a successful product for the BOP is to understand the specific needs of the local population to create a product meeting the local demand and creating a product delivering a superior value, which would transform non-consumers into consumers. The above discussion suggests that acquiring this knowledge about the BOP market can be challenging. To increase chances of success, businesses have to recognize the BOP as a customer group with specific needs and demands and have to get involved in market research to explore the demand or get involved in intense consumer–producer interactions to embed the needs of the local consumers in product development. Gathering this knowledge is essential to reduce the risk of non-acceptance by the BOP. In the following section, modes of involvement for multinationals are presented, which have the potential to facilitate innovation within the BOP or in collaboration with the BOP as well as approach the needs of these new customer groups. The discussion also provides both opportunities and challenges for the varying approaches.

Formal R&D is recognized as one mechanism, but formal R&D has a lag time in bringing product to market. However, the entry mode through FDI can have considerable impact on the ability of companies to access market knowledge supporting the development of BOP products (Narula and Dunning 2010; Reddy 2005). A greenfield site would ensure compliance with the general needs of a company but would require "time" to get accepted by the locals as well as the scientific community in the host country; this can delay the process of accessing local knowledge. Acquisition of an existing facility or company may provide quicker linkages to suppliers and markets. Another mechanism is to use R&D for specific products and technologies. In the context of FDI in R&D, establishing indigenous development units seem to be a way to relate product development to locally available technologies. However, a working relationship with the potential consumers has to be established to ensure the creation of a suitable product and reduce the risk of failure (Reddy 2005). Sehgal et al. (2010) suggest that it is not sufficient to alter existing designs to serve the BOP and that only a wholly new development process could provide creative freedom to incorporate the specific needs of the BOP. In this context, establishing corporate units to perform basic research and develop a specific knowledge base might be beneficial.

Another aspect is more direct interaction with potential customers. It can be challenging for an MNC to acquire the necessary knowledge to develop a potentially successful product as well as to ensure its acceptance. To become part of the local network providing an MNC with local knowledge, cooperation with intermediaries (such as NGOs), which serve either as facilitator of contacts with the local network or provide information, may be a suitable strategy for companies (Tasavori 2013). Generally, company/intermediary partnerships are not unproblematic given that some may differ in their goals. While a company has financial capital and has to make profit, a not-for-profit intermediary has social capital and is driven by the goal of development. The difference in how both organizations value their outcomes may be a constant strain on their relationship requiring MNCs to invest significant time and money into relationship management (Oetzel and Doh 2009). According to Tasavori (2013), for example, the cooperation between NGOs and MNCs is based upon commitment, trust and legitimacy where a company has to convince the NGO of its ambitions.

Commitment requires an initial involvement of the MNC; trust between the NGO and the MNC can be earned if short-term goals are not determining the actions of a company within the BOP context; in other words, legitimacy can be achieved by continuous involvement and fulfillment of promises. A Dutch home-appliance MNC, for example, developed together with an NGO's a smokeless stove to battle respiratory illness caused by stoves used inside the home (Tasavori 2013). Cooperating with a local NGO, the company gained access to BOP representatives, creating a viable solution through cooperation. The company gave up the intellectual property on the stove and entrepreneurs have received the designs for the smokeless stove free of charge. The company has established itself as a committed, reliable and trustworthy partner enhancing its ability to interact with the BOP and NGOs in further projects (compare with Tasavori 2013). This approach is closely related to inclusive innovation and indicates that MNCs involved in BOP are often focused on earning long-term gains. These gains include consolidating the general reputation of the company, the development of corporate social responsibility values, the establishment of the company as a BOP brand, the creation of customer loyalty and also market expansion beyond the target BOP (Agnihotri 2012; Ramani and Mukherjee 2014; Tiwari and Herstatt 2012b).

Establishing a cooperative understanding with an NGO allows MNCs to access its network, which may grant access to influential people (Oetzel and Doh 2009). This network helps companies identify entrepreneurial potential, which can be incorporated in their own R&D process and/or marketing. If the social or economic benefit of a product for the poor is substantial enough, an NGO may also be an effective partner to support marketing and distribution of a good. Establishing a system that is able to foster systematic frugal innovation on a large scale requires a combined effort from various interested parties such as government institutions, MNCs, NGOs as well as the BOP itself (Hietapuro 2011). Aside from business needs and goals, the attitude (for example, cognitive barriers) towards the BOP is likely to influence the mode of involvement by MNCs, the role of the BOP, as well as the future of the MNC–BOP relationship (Hill 2010).

13.5 CONCLUSIONS AND IMPLICATIONS

This chapter has addressed the development of new types of innovation spaces driven by the recognition of new types of consumers in an emerging economy. This new space is the BOP in India. Traditional FDI in R&D in India and other emerging-market economies focuses on adaptation of technologies and products according to the market conditions – therefore, access to the market is important followed by the significance of the abundance of cheap but skilled labor. In recent years, FDI in R&D is growing in high-technology industries such as pharmaceuticals, biotech, electronics and ICT in emerging economies with a goal of market expansion as well as augmentation of innovative processes for both the investing MNC and the host country (for example, joint-venture partners, suppliers). Multinationals involved in emerging economies are also interested in finding innovation spaces within the BOP given the size of the BOP as a potential source of ideas as illustrated by the literature with case examples.

Collaboration between MNC and BOP does not easily fit the traditional R&D models discussed to understand the internationalization of R&D by MNCs. The new process of MNC engagement innovation is a way to create strong partnerships among non-traditional collaborators such as the NGO and the MNC with goals that combine profit orientation with sustainability of livelihoods at the BOP. Such partnerships have to be facilitated and are not without risk. Trust and legitimacy are important in these interactions requiring companies to make commitments and to focus on long-term gains or sustainable outcomes. Tapping into an environment that fosters the creation of cheaper or frugal products with higher quality or are at least more resource efficient with significant impact in terms of inclusiveness is a challenging task for organizations. If the BOP continues to emerge as a strategic asset for the home country and the host, new forms of governance are expected to emerge to manage, protect and sustain this asset.

NOTE

1. The authors would like to acknowledge the support of their respective universities. Moreover, Professor Bagchi-Sen would like to acknowledge the support of the companies and foundations who support the Visiting Professor's Program at the School of Business, Economics, and Law, University of Gothenburg, as described in the acknowledgments to this book. Evangelos Bourelos would like to acknowledge the support of the Broman Foundation and of the Sten A. Olsson Foundation for Research and Culture.

REFERENCES

Agnihotri, A. (2012). 'Revisiting the debate over the bottom of the pyramid market'. *Journal of Macromarketing*, **32** (4), 417–23.

Aoyama, Y. (2012). 'Geography of economic governance: industrial dimensions of state-market relations'. In R. Lussà, J. Feliu and X. Paunero (eds), *Crisis económica e impactos territoriales – V Jornadas de Geografía Económica,* Girona: University de Girona, pp. 19–36.

Basant, R. and Mani, S. (2012). 'Foreign R&D centres in India: an analysis of their size, structure and implications'. Indian Institute of Management, Ahmedabad, Working Paper 2012-01-06. Available at: http://115.111.81.83:8080/xmlui/bitstream/handle/123456789/11380/2012-01-06Rakesh.pdf?sequence=1&isAllowed=y (accessed 27 July 2014).

Benito, G.R., Petersen, B., and Welch, L.S. (2009). 'Towards more realistic conceptualisations of foreign operation modes'. *Journal of International Business Studies*, **40** (9), 1455–70.

Bound, K., and Thornton, I.W. (2012). *Our Frugal Future: Lessons From India's Innovation System*. London: Nesta.

Buckley, P.J. (2002). 'Is the international business research agenda running out of steam?'. *Journal of International Business Studies*, **33**, 365–73.

Cantwell, J. (1995). 'The globalisation of technology: what remains of the product cycle model?'. *Cambridge Journal of Economics*, **19**, 155–74.

Chakrabarty, S. and Bass, E. (2013). 'Encouraging entrepreneurship: microfinance, knowledge support, and the costs of operating in institutional voids'. *Thunderbird International Business Review*, **55** (5). Article first published online 19 August.

Dhume, S. (2011). 'Unloved at any speed'. *Foreign Policy*, 7 October. Available at: http://www.foreignpolicy.com/articles/2011/10/07/tata_nano_unloved_at_any_speed?page=full (last accessed 10 January 2014)

Dodgson, M., Gann, D. M., and Phillips, N. (2013). 'Perspectives on innovation management'. In Dodgson, M., Gann, D.M., and Phillips, N. (eds), *The Oxford Handbook of Innovation Management*, Oxford University Press: New York, pp. 3–25.

Doh, J.P. and Teegen, P. (2002). 'Nongovernmental organizations as institutional actors in international business: theory and implications'. *International Business Review*, **11** (6), 665–84.

Dunning, J.H. (2001). 'The eclectic (OLI) paradigm of international production: past, present and future'. *International Journal of the Economics of Business*, **8** (2), 173–90.

Dunning, J. H., and Lundan, S.M. (2009). 'The internationalization of corporate R&D: a review of the evidence and some policy implications for home countries'. *Review of Policy Research*, **26** (12), 13–33.

Fletcher, R. (2005). 'International marketing at the bottom of the pyramid'. Available at: http://www.anzmac.org/conference_archive/2005/cd-site/pdfs/9-Marketing-Int-C-Cultural-Env/9-Fletcher.pdf (accessed 27 July 2014).

Gassmann, O., and Von Zedtwitz, M. (1999). 'New concepts and trends in international R&D organization'. *Research Policy*, **28** (2), 231–50.

George, G., McGahan, A.M. and Prabhu, J. (2012). 'Innovation for inclusive growth: towards a theoretical framework and a research agenda'. *Journal of Management Studies*, **49** (4), 661–83.

Guth, M. (2005). 'Innovation, social inclusion and coherent regional development: a new diamond for a socially inclusive innovation policy in regions'. *European Planning Studies*, **13** (2), 333–49.

Hall, B.H. (2011). 'The internationalization of R&D'. Available at: http://elsa.

berkeley.edu/pub/users/bhhall/papers/BHH10_RND_international_August.pdf (accessed 27 July 2014).
Hennart, J.F. (2010). 'Transaction cost theory and international business'. *Journal of Retailing*, **86** (3), 257–69.
Hietapuro, M. (2011). 'Partnerships in BOP business'. Available at: https://aaltodoc.aalto.fi/bitstream/handle/123456789/661/hse_ethesis_12529.pdf?sequence=1 (accessed 27 July 2014).
Hill, R.P. (2010). 'A naturological approach to marketing exchanges: implications for the bottom of the pyramid'. *Journal of Business Research*, **63** (6), 602–607.
Karippacheril, T.G., Nikayin, F., de Reuver, M. and Bouwman, H. (2013). 'Serving the poor: multisided mobile service platforms, openness, competition, collaboration and the struggle for leadership'. *Telecommunications Policy*, **37** (1), 24–34.
Karnani, A. (2006). 'Mirage at the bottom of the pyramid'. Available at: http://deepblue.lib.umich.edu/bitstream/handle/2027.42/57215/wp835?sequence=1 (accessed 27 July 2014).
Karnani, A. (2007). 'The mirage of marketing to the bottom of the pyramid'. *California Management Review*, **49** (4), 90–111.
Kedron, P. and Bagchi-Sen, S. (2012). 'Foreign direct investment in Europe by multinational pharmaceutical companies from India'. *Journal of Economic Geography*, **12** (4), 809–39.
Kolk, A., Rivera-Santos, M. and Rufin, C.R. (2012). 'Reviewing a decade of research on the "base/bottom of the pyramid" (BOP) concept'. *Business & Society*, **53** (3), 338–77.
Kuemmerle, W. (1997). 'Building effective R&D capabilities abroad'. *Harvard Business Review*, **75**, 61–72.
Lambell, R., Ramia, G., Nyland, C. and Michelotti, M. (2008). 'NGOs and international business research: progress, prospects and problems'. *International Journal of Management Reviews,* **10** (1), 75–92.
London, T. and Hart, S.L. (2004). 'Reinventing strategies for emerging markets: beyond the transnational model'. *Journal of International Business Studies*, **35** (5), 350–70.
Lundvall, B.A. (2009). 'Innovation as an interactive process: user-producer interaction to the national system of innovation'. Research paper. *African Journal of Science, Technology, Innovation and Development*, **1** (2–3), 10–34.
Luo, Y. and Tung, R.L. (2007). 'International expansion of emerging market enterprises: a springboard perspective'. *Journal of International Business Studies*, **38** (4), 481–98.
Mrinalini, N., Nath, P. and Sandhya, G.D. (2013). 'Foreign direct investment in R&D in India'. *Current Science*, **105** (6), 767–73.
Narula, R. and Dunning, J.H. (2010). 'Multinational enterprises, development and globalization: some clarifications and a research agenda'. *Oxford Development Studies*, **38** (3), 263–87.
The Organisation for Economic Co-operation and Development (OECD) (2012). 'Innovation and inclusive development'. Available at: www.oecd.org/sti/inno/oecd-inclusive-innovation.pdf (accessed 20 October 2014).
Oetzel, J. and Doh, J.P. (2009). 'MNEs and development: a review and reconceptualization'. *Journal of World Business*, **44** (2), 108–20.
Parthasarathy, B., Aoyama, Y. and Menon, N. (2014). 'Innovating for the bottom of the pyramid: case studies in healthcare from India'. Available at: http://www.

academia.edu/7167633/Innovating_for_the_Bottom_of_the_Pyramid_Case_Studies_in_Healthcare_from_India (accessed 11 August 2014).
Peng, M.W. (2004). 'Identifying the big question in international business research'. *Journal of International Business Studies*, **35**, 99–108.
Pervez, T., Maritz, A. and De Waal, A. (2013). 'Innovation and social entrepreneurship at the bottom of the pyramid – a conceptual framework'. *South African Journal of Economic and Management Sciences*, **16** (5), 54–66.
Polak, P. and Warwick, M. (2013). *The Business Solution to Poverty: Designing Products and Services for Three Billion New Customers*. San Francisco, CA: Berrett-Koehler Publishers.
Prahalad, C.K. (2004). *The Fortune at the Bottom of the Pyramid: Eradicating Poverty Through Profits*. Harlow: Financial Times/Prentice Hall.
Prahalad, C.K. (2009). *The Fortune at the Bottom of the Pyramid: Eradicating Poverty Through Profits*. Revised and updated 5th anniversary edition. Harlow: Financial Times/Prentice Hall.
Ramani, S.V. and Mukherjee, V. (2014). 'Can breakthrough innovations serve the poor (BOP) and create reputational (CSR) value? Indian case studies'. *Technovation*, **34** (5), 295–305.
Reddy, P. (2005). 'R&D-related FDI in developing countries: implications for host countries'. In *Globalization of R&D and Developing Countries: Proceedings of the Expert Meeting*, Geneva, 24–26 January, New York: United Nations, pp. 89–105.
Reddy, P. (2011). *Global Innovation In Emerging Economies*, vol. 17. London: Routledge.
Sehgal, V., Dehoff, K. and Panneer, G. (2010). 'The importance of frugal engineering'. *Strategy+Business*, **59**, 1–5.
Serapio, M.G. and Hayashi, T. (2004). *Internationalization of Research and Development and the Emergence of Global R&D Networks*. Oxford: JAI Press/Elsevier Science.
Seyfang, G. and Smith, A. (2007). 'Grassroots innovations for sustainable development: towards a new research and policy agenda'. *Environmental Politics*, **16** (4), 584–603.
Tasavori, M. (2013). 'The entry of multinational companies to the base of the pyramid: a network perspective'. EBS Working Papers WP2013-8, University of Essex, Colchester, UK. Available at: https://www.essex.ac.uk/ebs/research/working_papers/WP2013-8_Tasavori_multinational_companies.pdf (accessed 27 July 2014).
Tiwari, M. (2009). 'Does FDI reduce poverty? Case studies from India'. In Rugraff, E., Sánchez-Ancochea, D. and Sumner, A. (eds), *Transnational Corporations and Development Policy: Critical Perspectives*, London: Palgrave Macmillan, pp. 202–24.
Tiwari, R. and Herstatt, C. (2012a). 'Open global innovation networks as enablers of frugal innovation: propositions based on evidence from India'. Working Paper, Technologie- und Innovationsmanagement, Technische Universität Hamburg-Harburg, No. 72. Available at: http://www.econstor.eu/bitstream/10419/68261/1/733539696.pdf (accessed 27 July 2014).
Tiwari, R. and Herstatt, C. (2012b). 'Frugal innovations for the "unserved" customer: an assessment of India's attractiveness as a lead market for cost-effective products'. Working Paper, Technologie- und Innovationsmanagement, Technische Universität Hamburg-Harburg, No. 69. Available at: http://www.econstor.eu/bitstream/10419/55862/1/687849292.pdf (accessed 27 July 2014).

Utz, A. and Dahlman, C. (2007). 'Promoting inclusive innovation'. In Dutz, M.A. (ed.), *Unleashing India's Innovation: Toward Sustainable and Inclusive Growth*, Washington, DC: World Bank, pp. 105–28.

Vermeulen, P., Bertisen, J. and Geurts, J. (2008). 'Building dynamic capabilities for the base of the pyramid: a closer look at company practices'. In Kandachar, P. and Halme, M. (eds), *Sustainability Challenges and Solutions at the Base of the Pyramid: Business, Technology and the Poor*, Sheffield: Greenleaf Publishing, Chapter 21.

World Bank Open Data (n.d.). 'Foreign direct investment, net inflows (BoP, current US$)'. Available at: http://data.worldbank.org/indicator/BX.KLT.DINV.CD.WD?page=1 (accessed 24 July 2014).

Yaziji, M. and Doh, J. (2009). *NGOs and Corporations: Conflict and Collaboration*. Cambridge: Cambridge University Press.

14. Problem-framing in new innovation spaces: insights from software outsourcing

Rasmus Lema

14.1 INTRODUCTION

This chapter examines how emerging innovation spaces are created, by focusing upon problem-solving and problem-framing tasks in the Indian software industry. Over the last three decades, outsourcing has had a major influence on the international division of labour. It is clear that it has been an important reason for the enormous build-up of production capabilities in the developing world, in particular in the export platforms of Asia. While there is increasing acknowledgement that the globalization of research and development (R&D) and innovation is following production, it is not clear how this process unfolds and whether it extends to advanced innovation.

Until recently, the literature on outsourcing and global value chains maintained that only certain stages of the chain was outsourced to low-cost economies, mainly manufacturing and standardized services. Innovation activities, on the other hand, remained in so-called advanced economies (Mudambi, 2008). Outsourcing should merely strengthen suppliers' existing position in the global division of labour – where competitive advantage is defined primarily by low cost. Innovation activities were 'detached' from outsourced tasks and they remained centralized in and around lead firms and global buyers residing in advanced economies.

There is now increasing acknowledgement that outsourcing is changing in nature: firms are outsourcing activities, which they used to undertake in-house. They are farming out knowledge-intensive activities, including some innovation and R&D activities (Ernst, 2008; Jensen, 2009; Manning et al., 2008; Tate et al., 2009).

This chapter seeks insights by focusing on the global software-outsourcing industry and the supply platform in Bangalore (India), one of the most prominent cases of latecomer development in the global economy.

So far the literature on the software industry has tended to argue that the global division of labour is relatively stable (for example, Sharma, 2014). This is the key conclusion arising from studies that have been looking at 'hard indicators', mainly the distribution of patented intellectual property and formal R&D expenditure (Arora, 2006; Arora et al., 2008; Dossani, 2006). However, this conclusion is closely related to the division of tasks in distributed software development projects. Arora (2006: 400) describes the division of labour in outsourced software services as follows.

> At the risk of oversimplification, software-related activities generally fall into one of three categories: design, coding, or maintenance. Design, which translates approximately into R&D and product development, has the highest value added of the three activities. Coding and maintenance may be thought of as analogous to production in other industries and consequently entail lower-end tasks. . . . Most of the functions offshored (especially to India) involve production, while design has tended to remain local.

In other words, 'production' is outsourced, but 'design' is not. Core and strategically important innovative activities of OECD-based customers, R&D and product development, are typically perceived as 'non-globalized' and 'bound' to their home locations; they are thought to depend on localized and intricate linkages between firms and institutions in lead markets (see Wibe and Narula, 2002: 243). For example, Hoekstra (2006) argues that outsourcing relationships do not provide proximity to tacit knowledge and domain expertise because customers are at a physical and social distance from India. Furthermore, key constraints arise from the way outsourcing relationships are 'modularized' and structured.

According to D'Costa (2004: 57), 'No firm wants to co-locate critical projects overseas due to coordination and communication problems . . . Each project/product is decomposed into self-contained modules, each with varying demand on tacit knowledge, making it possible to co-locate certain modules in certain places.' Hence, because lead firms keep core competences in-house, the formation of innovative capabilities in India is constrained. 'Export services that are outsourced to India are likely to remain non-critical adjuncts to central functions' (D'Costa, 2003: 214). These were the conclusions derived by scholars studying software outsourcing ten years ago, so this chapter provides a critical discussion of whether this still holds true.

The chapter questions the widely held position – following Arora (2006) and D'Costa (2004) – that only lower-order activities are outsourced. To develop this understanding of emerging innovation spaces in India, the chapter adds to the literature on knowledge-intensive outsourcing (Jensen, 2012; Manning et al., 2008; Massini and Miozzo, 2012; Tate et al., 2009) and qualifies the argument that outsourcing relationships are unlikely to evolve beyond certain threshold levels because they do not provide supplier proximity to tacit knowledge and domain expertise.

The key contribution of the chapter lies in specifying the circumstances in which innovative tasks may (or may not) be relocated from Western countries to India, thereby creating new spaces for innovation at the Indian end of the outsourcing relationship. It develops an understanding of how innovation spaces are created in India, by focusing upon problem-solving and problem-framing tasks in software development. The next section presents the conceptual framework for the study, combines this with a review of the relevant literature and specifies the knowledge gaps and questions explored in this chapter.

14.2 OUTSOURCING AND INNOVATION: INSIGHTS FROM THE LITERATURE

The nature of innovation activities that are potentially 'transferred' to emerging economies through outsourcing is debated because it has traditionally been seen as a cost-driven phenomenon. There is widespread agreement that access to 'production' capability at low cost is the main driver of offshore outsourcing, at least initially. However, outsourcing may evolve from production as the experience may give rise to a deepening of the outsourcing relationship. It may therefore be important to pay attention to many the intermediate stages between pure production activities, which depends only on the *use* of existing/given knowledge, and pure research, which is only concerned with the *creation* of knowledge.

Schmitz and Strambach (2009) have recently emphasized the need to recognize not only innovation activities that are loosely connected to production (standalone innovation), but also those that are tightly connected to production (integrated innovation). In an outsourcing setting the latter could arise through the engagement of suppliers of manufacturing services in developing new products or processes. The key characteristic of this type of outsourcing is that it concentrates on the acquisition of a good or services but that some innovating by the supplier is required to provide this good or service. Despite the acknowledgement of the changing nature of outsourcing, most of the literature argues that dispersed innovation activities are of a second order.

Schmitz (2007) suggests that the notion of 'advanced' innovation in global value chains can be specified further. He argues that this type of innovation activity is 'problem framing'. He draws on the modularity literature, which shows that tasks that integrating technology and systems are critical steps in the innovation processes, even where the integrator (buyer) itself is a sub-system supplier in intermediate markets (Brusoni, 2005; Brusoni et al., 2001; Pavitt, 2005). The failure to retain the integrating step in the innovation process could result in a situation where the buying firm no longer possesses the capabilities to incorporate new knowledge and components effectively into its systems (Chesbrough, 2003: 191). For this reason, lead firms are much more readily prepared to outsource 'problem-solving' innovation such as the design and engineering activities associated with the development of a system component. The situation that arises is that buyer firms keep problem-framing activities in-house (or close to home) and only disperse problem-solving activities to lower-cost suppliers in new economic regions (Schmitz and Strambach, 2009).

The hypothesis identified in the literature is thus that the 'quality' of outsourced innovative activities is limited to 'problem solving' (Schmitz and Strambach, 2009), 'subsystem design' (Chesbrough, 2003) or simply 'routine low-end innovation' (Chen, 2008). Integrative capability in technical fields is strategic to buyer firms and that the strengthening of supplier capabilities in these realms is against their interests.

In manufacturing, this constellation is evident in the turnkey production network as found in Figure 14.1. In this type of network the problem framing tasks are undertaken by the brand name firm whereas the all remaining problem solving and production tasks are undertaken by the turn-key contractor. Adapted from Sturgeon (2002), there is a chain of three activities in turnkey production networks. Problem-framing includes activities like product R&D, product definition, functional design, form design and prototype fabrication. Problem-solving activities include

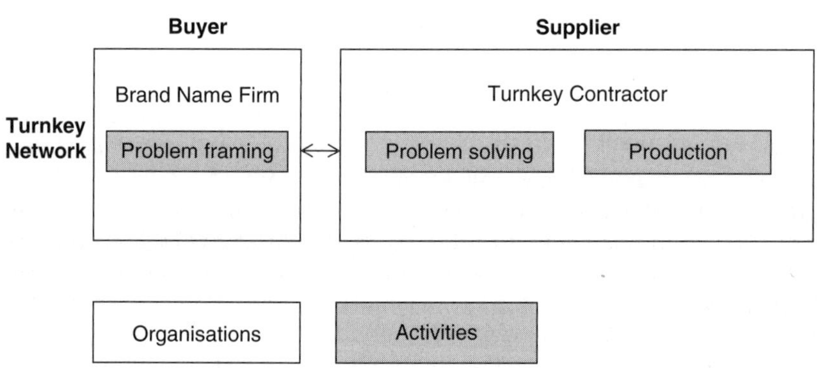

Source: Adapted from Lema (2012a), based upon Sturgeon (1997)

Figure 14.1 Problem-framing and solving in turnkey production networks

process R&D, design for manufacturing and testing. Production includes manufacturing, component purchase and assembly.

Even in manufacturing but especially for software, the question is whether problem-framing and problem-solving activities can be separated. Henderson and Clark (1990) distinguish between two different kinds of relevant knowledge. The first is knowledge of the system components and the second is knowledge about the linkage between them, that is, architectural knowledge. In many industries, component knowledge is provided by suppliers, whereas architectural knowledge is retained by lead firms that function as system integrators. While such structures are particularly pronounced in industries dealing in highly complex products, some scholars have suggested that coordinating and integrating specialized knowledge is an important feature of innovation processes in general (Pavitt 2005). System integrators must have architectural knowledge as well as some degree of component knowledge in order to cope with technological change and product-level interdependencies. They must 'know more than they make' (Brusoni et al., 2001) and this requires the presence of in-house staff that possess a 'higher level understanding'.

However, a number of scholars have pointed out that there are limits to decentralization through modularity (Chesbrough, 2003; Ernst, 2005). Research on the outsourcing of printed circuit board manufacturing confirms that the overall design parameter typically remains in the hands of the customer because the requirements depend upon the electronics product into which it is inserted. A part of the design process is transferred to competent suppliers who then gain some scope to draw on and further develop distinct in-house competences (Lee and Chen, 2000). Ernst (2005) has shown that there are limits to modularity in chip design because of cognitive complexity. Technology change in this sector was unpredictable and changed faster than the ability to codify.

Modular production and innovation networks are dependent on what Baldwin and Clark (2006) call a pinch point in the flow of activities, that is, a codified and standardized transfer (hand-off) of design parameters and specifications. However, the establishment of such a pinch point can be difficult and costly. If a lead firm wants to place a transaction in a particular location it must define and find a way to measure the transacted 'objects' (Baldwin, 2008). Significant costs can occur when the buyer undertakes product definition and specifies service-level agreement, process-approach, testing methods, and so on. Transactions between firms (externalization) are more likely to occur where these costs are low. By implication, the outsourcing of standardized product components is feasible and straightforward when firms can rely on general industry standards and on clear specifications. The ability to codify becomes more difficult as you proceed with functions that have knowledge-creating elements, such as design and engineering activities (D&E) or even R&D.

This has important implications for the division tasks between buyers and suppliers in outsourcing relationships in software. Dividing the value chain up into functions that are undertaken by separate firms (as in Figure 14.1) is relatively straightforward when lead firms have a strategy of 'controlling the creative heart of the value chain, while outsourcing all other activities' (Mudambi, 2008: 702). By contrast, 'linkage economies'[1] may make such splitting up more difficult when lead firms adopt a more deliberate strategy to outsource innovation.[2] In such cases lead firms may find it more efficient to bundle an increasingly large share of the value chain thread with supplier firms and work much more closely with suppliers on the highly creation tasks. This typically means firm transactions cannot be fully specified *ex ante*. The 'pinching' of the inter-firm link is likely to be pushed to the limit, but the parties may invest in a (temporary) 'transaction-free zone'. Such zones are 'physical, virtual, or social spaces where, by convention, a designated set of transfers occurs freely' (Baldwin, 2008: 181). Such zones are needed to facilitate complex, interdependent, and iterative transfers in the task network.

So far there are few studies that have examined modularity patterns in offshore software outsourcing empirically through case studies, even literature assumes that modularity dynamics pose threshold limits to the outsourcing of highly creative (problem-framing) work. The literature suggests more or less explicitly that this type of activity remain 'non-globalized', not only in software outsourcing (Arora et al., 2008; Dossani, 2006) but sometimes also more widely in outsourcing generally (Mudambi, 2008; Schmitz, 2007; Sturgeon, 2002). To explore this assumption, the next section examines the character and evolution of outsourcing relationships and focuses on the outsourcing (or not) of problem-framing activities.

14.3 CASE STUDIES OF OUTSOURCED SOFTWARE: PROBLEM-SOLVING AND PROBLEM-FRAMING

Despite the 'non-globalization of innovation' argument discussed in the introduction, recent literature suggests that more complex tasks are being outsourced from the rich world to emerging markets (Jensen, 2009; Manning et al., 2008; Tate et al., 2009). Following upon this recent research, this chapter seeks to understand how more complex activities are outsourced as well as how and why this occurs in software. To do so, the

Table 14.1 Chain of activities in software development

Problem framing	Requirement analysis and definition; high-level design
Problem solving	Low-level design
Production	Programming (coding and testing); deployment

Source: Lema (2012a)

empirical analyses draws on the empirical analysis put forth in the previous section.

This chapter is based on a larger piece of work, specified in Lema (2012a). Table 14.1 outlines our conceptualization of problem-framing, problem-solving and knowledge-using activities in software development. The problem-framing activities identify what a product or system should do and broadly how it should do it. It relates to the step in the software project lifecycle that deals with requirement definition and high-level design, which mainly occurs in the inception phase of projects.

Analysing software along these lines is done by analysing three different buyer–supplier relationships in the case studies[3] to maximize the comparative element of the chapter.

There is a primary software firm focused solely on the development of software systems. This is Digital Media Networks and the outsourcing of product development (OPD) to Aditi Technologies. There is a secondary software firm with an origin as an in-house software developer for a large manufacturing enterprise. This is Automobile IT and the outsourcing of custom application development (CAD) to MindTree Consulting. There is also an electronics firms producing and purchasing software for its hardware products. This is Telecommunication Corporation and the outsourcing of engineering services (ESO) to Sasken Communication Technologies. The buyer firms are all incorporated in Europe or the US and the supplier firms in India.

14.3.1 Digital Media Network and Aditi Technologies

Digital Media Networks is a privately funded US start-up company in the online digital media business. It was established in 2002 with the idea of a media service engine for legal digital content sharing. Aditi Technologies, was established in 1994 by entrepreneurs returning from the US. The chief executive officer (CEO) came from a position as a general manager of a division in Microsoft. During the 1990s the firm concentred on developing its own customer relationship management (CRM) product,

Talisma, while also supplying services, including technology support for independent software vendors (ISVs). In the early 2000s the CRM product business was spun off, so the firm could concentrate solely on outsourced product development. Today the firm has 650 employees and has acquired capabilities for end-to-end new product development.

The founder and CEO of Digital Media Networks previously had a career in Microsoft in which he was a senior sales leader. The firm was 'born open' with a complex business model and revenue-sharing agreements. As a business-to-business (B2B) company, this firm focuses its managerial resources on sales, network alliances and strategic management. A strong network of industry contacts helped the CEO to build the business and the various commercial and technical networks it entails. The media service engine was based on Microsoft technology and standards, content was provided by record label companies such as Sony, Universal and Warner, third-party providers such as PayPal provided critical components and outlets were provided by Microsoft Media Player and eBay. The company was the exclusive alliance partner for powering eBay's foray into the music download business.

The decision to outsource the entire development of the system to an offshore provider was an easy one – it was inherent in the business plan. This decision had been made for three primary reasons. First, it was believed that time to market for the flagship solution was crucial for the success of the firm. However, it was felt that it would not be possible to quickly build a team in-house with sufficient knowledge and experience as such a team was not easy to assemble in the US. Second, it was important to have the ability to ramp up and down effortlessly once the major phase of creation was complete and to ramp up again for the second release. Such flexibility could not be achieved with an in-house team. The third reason was the combination of rich experience and prices comparable or below US prices. Most important, the firm was attracted by the ability to get inputs to the project from an experienced product development firm.

As the very foundation of Digital Media's business, the solution provided by Aditi was mission critical. The initial requirement had been described in just an eight-page 'visioning document'. This document formed the basis for proposal building and preparatory activities. Requirements were then settled during a one-week meeting at the Digital Media office. Thus, Aditi was closely involved in the requirements-definition stage as pertaining to the technical solution. As explained by the CEO of Digital Media, some of the requirements came from the supplier's ability to envisage usage scenarios. When asked about whether there was a concern that the supplier would not understand the end-user scenario he replied:

> There were things they thought of that we missed. This Group at Aditi, there were many things they brought to the table that added value. I would like to think that we thought of most of the usage cases, but there were things they brought to the table that we hadn't even thought about. So I wouldn't say that they don't understand the consumer situation, again there were many cases were they did add value there. (Digital Media informant)

For Digital Media, the focus on sales and the management of a new business model were enabled by a far-reaching outsourcing strategy in the sphere of technology. It was decided very early to outsource the development of the core technology platform to Aditi. The availability of the supplier's R&D services allowed for an operational business model focused on customer-facing activities and management of alliance relationships. This was dependent on the supplier's depth of competences in the involved technology domains, which could aid technology decisions for the system as well as the ability to provide end-to-end solutions from vision to launch.

The radical outsourcing strategy was not adopted without problems. At the outset, the leadership in Digital Media had envisaged a business model with no in-house technical resources. This strategy needed to be revised. The main complicating factors were about communication between non-technical (buyer-side) and technical (supplier-side) people in the distributed work environment. The division of labour which was originally envisaged did not work. The buyer came to realize that a certain amount of overlap was needed. As the CEO explained, 'you need to have technical people on your side who completely understand the vision of the project' in order to effectively manage the relationship with the offshore provider. Therefore, an in-house technical team was gradually built to improve the work process for the release of the second version of the system.

14.3.2 Automobile IT and MindTree Consulting

For more than 30 years, until 1998, Auto IT was the in-house information technology (IT) division of a European auto manufacturer now specializing in trucks and buses. In 2001, the firm acquired two truck manufacturers (located in France and US) and the IT services divisions were consolidated in Auto IT, which had become a wholly-owned subsidiary. The IT organization was to play a new role, offering its services in the international marketplace for software development services. In the same year, the organization initiated a competitive sourcing programme and established relationships with suppliers in Poland and India, in order to reduce costs, speed up deliveries and learn from skilled partners. The outsourcing practice grew rapidly and the customer base expanded beyond the capacity of the organization; it was therefore clear that a strategy of internal competence transition was needed. This strategy had two main elements. First, Auto IT needed to establish a new role for the organization, one that was closer to the customer and with more of the deliveries managed by suppliers. Second, it needed internal employees – now perceived as 'high-cost employees' – to move up the value chain, 'out of the technical areas and over to the business side of things' (Auto IT informant).

A key element in the definition of requirements was the experience that was gained from collaborating with MindTree Consulting, a spin-off from Wipro. MindTree was established in 1999 by a former Wipro leader with the explicit aim of creating a knowledge-intensive software solutions company. The firm initially latched on to the US internet economy by providing e-business integration services on the enterprise side, but soon after inception the market took a downturn. As a survival strategy the slump was used to 'build processes for the future' in areas such as tools, methodologies and quality in the more traditional customized application development (CAD) space. Once this business line picked up, a key mechanism for the deepening of domain knowledge and related processes was to develop a strong knowledge management culture and system. Furthermore, MindTree was concerned with following in the footsteps of the established firms by becoming a 'global company' by instituting a strong presence in customer locations. Top management and founders drove this process as an opportunity for building deeper domain competences in the CAD segment as one of key vehicles for upgrading in this segment.

According to the CEO of Auto IT, the relationship with MindTree 'is the only true partnership of Auto IT'. Already in 2001, Auto IT had engaged MindTree to build and maintain a new global dealer management system (DMS) for its trucks division. The system was eventually rolled out in 18 countries and was perceived as 'mission critical'. MindTree's independent development of the system and the effective building of new skills showed Auto IT that increased outsourcing to capable suppliers could support a new growth strategy.

Over time, MindTree has become more closely involved in the outsourced projects, and the supplier is involved in complex tasks in the software development lifecycle. It no longer merely develops systems to Auto IT's specifications, but also participates in the development of those specifications by finding resolutions to user requests.

A good example was the development of a CRM sales tool for a leading trucks manufacturer. With external financing, this was a critical project with high visibility. The decision to engage MindTree in the end-to-end development of the system was rooted in a 'critical situation'. The packaged-legacy CRM system for pre-owned trucks was being phased out by the provider, and the customer urgently needed a new system in its place. However, the proposal initially developed by Auto IT, which depended on in-house resources for the critical phases of the project, had a budget and a schedule that was far beyond what the customer was willing to accept. After deliberations among the board, it was decided to challenge MindTree by giving them key responsibility for the project, in order to avoid the loss of an important business opportunity. However, there was also a more fundamental reason that was to do with the difficulty of transferring complex knowledge. As an informant in MindTree explained:

> They wanted to develop the system themselves and then involve us in the next phase of back-end integration. That was the initial plan they presented to the management. But [the executive vice president and head of Auto IT's 'region international'] felt that this was not right. He knew us very well. He said: 'You say that you will involve MindTree in Phase Two. But when it comes to Phase Two, you will come back and say that MindTree does not have the business knowledge of Phase One, so we cannot involve them. So don't make that mistake. Involve MindTree from the beginning.' That is when the whole plan changed. Later on they told us that it was one of the best decisions they had taken.

MindTree was able to draw on its experience of working on and developing CRM systems for customers in other industrial domains. However, MindTree used this 'generic knowledge' in this business-critical project within Auto. It was able to do so because of the close relationship between the two firms. A full-time MindTree manager is posted permanently on site, with access to the entire buyer organization. Key personnel in the supplier firm have accumulated customer-specific knowledge and competences incrementally, which has enabled them to add value and provide Auto IT with new ideas capabilities for innovation in new projects.

14.3.3 Telecommunication Corporation and Sasken Communication Technologies

In the early 1990s when Telecom Corp introduced its first GSM (Global System for Mobile Communications) handsets to the market, the company was able to undertake all processes in-house, even the design of its own chips. As an industrial conglomerate, Telecom Corp could internalize all stages of mobile phone development, including R&D, design, assembly and manufacturing. However, over time, this strategy was abandoned. Throughout the 1990s and continuing in the 2000s it sold off parts of the corporation to focus on key processes, using the newly formed firms as suppliers. During the 1990s, the value of purchases grew three times faster than the value of sales. During the 2000s, the firm consciously worked to reduce R&D spending and rely more on an external network of providers. A key driver of this process of externalization is the increasing complexity of technologies and supply chains, factors that make it impossible to undertake all innovation processes in-house. A distinction developed between elements and processes that were 'core' and 'context', respectively. The latter included what the firm referred to as 'commodity R&D and technology', which was now acquired in the market.

A fraction of this contextual R&D was provided by Indian firm Sasken Communication Technologies, a firm specialized in IP development and outsourced engineering services for the handset industry. Established in 1989, Sasken was one of the firms hit severally by the technology slump of the early 2000s due to its dependence on the US 'dot com' technology sector. The slump spurred a refocusing of the firm from the personal computer (PC) domain to the mobile communications domain. According to informants, the downturn was used to make investments for the future in this area. The relationship with Telecom Corp reflects the firm's new profile as a pure player in this area.

For instance, Telecom Corp made some use of video-application and codec licensing from Sasken. These are subcomponents and commodity inputs. As stated by an informant in Telecom Corp: 'There has been some licensing of certain application and features, but they are not really key components.'

The relationship between the two firms was strengthened in 2005 when the venture capital arm of Telecom Corp made a US$3 million investment in Sasken. Despite this, Sasken was unable to license out or work on more critical technology and processes for Telecom Corp. Sasken had developed core applications such as an integrated multimedia suite, but the supplier was unable to sell this to Telecom Corp (Telecom Corp informant):

> The problem for Sasken is that multimedia happens to be one of the key areas for [Telecom Corp]. . . . In order for Sasken to sell their subsystem it would have required that [Telecom Corp] makes a decision to withdraw its own in-house developed subsystem and replace it with Sasken's and start paying money to Sasken for the licensing and the further development. The control over that subsystem would not have been inside [Telecom Corp]. . . . There are certain areas there in which [Telecom Corp] would like to keep the control in its own hands. This multimedia subsystem and multimedia applications and services are those things that are not likely to be outsourced or licensed from outside.

Sasken was not the only Indian firm that supplied Telecom Corp with outsourced engineering services. Wipro, a major Indian service provider, was a key source of outsourcing and staff augmentation services for particular

Telecom Corp projects. Wipro was a part of the R&D supply chain in a major way. However, certain core hardware and software design and testing services (including radio frequency testing) were not outsourced to this supplier. As an informant stated, there are certain types of process and knowledge that Telecom Corp keeps under European control. The reason was a concern within Telecom Corp about dependence on this large firm for critical resources. There were certain types of knowledge that it did not want to put in the hands of this supplier. Rather it sourced these services from a small number of domestic firms. Most of these adopted 'follow sourcing' strategies, and went global in order to service Telecom Corp in new markets such as China and India. This led to Sasken becoming part of the innovation chain, but there were clear limits to the involvement.

Telecom Corp had a very close relationship with Wireless Hightech, a European supplier of design and radio frequency testing services. Leading managers in Wireless Hightech had an employment history in Telecom Corp. However, Wireless was not globally oriented and did not have the size to venture abroad as was required by Telecom Corp. For this reason, Wireless was put under pressure to merge with Sasken in order to service Telecom Corp in Europe and globally (in India and Mexico). As a result, Sasken acquired Wireless in 2006. This act of supply-chain coordination exercised by Telecom Corp was initiated for two primary reasons. The first was to ensure that the particular engineering service capabilities of Wireless could be scaled up globally. The second was to create more balance in the supply network and to develop a degree of control over certain R&D services outsourced to Indian organizations. On the other hand, this reconfiguration of the supply chain provided Sasken with an opportunity to move into new competence areas, such as advanced hardware testing, which had previously been out of bounds for Indian suppliers. However, it did not enable the firm to move into mission-critical R&D such as high-level design services or a licensing implementation service for key components such as multimedia applications.

14.3.4 Analysis and Insights from the Case Studies

A series of points from the analysis and insight are discussed here. A first point is that outsourcing of problem-framing activities does occur but is confined to software services. In the software services industries, represented by Digital Media Network and Auto IT, suppliers are now often invited to participate in requirement-definition – problem-framing activities – in a substantial way. In the primary software industry of Digital Media networks it relates to the identification of user needs (for example, from market and customer surveys) and the capturing of these in the definition of functional specifications. The same is true for the secondary software industry of Auto IT, but in this setting there is a much more direct relationship with users. The insights from both cases run counter to the extant hypothesis that such problem-framing activities are kept at home in buyer organisations.

This finding is specific to the software buyer segments as opposed to the electronics hardware segment. As the Telecom Corp case illustrated, electronics and telecom firms mainly outsource problem-solving and innovation support activities, thus confirming largely to the extant hypothesis identified in earlier sections. Indeed, the analysis of Telecom Corp and Sasken illustrate that engineering services provided in India tend to feed into highly coordinated networks and innovation processes in which these Indian service providers play a specialized and bounded role. In Telecom Corp problem framing is associated with overall product architecture and systems integrations. The firm provides carefully defined and limited spaces in which suppliers can operate.

The second point is that certain elements of problem-framing are becoming less strategic to buyers of software services. Outsourced problems helped certain buyers to tackle particular challenges define more narrow core competences for their own organizations. This does not mean that the distinction between strategic and non-strategic activities has vanished. It has shifted to somewhere else. The key innovation processes – those that provide the most value in new business models and ensure future competiveness – are becoming non-technical. Instead of focusing on product and systems development, managers focus on the customer interface. Internal resources are deployed in the areas that enhance user-knowledge and sales capability; they are increasingly focused on managing external relationships and on capturing rent from new business models.

This means that new upper boundaries emerge for what the buyers are prepared to outsource. This was clearly expressed by buyers. As an informant in Auto IT stated: 'We do not want to bring in someone else to take the layer between us and the customer.' Controlling access to the customer is increasingly vital. Even though suppliers may contribute substantially to functional requirements, the lead firm are strongly positioned as mediators and relationship anchor-points forward and backwards in the chain. Thus, the question arises for firm management, namely whether firms are willing to let go of component knowledge (problem-solving), while seeking to retain architectural knowledge (problem-framing). In this scenario, architectural knowledge is what matters and knowledge spillover arising from buyer–supplier interactions is a dangerous threat. However, when the rent-generating processes move forwards towards the user, the

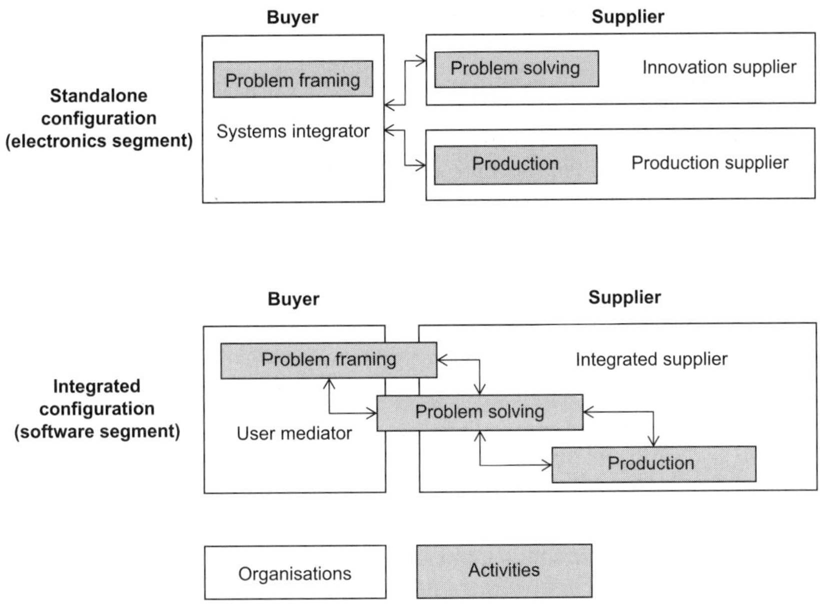

Source: Lema (2012a)

Figure 14.2 Problem-framing in standalone and integrated innovation outsourcing

strategic importance of architectural knowledge is reduced in the services segment.

The third point is that problem-framing activities and system development activities are often outsourced in a bundle. Key differences across the case studies relate to the bundling of value chain tasks. Where innovative services are bought on a standalone basis, they do generally not extend beyond problem-solving. Where production and innovation tasks are integrated there is a wider scope for or incentive to draw the supplier into problem-framing elements of the development process. Based on the analysis of the case studies, Figure 14.2 illustrates the difference between the standalone and the integrated supplier setting in this regard.

Standalone innovative activities are typically undertaken within the realm of new product development. It is therefore easy to assume that these are 'most proximate' to problem framing. However, a loose connectedness as detected in software outsourcing by electronics firms means that highly differentiated roles – for example, systems integration vs modular component provision – can easily be assigned to separate organizations. Typically, there are relatively modest interactive requirements despite a high degree of system interdependency. The interface between the software component and the overall product is specified by the overall product design and the technical standards. The buyer is overseeing the design of the overall product (for example, a chip or wireless device) and defines the functional requirements of the component. These specify the behaviour of the component and the interface (external design). The supplier is left with responsibility for non-functional requirements (such as performance, security and reliability) and internal design.

Conversely, because integrated innovation activities are undertaken in tight connection with production (that is, implementation) it is easy to assume that these are the 'furthest away' from problem-framing. However, this is not the case. This is because some problem-solving activities are difficult to codify in the primary and secondary software industry. If buyers want to outsource problem-solving activities, they typically need to open up for elements of the problem-framing processes. Buyer firms expanded the outsourced value-chain thread from implementation activities all the way into the realm of problem-framing.

The modular view on outsourcing suggests that while specialized component knowledge can be acquired externally, lead firms are likely to retain architectural knowledge and integrative capacity in-house (Baldwin and Clark, 2006; Pavitt, 2005). However, the analysis above suggests that this point may have been exaggerated when it comes to software. Or, to be more precise, it suggests that the modular view only applies in certain segments of the software outsourcing industry where integrated suppliers provide a full range of innovation and development functions.

The fourth point is that limits to codification are a driver of outsourcing of problem-framing activities. It is not always possible to draw a clear dividing line between problem-solving and problem-framing.[4] It is this limit to codifiability, and the resulting needs for buyer–supplier interaction, that explains why co-framing of requirements was widespread in the tightly connected relationships that were studied in this research. Sometimes requirements and specifications have not been fully fleshed out at the point of engagements and sometimes they are virtually impossible to work out at this point. More importantly, they are virtually impossible to relay to a supplier firm.

Business analysts and software architects construct the high-level design and the specifications at this stage. They need to write these specifications in a highly detailed form if the buyer intends to transfer them to a supplier who will take over during the coding phase. This process of codification of requirement is what Baldwin and Clark (2006) refer to as the costly process of 'pinching'. However, these costs can be reduced and externalized by involving the supplier in the elaboration phase. Because the supplier firm performs high-level design activities as well as execution, it increases the efficiency of each of these activities. The supplier relies on the 'cognitive proximity' of onsite staff and on highly developed methods for the within-firm 'transfer' of these specifications to the execution team in the offshore development centre.

In highly evolved outsourcing relationships, the buyer needs to draw supplier staff into the architecture and sometimes even the 'vision' of the project. Hence they take part in the inception phase, and activities in this phase are very difficult to codify. A pinch-point is virtually impossible to establish and the inter-firm connection becomes relational. They create limited transaction-free zones by placing supplier staff onsite, thereby reducing the problems that arise in the organizational-geographical nexus of advanced offshore outsourcing. These zones typically exist for a limited period; but occasionally they are created on a permanent basis so that the suppliers can obtain a deep level of understanding of the vision and context of consecutive projects. They are drawing suppliers into the 'creative heart of the value chain', not only to externalize the cost of pinching but also to leverage supplier assets in higher-order activities.

14.4 CONCLUSIONS AND IMPLICATIONS

There are now several strands of literature that stress the centrality of outsourcing in the debate about the global distribution of innovation activities (Jensen, 2009; Manning et al., 2008; Maskell et al., 2007). Yet, the more precise extent and importance of outsourcing is a matter of controversy, not least because of conceptual ambiguity. The challenge confronting the researcher is to identify particular hypotheses, provide definitions and find ways to examine them. This is what this chapter has sought to do. The aim of the chapter was to explore the widely held hypothesis that lead firms keep advanced innovation in-house (or closes to home) while dispersing only tasks of a second order. It has drawn on evidence from the global software outsourcing industry that connects lead firms in OECD countries with suppliers in Bangalore, India. While recent literature suggests that services outsourcing is becoming more knowledge-intensive, there is controversy about how deep these relationships and complexity goes and why.

The framework proposes to equate advanced innovation with those elements of software development that are 'problem-framing'; these elements define what the software should do and how it should do it. One of the main results of the case study research was that some buyers are beginning to open up the problem-framing process, relying on key inputs from Indian providers of software services.

The case studies showed how IT departments and independent software vendors build on their experience with the outsourcing of software design activities to take a step further and give suppliers the responsibility in the entire chain of software development activities, sometimes including those that define the system or product. Outsourced activities are no longer merely non-critical adjuncts to central functions. Indian firms even become involved in specifying (framing) and solving 'mission critical' problems.

This chapter has also cast light on the issue of why and how innovative tasks are shifting to the supply base, including critical observations about salient conceptual and causal connections that are central to the literature on outsourcing and innovation.

The literature stresses that offshore outsourcing has emerged as a cost-driven phenomenon, and that the scope innovation is constrained by the prominence of activities with high labour requirements (Mudambi, 2008). It is commonly assumed that labour-intensive activities have fixed or unfavourable learning exponentials and that such activities do not provide access to crucial domain expertise and tacit knowledge which 'sticks' to customer locations. The cases in this chapter confirmed that 'routine' production tasks are important, but this did not limit the scope for further capability building as discussed in the introduction. There is considerable scope for innovation as an incremental extension of routine outsourcing low-cost countries.

Three main findings relate the specific case of software outsourcing from US and Europe to India to the broader issue of the development and flows to specific innovation spaces.

One finding is that in a rapidly changing world it is not feasible to assign 'fixed values' to categories such as 'strategic' and 'non-strategic'. Architectural knowledge is only strategic in certain circumstances and it is a fallacy to view the limits to outsourcing as given. The chapter has suggested that that supplier firms have moved into the chamber of 'advanced' innovation activity – not despite the interest of buyers but often because of the interests of buyers.

Second, it appears that the constraints of modularity in products and organization may have been over-emphasized. Despite, the initial focus on 'low end' tasks and less critical projects, outsourcing relationships seems to have provided 'sufficient' knowledge and domain expertise for the subsequent outsourcing of higher-end tasks and more critical projects. The inability to fully codify tasks often means that even if systems have a modular architecture, face-to-face interaction is necessary. Iterative information flows are forged with the presence of onsite personnel (actual proximity) supplemented with new technologies that enable frequent and high-bandwidth communication across geographical distance (virtual proximity).

Finally, linkage economies can be achieved through the integration of production and design tasks in the outsourced value chain thread. 'Integrated innovation outsourcing' reduces the substantial coordination costs associated with up-front investments undertaken during the initial phases of software projects. Buyers reduce and externalize the substantial tasks of detailed codification and communication by drawing supplier closer to the vision of the project. There are cumulative dynamics arising from initial outsourcing of mundane tasks, which may lead to further rounds of more advanced outsourcing of innovation.

These three findings suggest that the roads used to redistribute increasingly deep innovative activities are paved by global production networks and lead firms in these networks. But redistribution does not occur automatically. It requires active efforts in supplier firms, including internal investments, project organisation or learning and external acquisition of knowledge to meet new demands.

A crucial question is how this deepening will affect the economic power balance between the current demand-bases (the loci of buyers and lead firms in OECD countries) and the new innovation spaces in China and India. Will the further outsourcing of innovative tasks create mutually beneficial relationships whereby buyers shift to higher value activities, shedding more and more complex activities to Indian suppliers, but keeping them in a complementary and subordinate function? Or will it create relationships that entail risks for the buyers who may eventually lose their competitive edge in problem-framing and customer-related activities?

NOTES

1. Linkage economies arise from connections between the production functions associated with different activities in the value chain. Mudambi (2008) argues that if linkage economies are high, lead firms are likely to opt for offshore foreign direct investment (FDI) rather than outsourcing. Information flows with highly complex knowledge are easier to forge between units within a single firm than between disparate enterprises.
2. For example, if these lead firms adopt more or less 'open' business models (Lema, 2010).
3. These cases are drawn from a larger study of 36 'critical events' for attainment of innovation capability in ten Indian software firms. A major task for the analysis of these events was the analysis of external sources of innovation. It is the retracing of customer linkages that forms the basis of the material presented in this chapter. Interviews were made in the buyer firms in order to understand driving forces and circumstances for

outsourcing of different types of activities across the software value chain. In total, more than 100 interviews were conducted covering both the supply and demand side using semi-structured questionnaires. Most buyer firms were therefore interviewed by phone. However, European buyers were interviewed face to face. Names of buyer firms (but not supplier firms) are anonymous as to adhere to non-disclosure agreements. Further details on methodology are provided in Lema (2012a, 2012b).

4. The distinction between problem-solving and problem-framing is not in the buyers' minds. Their language was vague referring to (in CAD) 'the tasks of high-cost employees' and 'the business side of things'. In most cases they want a specific and often pressing problem solved, and this then becomes a route to 'value-chain creep' in which suppliers enter the loci of problem-framing activity.

REFERENCES

Arora, A., 2006. 'The emerging offshore software industries and the US economy'. In Collins, S.M., Brainard, L. (eds), *Brooking Trade Forum 2005: Offshoring white-collar work*. Washington, DC: Brookings Institution Press, pp. 399–409.

Arora, A., Forman, C., Yoon, J., 2008. 'Software'. In Macher, J.T., Mowery, D.C. (eds), *Innovation in Global Industries: U.S. Firms Competing in a New World*. Washington, DC: National Academies Press, pp. 53–100.

Baldwin, C.Y., 2008. 'Where do transactions come from? Modularity, transactions, and the boundaries of firms'. *Industrial and Corporate Change* **17**, 155–95.

Baldwin, C.Y., Clark, K.B., 2006. *Where Do Transactions Come From? A Network Design Perspective on the Theory of the Firm*. Boston, MA: Harvard Business School Press.

Brusoni, S., 2005. 'The limits to specialization: problem solving and coordination in "modular networks"'. *Organization Studies* **26**, 1885–907.

Brusoni, S., Prencipe, A., Pavitt, K., 2001. 'Knowledge specialization, organizational coupling, and the boundaries of the firm: why do firms know more than they make?'. *Administrative Science Quarterly* **46**, 597–621.

Chen, Y.-C., 2008. 'Why do multinational corporations locate their advanced R&D centres in Beijing?'. *Journal of Development Studies* **44**, 622–44.

Chesbrough, H.W., 2003. 'Towards a dynamics of modularity'. In Prencipe, A., Davies, A., Hobday, M. (eds), *The Business of Systems Integration*. Oxford: Oxford University Press, pp. 174–99.

D'Costa, A.P., 2003. 'Uneven and combined development: understanding India's software exports'. *World Development* **31**, 211–26.

D'Costa, A.P., 2004. 'Export growth and path-dependence: the locking-in of innovations in the software industry'. In D'Costa, A.P., Sridharan, E. (eds), *India In The Global Software Industry: Innovation, Firm Strategies and Development*. Basingstoke: Palgrave Macmillan, pp. 51–82.

Dossani, R., 2006. 'Globalization and the offshoring of services: the case of India'. In Collins, S.M., Brainard, L. (eds), *Brookings Trade Forum 2005: Offshoring White-collar Work*. Washington, DC: Brookings Institution Press, pp. 241–77.

Ernst, D., 2005. 'Limits to modularity: reflections on recent developments in chip design'. *Industry & Innovation* **12**, 303–35.

Ernst, D., 2008. 'Innovation offshoring and outsourcing: what are the implications

for industrial policy?'. *International Journal of Technological Learning, Innovation and Development* 1, 309–29.
Hoekstra, B., 2006. 'Innovation in R&D'. In NASSCOM (ed.), *Building an Ecosystem for IT Innovation in India.* New Delhi: National Association of Software and Service Companies, pp. 24–9.
Jensen, P.D.Ø., 2009. *Offshoring of Advanced and High-value Technical Services: Antecedents, Process Dynamics and Firm-level Impacts.* Copenhagen: Department of Intercultural Communication and Management Copenhagen Business School.
Jensen, P.D.Ø., 2012. 'A passage to India: a dual case study of activities, processes and resources in offshore outsourcing of advanced services'. *Journal of World Business* **47**, 311–26.
Lee, J.R., Chen, J.S., 2000. 'Dynamic synergy creation with multiple business activities: toward a competence-based business model for contract manufacturers'. In Sanchez, R., Heene, A. (eds), *Theory Development for Competence-based Management, Advances in Applied Business Strategy.* Stanford, CT: Jai Press, pp. 209–28.
Lema, R., 2010. 'Adoption of open business models in the West and innovation in India's software industry'. *IDS Research Reports 2010*, 1–144.
Lema, R., 2012a. *Global Redistribution of Innovation Activities: Outsourcing of Software Services and Innovation Capability in Bangalore.* Saarbrücken, Germany: Lambert Academic Publishing.
Lema, R., 2012b. 'Outsourcing and supplier learning'. *Insights From The Indian Software Industry* **6**, 285–311.
Manning, S., Massini, S., Lewin, A.Y., 2008. 'A dynamic perspective on next-generation offshoring: the global sourcing of science and engineering talent'. *Academy of Management Perspectives* **22**, 35–54.
Maskell, P., Pedersen, T., Petersen, B., Dick-Nielsen, J., 2007. 'Learning paths to offshore outsourcing: from cost reduction to knowledge seeking'. *Industry and Innovation* **14**, 239–57.
Massini, S., Miozzo, M., 2012. 'Outsourcing and offshoring of business services: challenges to theory, management and geography of innovation'. *Regional Studies* **46**, 1219–42.
Mudambi, R., 2008. 'Location, control and innovation in knowledge-intensive industries'. *Journal of Economic Geography* **8**, 699–725.
Pavitt, K., 2005. 'Innovation processes'. In Fagerberg, J., Mowery, D.C., Nelson, R.R. (eds), *The Oxford Handbook of Innovation.* Oxford and New York: Oxford University Press, pp. 86–114.
Schmitz, H., 2007. 'Transitions and trajectories in the build up of innovation capabilities: insights from the global value chain approach'. *Asian Journal of Technology Innovation* **15**, 151–8.
Schmitz, H., Strambach, S., 2009. 'The organisational decomposition of innovation and global distribution of innovative activities: insights and research agenda'. *International Journal of Technological Learning, Innovation and Development* **2**, 231–49.
Sharma, D.C., 2014. 'Indian IT outsourcing industry: future threats and challenges'. *Futures* **56**, 73–80.
Sturgeon, T.J., 2002. 'Modular production networks: a new American model of industrial organization'. *Industrial and Corporate Change* **11**, 451–96.
Tate, W.L., Ellram, L.M., Bals, L., Hartmann, E., 2009. 'Offshore outsourcing

of services: an evolutionary perspective'. *International Journal of Production Economics* **120**, 512–24.

Wibe, M.D., Narula, R., 2002. 'Interactive learning and non-globalisation: knowledge creation by Norwegian software firms'. *International Journal of Entrepreneurship and Innovation Management* **2** (2–3), 224–48.

15. The internationalization of innovation: off-shoring home-base innovative processes in software to a host-nation innovation system

Olof Zaring

15.1 INTRODUCTION

This chapter addresses the innovation space of Sri Lanka, set in the context of internationalization of outsourcing.[1] This chapter provides a detailed description of one company over time, and complements Chapter 14. The perspective here is how, over time, the firm interacts with the national innovation system in the host nation, in such a way as to change the location of innovative activities for reasons having to do with financing, markets and crisis. Generally in the literature, the globalization of innovative activity is associated with growth of multinational (MNE) activity and in particular foreign direct investment (FDI) since 1945 (Narula and Zanfei, 2005). However, the degree and impact of internationalization differs between different regions of origin, where firms from the European Union (EU) obtain a larger share of patents from their foreign subsidiaries than their Japanese or US counterparts (Cantwell and Janne, 2000).

At the sectoral level, firms tend to internationalize production activities at a faster pace compared to innovative or research and development (R&D) activity (Benito et al., 2003). This is because it is believed that systems of innovation are embedded in the MNE's home networks (Cantwell et. al., 2010; Cantwell and Zhang, 2011). In recent years, internationalization of innovative activities has receivied a lot of attention given the potential of large Asian markets to provide intelligence toward the development products, processes, and services, especially in knowledge- and technology-intensive industries such as information and communication technology (ICT) and related services.

Until recently, MNEs constituted the main object of study in this literature. The main explanatory factor for the process of internationalization of innovative activity was the requirement for firms to adapt their products and production processes to different market conditions across locations. This was often accomplished by locating R&D activities in foreign locations. In the literature this phenomenon is referred to as asset-exploiting R&D (Dunning and Narula, 1995; Kuemmerle, 1999). Asset-augmenting R&D is another explanatory factor for the internationalization of innovation (Andersson, et. al., 2002; Cantwell and Zhang, 2011; Dunning and Narula, 1995; Makino et. al., 2002; Wesson, 2005). Firms secure access to the technologies or competences that complement the assets available in their home, for example, their nation of origin or location. The underlying motivation for these two explanatory factors is that technological specialization is seen as a national trait despite economic globalization, and firms consequently may need to access foreign national technological specialties by moving R&D activities abroad. In this latter case of internationalization of innovation, home locations are unable to provide what firms may require to stay viable in their industry. Technological advances at the industry level may very well move faster than the rate at which national innovation systems can adapt. Consequently, firms need to acquire the necessary technology from other sources, such as other innovation systems located abroad, either by importing them or by establishing asset-augmenting R&D operations in foreign locations. Thereby, firms may gain access to the resources they require in addition to the advantages of moving closer to markets and foreign production assets.

Such flows will impact the development of innovation spaces. The internationalization of innovation raises issues connected to the potential for spill-overs from MNEs to the host-country innovation systems and the host-country economy as well as spill-overs between the MNE's foreign subsidiary and the source-country establishments. On the one hand, there is a question whether there is a loss of competitiveness in the home base when the firm moves its access to national innovation system abroad. On the other hand, the host nation's innovation systems may increase their competiveness as a consequence of foreign firms' establishment of R&D operations. The research evidence indicates that spill-overs to host countries are limited, and by implication, the home-base is able to maintain its competitiveness (Castellani and Zanfei, 2003). The explanation may be that the low adaptation rates in national innovation systems in general, and limited absorptive capacity in many host countries works towards preserving home countries' competitiveness.

Therefore, a related issue – specifically addressed in this chapter – refers to the relative competiveness between sub-components of MNEs where the strengths of the home-based parts may not transfer easily to foreign subsidiaries. The absorptive capacity of the foreign subsidiary may be insufficient due to low absorptive capacity in the host nation's innovation system. Factors internal to the firm that tend to decrease the pace of international spill-over of innovation are requirements such as internal cohesion of value chains in MNEs as well as the quality level of the infrastructure for R&D activity in the host nation (Narula, 2002; Blanc and Sierra, 1999). One factor that is expected to be of considerable importance is the strategic motivations of firms that internationalize their R&D activities (compare with Dunning and Narula, 1995). Based on the framing found in the above research, this chapter addresses the following two research questions: (i) What are the drivers of change and pattern in firm innovativeness during the internationalization of R&D activity in an MNE? (ii) Do home-base innovative capabilities spill over to the MNE's subsidiary in the host nation following the initial attempts at internationalization of R&D activity? In addition, does any spill-over influence the division of labour within the MNE's value chain?

To investigate these issues, a case study of a Swedish high-tech MNE, which has moved to outsource software to Sri Lanka and other countries, is used.

15.2 BACKGROUND OF THE IT INDUSTRY AND COMPANY

The knowledge base of the IT industry is characterized by it being persistently divided in two segments: mass-produced and customized software. The custom segment, which is of particular interest in this chapter, produces context-specific software where user–producer interaction is the key to product innovation and firm competitiveness (Athreye and Cantwell, 2007). The IT industry has been internationalized through consecutive waves of business process outsourcing beginning during the 1970s (Arora and Gambardella, 2005). These waves initiated processes where sectorial host-nation innovation systems emerged, eventually resulting in the establishment of a local competitive industry also being competitive in R&D-intensive custom software production (Athreye, 1998). The necessary capability building in host nations seems to involves public and private investments in research and advanced training, as well as in local infrastructure (Nelson and Malerba, 2011).

This case study is about the internationalization of innovation processes of the custom software company XYZ[2] with its home base in Europe. The case study focuses on the process when this company, which was already an MNE, establishes an R&D subsidiary in Sri Lanka as a host nation. The aim is to study the effects on the company's innovativeness chain following on the decision to conduct R&D in South Asia. Hence the case study follows the company during the 1983–2010 period. Based upon a descriptive case study built from secondary data, public data and interviews, this chapter proposes that the company went through three distinct phases during this time span when it pursued different corporate strategies and approaches to conduct its R&D activities culminating in the splitting of the R&D value chain through the move to South Asia. In the process, it also moved from an entrepreneurial venture in Sweden to a multinational enterprise working primarily in Sweden but internationalized to a multinational enterprise where significant innovative activities for software were located in Sri Lanka.

XYZ was founded in 1983 by a group of engineering students still living in a hostel. They were the original inventors and entrepreneurs and remained as major owners and top managers in the company as it grew into a multinational enterprise during the ensuing 20 years. Originally, the company provided software consultancy services, particularly in relational database technology. The company entered the market as a software consultancy firm to the energy utilities industry, and particularly for maintenance systems. XYZ subsequently established its own R&D department in Europe and launched its first integrated software product in 1990: a software system for enterprise resource planning (ERP).

The company has been active in this industry (that is, business software) since that point in time and remains a global supplier of ERP-systems. Today XYZ internally develops and supplies ERP systems to medium-size and large enterprises in a broad range of industries, globally. However, the R&D activity is today distributed across host nations in the main geographical market segments and is no longer carried out exclusively in Sweden (the home base of the company). The company was, and still is, an actor in the same market segment as ERP system providers SAP and Oracle. XYZ still delivers consultancy services but they are now directly related to the installation and use of the offered software system. Today XYZ receives earnings from three main sources: software license fees, software consultancy and systems support contracts to customers.

As explained above XYZ expanded significantly through the acquisition of software companies in national market during the late 1990s as to speed growth and by acquiring local R&D, marketing and service capacity. At that time XYZ became the fifth company in size in its global market segment. The growth strategy eventually proved costly as the market for ERP systems weakened around the year 2000. The company shifted its strategic focus and tried to balance cost reductions in R&D with continued growth through the use of partnerships with other software companies.

The following case description is divided in three phases: (1) the founding phase: building the domestic R&D base in the early 1980s to the early 1990s; (2) the expansion phase: R&D internationalization through direct investments in the early 1990s to the late 2000; and (3) the phase of consolidation, adaptation, internationalization of innovative activities, 2000–10.

Our study is focused on the third phase, when the company moved a considerable part of its R&D capacity to South Asia. As part of the case study, interviews were conducted by the author with the top management of XYZ active in the period 1983–2010, including all CEOs and personnel responsible for R&D.

15.2.1 The Founding Phase

In 1983, the XYZ software company was started by five engineering students at a technical university. The university was located in Linköping, a city that was incidentally the home of the national aerospace industry and also to XYZ's two main domestic competitors. During its first years, the company was financed internally and through bank lending. Only when the company established its international subsidiary in Poland in 1991 did it begin to raise capital from investors.

In 1983, the offering consisted solely of software-related services. The main services then provided were: software services, such as systems development; technical and economic documentation services for software systems; and project management services for implementing software solutions to management problems.

During its first years of operations in the 1980s, XYZ developers in Sweden accumulated specific expertise in relational database technology and linked this to knowledge on the management of preventive maintenance, which it acquired in connection with assignments in the nuclear power industry. This resulted in the development of 'XYZ Maintenance', the first software product of XYZ, which was launched in 1986.

In 1985, XYZ decided to diversify further, and acquired two other IT companies and changed the product offering to also including an in-house R&D and manufacturing capacity for computer hardware. The major customer then was Nokia, at the time in the process of developing the 'Luxor' personal computer with the company as a hardware supplier. During 1986 Nokia decided to terminate the development work on Luxor computers. This hit one division of XYZ especially hard since its operations were to a large extent based on the Luxor computer. This turn of events also forced the first restructuring of the company. This resulted in a narrower strategic focus with two business areas in 1987, each aimed at a specific customer category for XYZ's software-related products. The business areas were banking and finance, and industrial manufacturing.

Operations not associated with these areas were liquidated. The strategic focus on the two business areas, and with increased recruitment and investments in product development, resulted in a stronger market position during the following years. The company continued to expand in 1988. Net sales increased from €3.0 to €3.7 million, the number of employees rose from 32 to 43. It also developed and launched three new software modules: an accounting module, a production management module and a reporting module. The new modules now meant that it had a complete Oracle[3]-based solution for industrial manufacturing companies in 1988.

During 1989 and 1990 XYZ incorporated the modules into one integrated platform called System5. The modules of this platform were eventually converted to relational database technology, which transformed System5 into one of the most modern systems in its market at the time. All development work was made at the R&D department of Goodsoft in Sweden. Once XYZ began offering an integrated product it invested heavily in R&D. The R&D spending increased rapidly with the decision made to convert System5 to System6 in 1994. The ratio between R&D spending and net sales has, however, been rather stable between 18 and 21 per cent during all phases described in this case study.

System5 was further developed during 1991 and 1992. In 1992, XYZ expanded its development organization for System5 to this reach the international market. This is the first time XYZ declared any international ambitions, which were emphasized through the establishment of a subsidiary in Poland.

15.2.2 The Expansion Phase

During the second half of the 1990s, XYZ invested heavily in a new international growth strategy, primarily a geographical expansion one. It was also the fastest-growing supplier of ERP systems during the years 1997–99, globally. The company's net sales rose from €24 million in 1995 to €315 million in 2001. At the end of 2000 it had a presence in 43 countries and could be considered a multinational enterprise from a Western country.

XYZ pursued this expansion primarily through acquisitions and partnerships with established software companies in each targeted international market, especially in the US. Several issues of new shares were made in order to finance the expansion. This was accomplished by XYZ acquiring software developers and suppliers of business applications, incorporating parts of its software as new features in System5 and System6. Another part of the strategy was to also acquire access to the existing customer base of those companies as well as local market and language knowledge. The average number of employees increased dramatically during the years

1997–2000. XYZ also invested heavily in R&D during this phase. The cost of R&D increased from €12 million in 1997 to €55 million in 2001, the main part of this spending took place in Sweden, while R&D in host nations was restricted to market adaptation and integration of modules from acquired companies.

While most competitors pursued a strategy where consultancy was an integral part of their offering, XYZ followed another path. Beginning in the US, it partnered with companies that would sell System6 and provide the services themselves. This was originally due to the difficulty in terms of cost and time for a small Swedish company of entering the US market. Another reason was to avoid problems and avoid competing with local partners when it should have its own consultancy operation in a country. It also proved to be a way of quickly expanding internationally at a relatively low cost. This strategy meant less revenue from services compared with competitors; on the other hand it meant that more software licenses were sold to customers resulting in a future support fees being generated.

In 1994, XYZ began the project to transfer System4 to an object-oriented technology to preserve the competitiveness of its product offering. The object-oriented technology enabled XYZ to construct near tailor-made solutions by different combinations of objects or modules to suit the individual customer. This also enabled it to provide special combinations of modules in systems tailored for specific industries. The following years were spent developing the system, and new features, such as support for mobile phones, was added. System6 was launched in 1997 entirely programmed in object-oriented technology. In the same year XYZ launched web clients for System6. XYZ has launched a new version of System6 annually since 1997 until today, adding incremental improvements, new modules, improved usability and functionality, and widened compatibility.

XYZ also began to allow the integration of System6 in other vendors' systems: it aimed at becoming a specialized supplier to other actors in the software industry, which then sells and installs to end customers. This model is radically different from selling and implementing ERP systems directly to end customers. This strategy was built on the competences held in the company's Swedish home-base R&D department.

During this period of expansion, the market for ERP systems grew rapidly between 1995 and 1998. During 1998 the rate of market growth decreased and continued to decline during 1999. The lowered growth was probably caused by companies focusing their IT resources on the Y2K problem rather than on investing in new systems.

Financing was also important to the developments. In 1996, XYZ issued shares to 12 European institutions. This was XYZ's first issue of shares and the new capital was needed to finance the introduction of its products abroad, primarily in Western Europe, Asia and the US. It was also the first step towards entering the stock market. Prior to this stock issue, XYZ was entirely owned by its managers and employees. The issue of shares generated €12.5 million, and in the annual report for 1996, the company announced intentions to enter the Stockholm Stock Exchange within the first six months of 1997. At the same time it would make a new issue of shares worth €25 million.

However, XYZ did not then manage to meet the requirements of the Stockholm Stock Exchange. Its low profitability, high R&D costs and the fact that its entry onto the international market lagged behind its major competitors, were the reasons for the Stockholm Stock Exchange denying market entry. Early in 1998, XYZ again announced its intention to enter the stock market the same year and managed to do so. In the ensuing years a series of issues of shares raised a total of €200 million.

15.2.3 Phase of Consolidation, Adaptation, Internationalization of Innovative Activities

During this final phase (2000–2010), XYZ went through many changes. These changes were driven by difficulties in maintaining profitability and growth caused by the burst of the stock market bubble in the year 2000, which effectively ended customers' willingness to invest in ERP systems. The stock market crisis hit the IT industry in early 2001 and this was the major reason for XYZ's crisis as well as the software industry's difficulties at that time.

Even though XYZ managed to maintain its market share, profitability plummeted and the company reported a large financial loss for that year. The stock market crash prevented the company from raising more capital, and as the company's organization was poised for continuous growth, the organization proved over-dimensioned for a situation of negative growth in the market. The pronounced growth strategy had already officially decreased in pace in December 2000 when the board of directors decided to strengthen the company's financial position. The intention was to give priority to generating an increased cash flow and increase profitability and attempt to maintain growth and R&D spending. The future expansion was to be conducted from a more stable financial position. The specific target was to lower costs by €20 million during 2001. The result was a €34 million cost reduction by the end of that year. The cost reductions continued during 2002 with €30 million and an additional SEK30 million for 2003.

During 2001, XYZ further emphasized its strategy in regard to creating new alliances and partnerships. Contrary to many of its competitors, it expected that partners would increasingly manage marketing, sales and implementation. Its partners had already managed implementation and installation of the System6 ERP software, but it needed additional partners to further increase sales. At the beginning of 2001, partners accounted for 10 per cent of the sales.

Subsequently in 2001, XYZ started to reduce costs in a programme to achieve balance in the company's finances. Up until that point in time the company had mostly been able to balance its costs through increased revenues from the growth in company sales resulting from the growth strategy. As the market weakened, the revenues no longer increased at the necessary pace. XYZ therefore started to reduce operating costs with the aim to reach profitability during 2001.

The aim for the cost-reduction programme was to reduce the work force, starting by reducing the consultancy personnel in Europe. Reduction of personnel in the US followed later in that year. The programme had been implemented in mid-2001 and showed a promising positive result for the second quarter of 2001 at €2 million. However, the crisis worsened and the cost-reduction programme proved insufficient.

XYZ continued to try to balance the reduction of costs with expansion during 2002. The company closed offices and reduced personnel and found itself in difficulties when trying to preserve the technological competiveness of its product offering. The competiveness was built on the strong R&D base in Sweden. However, the Swedish operation was costly.

A weak financial result in 2002 forced the company to further cost reductions in 2003. XYZ cut personnel by an additional 350 people, primarily within R&D, and primarily in Sweden. At the same time many of those activities were moved to a new R&D operation in Sri Lanka where a substantial part of all R&D work was to be undertaken. Part of the development work was still performed in Sweden but the aim was to develop software in joint teams where Swedish and South Asian engineers cooperated in software projects. The ambition was to transfer technical knowledge from Sweden to South Asia to lower costs without losing the competitive edge, and also still preserving core technological competences in the home-base nation. The innovation system in South Asia at this point in time had the capacity to absorb the knowledge transferred from the home country. This was due to a local supply of Western and locally trained software engineers being readily available.

The expected benefits of the shift in corporate strategy were to reduce the development costs: salaries were one tenth of the Swedish level and R&D work could also be conducted around the clock. The volume of R&D work could be maintained at a reduced cost. At this point the company's financial situation stabilized and it showed positive financial results.

15.3 CONCLUSIONS AND IMPLICATIONS

The case study illustrates how the internationalization of innovation processes can be used to maintain the competitive edge of a company in the software industry. A move of R&D activity to a host nation with an innovation system that augments the innovative capabilities of the home-base innovation system can be successful. In this case, the shift to South Asia was driven by crises, associated with the IT bubble and subsequent need to radically reduce costs while maintaining software development.

The three phases can be summarized as follows. The founding phase was building the domestic R&D base in the early 1980s to the early 1990s. Entrepreneurs – engineering students – started XYZ, which originally had a diversified product offering, developed internally, but that did not succeed. It then proceeded to narrow the product focus to two areas and managed to obtain capital to finance the necessary R&D to develop and maintain a viable integrated software product, known as System5, at the end of the first phase. The company can be characterized as: focusing on consultancy and custom software development conducted in an internal R&D department; using relational database technology at the technology frontline in Sweden; using self-financing as source of capital; providing for clients, especially software for support maintenance systems in power industry; and developing informal personal networks part of organizational structure of the R&D department. Knowledge was obtained through strong personal connection to the local university's software engineering department.

The second expansion phase was R&D internationalization through direct investments in the early 1990s to late 2000. This phase was characterized by the company pursuing a quite successful global expansion strategy built on having an ambitious and expensive R&D operation in the Swedish home base. The strategy was to prioritize growth over profitability in order to attain economics of scale and market penetration. The strategy included the use of alliances with other software companies as a way of supporting the expansion. The company became a multinational enterprise, with sales and R&D abroad, in 43 countries.

The third phase of consolidation, adaptation and internationalization of innovative activities took place from 2000 to 2010. XYZ was now hit by the crisis that was spurred by the burst of the stock market bubble for the IT industry in 2000. The company responded by cutting costs by moving development to Sri Lanka while trying to maintain its global expansion strategy and adopting a complementary global R&D strategy to augment and preserve the technological edge of the Swedish R&D home base. The main characteristics here were: (1) cutting costs was the original strategic driver of the decision to internationalize the R&D processes. The cost advantage was realized but only because the innovativeness was maintained in the host location in South Asia; (2) the original innovation capability of the company was preserved. However, the home base was partly reduced – but not eliminated – by transferring knowledge to South Asia, where innovation systems had the capacity to supply software engineers with sufficient training; (3) the eventual success of the internationalization process was attributed to the fact that the training of software engineers internationally shared a basic similarity: the same software tools were taught to engineers in both the home as well as the host location, during their advanced education; and (4) a rebound in innovation focus of the company can be seen, following internationalization of R&D activity due to the successful transfer of main parts of the innovative capacity of the company to a host nation, which could augment the company's innovation processes.

What are, then, the drivers of change and pattern in firm innovativeness during the internationalization of R&D activity in this MNE? The perception of reducing the cost of R&D work may be an important strategic driver when IT firms seek to internationalize innovative activity. However, the cost advantage may only be achieved if the process of transferring R&D activity is successful. This case study points to additional important factors that are important to the success of the transfer process. The first factor coincides with industry traits of the software industry. The second factor concerns characteristics of the workforce in the host nation, and specifically how the workforce is supported by the innovation system in the host nation.

If the transfer of innovative processes in MNEs is to be successful, industry traits may have to be considered: the case study suggests that this may be specific to the software industry where the knowledge base is relatively standardized. This may be an explanation for why Chapter 14 can find that advanced innovation can indeed be outsourced in software. As this case concerns the software industry it is important to note that the software engineers of the host as well as the home nation were independently trained in the use of the same standardized development tools. Hence, this case study points to the importance of the quality of the host-nation's advanced education in software engineering, and especially that it needs to train students the industry standards (compare with Athreye and Cantwell, 2007). Another factor of importance is complementary language skills of the workforce host and home nation, which seem to facilitate the transfer over of R&D activity. In this case the education systems of the host as well as the home nation produced students with a high professional proficiency in English (compare with Nelson and Malerba, 2011).

The relative competitiveness between R&D components of MNEs, where the strengths of the home-based parts may be transferred to foreign subsidiaries, may augment the total innovative capability of the MNE (compare with Cantwell and Zhang, 2011). The increased innovative capability contributes to firm competitiveness primarily through cost reduction for the same R&D activity in this case. Software companies need to be careful in monitoring the costs compared to the results achieved following a process of internationalization. Software firms need also to analyse the absorptive capacity of any proposed foreign R&D activity to ascertain that specific complementary supportive traits in the host-nation's sectoral innovation system are adequate.

NOTES

1. I would like to acknowledge and thank the Sten A. Olsson Foundation for Research and Culture for supporting this research. This chapter has been written in context of the research programme Radical Innovations for the Enhancement of the Swedish Economy, running at the Institute of Innovation and Entrepreneurship, with Professor Maureen McKelvey as principal investigator.
2. The names of the companies have been changed to protect confidentiality.

REFERENCES

Andersson, U., Forsgren, M., Holm, U., 2002. 'The strategic impact of external networks: subsidiary performance and competence development in the multinational corporation', *Strategic Management Journal* **23**, 979–96.

Arora, A., Gambardella, A., 2005. *From Underdogs to Tigers: The Rise and Growth of the Software Industry in Brazil, China, India, Ireland, and Israel*, Oxford: Oxford University Press.

Athreye, S., 1998. 'On markets in knowledge', *Journal of Management and Governance* **1**, 231–53.

Athreye, S., Cantwell, J., 2007. 'Creating competition? Globalisation and the emergence of new technology producers', *Research Policy* **36**, 209–26.

Benito, G., Grogaard, B., Narula, R., 2003. 'Environmental influences on MNE subsidiary roles: Economic integration and the Nordic countries', *Journal of International Business Studies* 34: 443–56.

Blanc, H., Sierra, C., 1999. 'The internationalization of R&D by multinationals: a trade-off between external and internal proximity', *Cambridge Journal of Economics* 23: 187–206.

Cantwell, J., Dunning, J. H., Lundan, S. M., 2010. 'An evolutionary approach to understanding international business activity: the co-evolution of MNEs and the institutional environment', *Journal of International Business Studies* 41: 567–86.

Cantwell, J., Janne, O., 2000. 'The role of multinational corporations and national states in the globalization of innovatory capacity: the European perspective', *Technology Analysis & Strategic Management* **12** (2), 243–62.

Cantwell, J., Zhang, Y., 2011. 'Innovation and location in the multinational firm', *International Journal of Technology Management* **54**, 116–32.

Castellani, D., Zanfei, A., 2003. 'Technology gaps, absorptive capacity and the impact of inward investments on productivity of European firms', *Economics of Innovation and New Technology* **12** (6), 555–76.

Dunning, J.H., Narula, R., 1995. 'The R&D activities of foreign firms in the US', *International Studies in Management and Organisation* **25**, 39–73.

Kuemmerle, W., 1999. 'Foreign direct investment in research in the pharmaceutical and electronic industries: results from a survey of multinational firms', *Research Policy* **28**, 179–93.

Makino, S., Lau, C.-M., Yeh, R.-S., 2002. 'Asset exploitation versus asset seeking: implications for location choice of foreign direct investment from newly industrialised economies', *Journal of International Business Studies* **33**, 403–21.

Narula, R., 2002. 'Innovation systems and inertia in R&D location: Norwegian firms and the role of systemic lock-in', *Research Policy* **31**, 795–816.

Narula, R., Zanfei Z., 2005. 'Globalisation of innovation', in J. Fagerberg, D. Mowery, R.R. Nelson (eds), *Handbook of Innovation*, Oxford: Oxford University Press, pp. 318–45.

Nelson, R. R., Malerba, F., 2011. 'Learning and catching up in different sectoral systems: evidence from six industries', *Industrial and Corporate Change* **20** (6), 1645–75.

Wesson, T., 2005. *Foreign Direct Investment and Competitive Advantage*, Northampton, MA, USA and Cheltenham, UK: Edward Elgar.

16. Tapping into India's innovation potential: consequences for a Western multinational enterprise and the case of SKF

Katarina Lagerström and Roger Schweizer

16.1 INTRODUCTION

This chapter addresses how a European multinational corporation (MNC) accesses, but is also changed by, moving into research and development (R&D) to support production and new products, within innovation space in India. Today, India receives a sizeable and increasing chunk of foreign direct investment (FDI) in R&D by MNCs (Basant and Mani, 2012).[1] The number of innovations emanating from India is rising (Kumar and Puranam, 2011). Since the liberalization of India's economy in 1991, not only have FDI inflows to India exploded from US$75 million to US$36 500 million 20 years later, but also, since 2004, there is a remarkable trend of foreign companies outsourcing R&D activities to their Indian subsidiaries as well as to other Indian R&D firms (Basant and Mani, 2012; Reddy, 2011). The R&D units not only support adaptations and further developments of products and services developed for the mature markets (Dossani and Kenney, 2007; Lewin et al., 2009), but they also develop products and services adapted to the local market to a lower cost, perhaps eventually to be able to reach the bottom-of-the-pyramid segment (Prahalad, 2004). The major concern is then not to strip the product or service of "unnecessary and costly" features, but rather to develop a product or service with the right features and to the right cost to fit in emerging markets; that is, frugal engineering.

In 2010, the total number of foreign R&D centers operating from India was more than 600 (Basant and Mani, 2012); most of them were established after the last millennium (TIFAC, 2006). Studies show that, since then, there has been a continuous increase in the number of foreign R&D units. Supporting that view is the increasing number of patents filled and granted for foreign applicants (TAFTIE, 2011). Of specific interest for this chapter is that, although since the 1980s MNCs have set up R&D facilities in India (Basant and Mani, 2012), R&D investments have not only accelerated in pace and grown in size, but also changed in strategic intent. Whereas previously many of the MNCs had their R&D setup as a support to their local units, today firms are increasingly opening up independent R&D centers for taking up R&D activities in new and emerging high-tech research, that is, innovating for the global market.

The strategic shift sketched above has been documented in the literature with a focus on describing and conceptualizing the phenomenon along with emphases on clarifying why the change has occurred in the perception of MNCs in India and its innovation potential, as indicated by R&D. Explanations so far have thus predominately emphasized changes in MNCs' global industries (for example, heightened technology intensity of innovation processes) or developments in the Indian context (for example, increased engineer competences). Drawing on the literature on subsidiary evolution (for example, Birkinshaw and Hood, 1998; Birkinshaw and Pedersen, 2009; Birkinshaw et al., 2005), this chapter argues that such a narrow focus on global and environmental determinism does not adequately explain why Indian R&D subunits have evolved from local centers supporting local production to increasingly global centers (TAFTIE, 2011). Rather, we argue that such an evolution can also be understood as the interplay between three interrelated factors, namely (1) headquarters assignment, (2) subsidiary choice, and (3) local environment adaptation.

Accordingly, this chapter aims to contribute to the ongoing discourse on India's changing locus in MNCs' global innovation networks by adding an MNC internal perspective (that is, headquarters assignment and subsidiary choice) that provides another view on how MNCs perceive and react towards the potential in India. We employ a single case study approach – studying how a Swedish MNC (SKF) has over time changed its perception of India and its potential. We draw on empirical evidence collected during in-depth interviews at the MNC's headquarters as well as at a global R&D center.

The chapter continues with a brief description of the recent changes in innovation spaces in Asia, in relation to MNCs. Thereafter, the concept of subsidiary role evolution is discussed, followed by methodological considerations. We then continue with the case and end the chapter with a concluding discussion.

16.2 LITERATURE REVIEW

16.2.1 India's Innovation Potential: From Knowledge Exploiting to Knowledge Augmenting

The relation of R&D units of MNC to India has changed over time. From the 1980s to the millennium shift, MNCs predominately set up R&D entities in India – often in close geographic proximity to existing manufacturing units – to facilitate modification of products and technologies to local requirements (Mrinalini and Wakdikar, 2008). Basant and Mani (2012) label this approach as a "knowledge exploiting" strategy. Hence, MNCs' Indian R&D FDI activities during this period can be described as a strategic reaction in order to leverage the "Indian opportunity" that was perceived as a large and growing market for existing products. To be successful in the Indian market, they had to adapt their products at least partially to local needs. The main *raison d'être* for Indian R&D centers, therefore, was to leverage local knowledge, especially gaining an understanding of the local customer, and then to adapt the existing products originally created for developed markets (TAFTIE; 2011; Mrinalini and Wakdikar, 2008).

Today, in contrast, R&D investments by MNCs in India are increasingly the result of attempts to attain or retain global competitiveness (Mrinalini and Wakdikar, 2008). Due to sharpened technology intensity, complexity of innovative processes, declines in product life cycles, and thereto an enhanced need to reduce costs and increase the speed to market, MNCs experience that when their existing R&D capabilities are predominately located in developed countries, primarily the home country, then R&D is inadequate, too expensive, or both (Basant and Mani, 2012). Mitra (2007), for example, shows that human-resource-related R&D costs in general (even after accounting for the hidden costs of managing offshore R&D units) are 30 to 40 percent lower in India compared to the US. Hence, despite rapidly increasing salaries for, say, engineers, India still promises a cost advantage (Growth Analysis, 2013). Indian engineers are also attractive (Lampel and Bhalla, 2008), due to their competence and skills. Furthermore, Indian engineers, brought up in an environment characterized by fierce competition and limited resources, have seemingly inherent cost and resource consciousness.

This heritage is essential for frugal engineering,[2] the overarching philosophy believed to be the key to success in emerging markets, as it recognizes that merely removing features from products developed for mature markets to sell them at a lower price in emerging markets is a losing game (Segahl et al., 2010). The prices are still too high for emerging-market customers, and the unique needs of these customers may not be met by these products, either (Segahl et al., 2010). Hence, in order to leverage the so-called "bottom of the pyramid" (Prahalad, 2004), constituting the Indian market opportunity, the MNCs need to change their way of thinking about product design and engineering by maximizing value for the Indian customer while minimizing nonessential costs. Furthermore, products developed with such an approach for emerging markets are starting to compete with products in developed markets (Prahalad, 2004).

Obviously, this shift over time of MNCs' view on India is a rugged and simplified depiction of reality, and different views are prevailing among researchers. For example, He (2007) argues that the main reasons are resource-seeking along with market-seeking, when establishing Indian R&D centers, while Schmiele and Mangelsdorf (2009) argue that knowledge-exploiting strategies are still the dominant reason. Then again, Basant and Mani (2012) point out that the development of new technologies for different markets is more important than modifying existing technologies for local markets and/or providing support to local manufacturing.

To summarize, despite certain disagreements regarding the pace and degree of change, whereas MNCs' early Indian R&D strategies were a result of their market-seeking behavior, today, increasingly, the choice to set up R&D centers is guided by resource-seeking behavior or, in the words of Basant and Mani (2012), by "knowledge augmenting" strategies. By tapping into India's scientific and technological resource pool – not only by recruiting skilled Indian personnel, but also through collaborations with Indian universities, contract R&D units, and information technology (IT) consultancy firms, for example – MNCs reduce costs and access new knowledge (TAFTIE, 2011; Growth Analysis, 2013). Hence, MNCs' perception about India appears to have changed, but this shift has so far predominately been explained by highlighting changes in the global and Indian context emphasizing local environmental determinism (compare with Birkinshaw and Hood, 1998). The next section augments this picture with another view, by drawing upon insights made in the subsidiary evolution literature.

16.2.2 Subsidiary Evolution

Another part of the story relates to the choices of subsidiaries in India, which are defined as value-adding entities in a host country (Birkinshaw and Hood, 1998). Research on MNCs' subsidiaries focuses on understanding the various roles or mandates that subsidiaries take in the global organization (Dörrenbächer and Gammelgaard, 2010). Many different classifications exist. Birkinshaw and Morrison (1995), drawing on literature, for example, distinguish between three roles: (1) local implementers with limited geographic scope adapting global products to local needs; (2) specialized contributors with expertise in certain specific activities highly integrated in the MNC's global network; and (3) world mandates with worldwide/regional responsibility for a product line or an entire business. The role of a subsidiary is not static, but can change and evolve over time through the accumulation of resources and development of capabilities (Eckert and Rossmeissl, 2007; Prahalad and Doz, 1981).

Birkinshaw and Hood (1998) summarize the comprehensive literature on MNC subsidiary management and identify three interacting mechanisms that determine a subsidiary's role at any given point of time. First, the subsidiary's role is determined by decisions made by headquarters' manager decisions regarding the allocation of activities to subsidiaries. Birkinshaw and Hood (1998) argue that this is the predominant driver for a subsidiary's role evolution in the early stages, when the level of local resources and capabilities is not yet advanced. Indeed, a subsidiary's charter or the visible manifestation of a subsidiary's role is often changed by a parent-driven investment (compare with Galunic and Eisenhardt, 1996). As pointed out by Birkinshaw and Hood (1998), headquarter decisions based on the aspiration to locate its various functions in the optimal location are often preceded by negotiation and discussion between the parent and the subsidiary, which in turn is followed by a period of resource and capability enhancement by the subsidiary in order to fulfill the new charter. Secondly, a subsidiary's role is further influenced by decisions made by the management of the subsidiary itself regarding the activities undertaken; that is, by subsidiary choice. Subsidiary-driven charter extension is often preceded by a lengthy and slow process of competence building driven by the subsidiary management's entrepreneurial capability (Birkinshaw, 1997). In situations in which the subsidiary's charter extensions are hindered by the MNC's immune system (Birkinshaw and Ridderstråle, 1999), suppressing the subsidiary's initiative taking, subsidiary management builds the required capabilities first and then seeks charter extension once all the needed capabilities are in place (Birkinshaw and Hood, 1998). Finally, the role of a subsidiary is also to be understood as a function of the constraints and opportunities in the local market context (Birkinshaw and Hood, 1998; Cantwell and Mudambi, 2005).

Thus, the development of the subsidiary's role can be described as a triangular showdown between headquarters, subsidiary, and local market (Chen et al., 2012), which is an analysis used in this chapter.

16.3 METHODOLOGY

The purpose of this chapter is to broaden the ongoing discourse on India's changing locus in MNCs' global innovation networks by adding an MNC internal perspective (that is, by emphasizing the impact of and interaction between headquarters assignment and subsidiary choice) to how MNCs perceive and react towards India. Due to the exploratory nature of the question and the necessary level of detail of the data required to understand these processes, a single case study is chosen (Easton, 2010; Yin, 1994). Indeed, case studies provide a unique means of "sharpening existing theory by pointing at gaps and beginning to fill them" (Siggelkow, 2007, p. 21).

Drawing on Eisenhardt and Graebner's (2007) thoughts about theoretical sampling and Merriam's (1998) discussion on convenience sampling, we study the R&D activities of a Swedish industrial MNC, by specifically focusing on one location in India and how the strategic intent of the local unit has changed over a ten-year period. The company studied is SKF. During this time period, SKF's Indian R&D unit went through a dramatic transformation, from being relatively small with the main task to assist local production by locally adapting existing products originally developed for mature markets, to becoming one of two global R&D centers that today perform advanced R&D, for the Indian market along with the global market.

Both primary and secondary data were collected (Merriam, 1998). The most important source of information was interviews with five global top managers (including the chief executive officer (CEO) and the senior vice president for R&D) and five interviews in India (including the country managing director and the managing director of the R&D unit). The interviews lasted approximately 75 minutes each. All interviews were taped and transcribed. The interview question was simply a request that the interviewees tell their story about how SKFs' perception of India and its innovation potential have changed over time and how this in turn has influenced the R&D activities undertaken in the country and vice versa. Secondary information was collected from media articles covering the development of the MNC and internal company documents. We also had the opportunity to visit the local R&D center twice. Hence, the study was retrospective in nature. The use of triangulation and multiple interviews about the same process reduce the risk of post-rationalization of previous actions, thoughts, and decisions as well as problems related to memory (compare with Easton 2010; Yin 1994).

All data was used to generate a chronological story of the development of the MNCs in relation to the Indian R&D site from 2000 to 2013.

Important to note is that the collected data does not cover the total process; rather, evidence of various main events was composed. Hence, the MNC internal processes are explained with a variance model (compare with Langley, 1999), attempting to clarify how headquarters and subsidiary aspects complement the prevailing discourse dominated by local adaptation explanations when discussing the emergence of India as an important hub within MNCs' innovation networks.

16.4 THE CASE STUDY OF SKF

The SKF Group is the leading global supplier of products, solutions, and services within its five technology platforms: roller bearings, seals, mechatronics, services, and lubrication systems. Founded in 1907, SKF rapidly grew into a global company. As early as 1920, the company was well-established in Europe, North and Latin America, Asia, and Africa. Today, SKF is represented in more than 130 countries and has more than 100 manufacturing sites and sales units supported by about 15000 distributors' locations. Mainly, SKF does business through its distinct business areas: SKF Industrial Market, Strategic Industries, SKF Industrial Market, Regional Sales and Service, and SKF Automotive.

Each business area works across the entire asset life cycle for the different industries and develops and delivers products, solutions, and services to original equipment manufacturers (OEMs) and end users. In India, SKF started trading operations in Kolkata in 1923, and since then operations have been consolidated into SKF India Limited and SKF Technologies Limited. In India, SKF has six manufacturing sites, two located in Pune, two in Bangalore, Ahmedabad, and Haridwar.

In Bangalore, SKF's technical center expanded from a local R&D site with the charter to support the manufacturing sites with the development of products for the Indian market. The center was inaugurated as a global R&D center in 2011 with the changed charter to not only drive innovation by developing products and solutions for local customers, but also for global customers. The new charter implied that the center had a global responsibility to test and provide advanced technical knowledge for design, process, product, and validation.

16.4.1 How SKF Developed R&D Activities in India

In 2000, SKF's R&D activities in India were located at the various manufacturing units and were mainly meant to provide support for limited product development for the local market. However, starting in 2001, a new perception of India and its potential for R&D started to emerge – a change that went hand in hand with SKF's overall aspiration to create global product development and engineering organization. As mentioned by the CEO, in 2002 it became more and more obvious that India and the demands from Indian customers had changed for good. No longer could MNCs sell "old" products, by which they meant products developed for and grown outdated in Western markets. The CEO summarizes the changes in India as follows: "India has changed from being an importer of knowledge to being a developer of knowledge. We work together with customers, developing products with them, specifically for that market."

This led to a localization of management. It was realized by SKF that, in order to take part in the growth in India, the activities had to be headed by Indians who understood the context, the customers, and the market. As explained by the CEO: "In 2002, we made a conscious decision – if we want to develop SKF in India, we need to go for local managers." Before that, like most other foreign MNCs in India, the local unit was headed by expatriates usually staying for three years. Since they, in the first year, had to become acquainted with the Indian context and, in the last year, were concerned about their subsequent job, these expat managers really only focused on the Indian activities for one year.

By appointing an Indian as the managing director, SKF sent an important signal. The CEO commented: "We gave the authority to the locals. We gave them the keys. We, however, also expected them to come to us and tell us what was needed – how we should assist them – in order to be successful in India." However, the CEO clearly stresses that the Indian manager and unit is part of a global company:

> We don't run India differently from other operations. What is different is that the need for change is faster. That is why it is absolutely vital that management is close to the market; that is why I visit them regularly. We also benefit from having an Indian on the Board. If you are not exposed to India, you won't respond in the right way.

Local management as above was followed by the development of global product development for two-wheelers. The next step taken in committing more to India was that in November 2004 SKF launched its Application Development Center (ADC) for two-wheelers and the automotive market in order to bring its engineering solutions closer to its customers. The reason for setting up the ADC in Bangalore was that, whereas India is a high-growth area for the two-wheeler and automotive segment, Bangalore is the IT hub of the country. The CEO explained:

> Out of the 24 million two-wheelers produced globally every year, 20 million come from Asia. This almost amounts to 83 percent of the total production. [With the ADC] SKF aims to strengthen its role as a global partner to its customers in all segments of the industry as well as its commitment to the Asian market and India in particular.

The focus of this center (ADC) was on the development of completely new technologies as well as on optimization of the existing product range. Thereby, it should support the two-wheeler customers of SKF worldwide with competencies and services.

Moreover, there was the formation of a global test center in India. Five years later, in November 2009, SKF inaugurated a Global Testing Centre at the onset of the Two-Wheeler, Automotive, and Sealing Solutions businesses in Bangalore. As pointed out by the CEO: "We are putting a knowledge infrastructure in place in India that allows us to provide state of the art technology solutions. The new testing center will significantly reduce lead time for our customers and ensure a speedy response to their requirements for new products." The *raison d'être* for the Global Testing Centre was to conduct testing in close proximity to SKF's Indian customers for design, process, and supplier validation, thereby entailing greater support for SKF's processes in a faster and more cost-effective manner by applying global test standards that adapt to local customer specific requirements.

16.4.2 Formation of the Global Technical Centre in Bangalore

In December 2011, SKF inaugurated its Global Technical Centre India (GTCI) in Bangalore, incorporating the existing Global Testing Centre and the ADC discussed above. The new center also houses a metallurgy and chemistry laboratory, thus marking the next development phase. The opening of GTCI can be understood as part of SKF's technology strategy to develop a global product development and engineering organization. As mentioned by the SKF Country Head in India: "With the launch of GTCI, our customers in India will benefit from closer-to-home availability of SKF's global knowledge and expertise to address the growing need in India for newer technology development at a faster pace." Indeed, local knowledge is essential especially when changing design rules (for example, developing a new product for the Indian market) but also when trying to adapt an existing design in order to reduce the costs.

The trigger for investing heavily in the GTCI came from a clear vision by SKF's CEO that he had communicated since 2007. The CEO recalled that he felt he heard the local voices loud and clear:

> We had to give them a technology center. You need to be able to develop things locally . . . you can't expect that you can develop things in Europe and sell them in India. If we hadn't had a technology center in India, we wouldn't have taken part in the market growth; we wouldn't have the right resources to meet our customers' demand.

These decisions were thus guided by the CEO's strong vision that, in order to develop a global R&D network, SKF needed to increase its Indian R&D capabilities. Thus, the senior vice-president for R&D and the head of technology travelled to India in 2010. He recollected:

> I remember meeting our head of India in Bangalore, and I said we will start our GTCI activities in India with 20 people. He said we need 20 people in house and 20 people by subcontracting, giving us the best of two worlds. Confidential things and IP we do; others we subcontract to an external partner.

Consequently, SKF began to collaborate with an external consultancy firm, thereby kick-starting the plan to increase efforts to not only exploit, but also to augment India's innovation potential. The senior vice-president explained the choice of the consultancy firm:

> We looked at some top universities and institutes; we visited our own sites and a number of potential engineering consultancies [among others the eventually chosen partner]. We were extremely impressed with what they could do . . .] not just by the quality and quantity of their engineers, but their processes and methodologies, how they managed and dealt with the tasks very professionally.

Moreover, SKF had already had positive experiences working with the chosen consultancy firm, since another of SKF's units had used the same firm. The collaboration with the consultancy firm worked well, and the firm is still an important partner. This partnership was also important as a trigger for further development of in-house R&D capabilities. Whereas initially, as mentioned, the need to build local R&D competence was driven by the Indian market's growth, the venture with the consultancy firm showed that "this whole activity of sending work to India works," as expressed by the GTCI head. Thus SKF became even more determined to build up Indian R&D capabilities in-house, and as the GTCI head explains: "India's market growth, SKF India's customers, and the venture with the consultancy firm created a positive story for having a captive center in India." Another positive aspect was that they could solve previously challenging confidentiality and IP issues when working together with local customers.

Not everybody was pleased with the creation of the GTCI, however.

There was some initial resistance and doubts as the senior vice president recalled:

> There was probably some resistance from people in Europe; 'my job just went to India.' We told them that this was not the case; rather, it was to get extra work done. We are expanding in Asia, we have to be able to work in Asia and train people there.

Moreover, despite the positive experiences, parts of the global organization still doubted if tasks given to GTCI indeed would be delivered on time and to the right quality. Hence, an important early undertaking of the GTCI management was to build credibility among other parts of the global organization. The head of GTCI explained: "Proving credibility takes time, one should have patience . . . You push them enough to get them to give you work and then you leave them to judge." The global management, which had understood that India needed a critical mass of projects, supported GTCI heavily, and the message to leverage the R&D capability in Bangalore was driven through each committee at various levels. Then again, GTCI also needed to develop a robust system for customer feedback in order to learn and develop. As pointed out by the head of GTCI, "When you prove your worth by robust processes – repeatability is guaranteed – you get more. Recommendation cannot be a long-term engagement. If that's the route they take, they will always try to find fault." This also led to human resource development. Appropriate training – through in-house video-based learning and three months' training staying abroad at other SKF units – were seen as important mechanisms to reach credibility.

The head of GTCI provided a perspective on culture and communication:

> The Indian engineers, when the relationship is still being established, do not openly express their opinions on new and better ways of doing a task to their global colleagues. This is basically due to the Indian culture of being obedient and polite. As the relationship progresses, rapport gets built and these engineers become more confident and comfortable with their global colleagues and lose their initial shyness. They start expressing their opinions. We are only now entering that phase when they start to give suggestions. The global colleagues find this as an excellent value proposition. We have knowledge-sharing sessions as part of the appraisal process.

Common training was thus important for creating relationships between Indian engineers and other engineers to enable knowledge sharing from GTCI to other parts of the global organization and vice versa. The relationship between the GTCI and the other SKF units in India also had to be developed. The head of GTCI expressed it as follows:

> One of the value propositions of GTCI to our SKF colleagues in India is to bring global SKF knowledge home, which will enable them to please their customers. One of the key elements towards achieving this is firstly for GTCI to build credibility through acquisition of knowledge, and secondly to work with them to learn how GTCI can be leveraged appropriately.

Over time, the trend was very positive; whereas in 2010, GTCI's utilization rate was 75 percent, in 2012, it was 94 percent. Similarly, in 2010, GTCI had 2.2 percent dissatisfied stakeholders and 0 percent in 2012. The GTCI head pointed out that the unit now was accepted as a credible and reliable engineering community. Thereby, expectations have gone up and tolerance for exceptions down as acknowledged by the SVP: "By now, India is doing pretty much the same as Europe and is developing rapidly."

The senior vice president further stated that one important decision taken at SKF headquarters was that the new global R&D unit in India should have all R&D activities (historically, in Europe, they were located at different units). "If we hadn't stepped in centrally, we probably would have ended up with small setups at existing R&D factories also in India." The head of GTCI agreed that having all R&D activities under the same roof was a huge advantage, since this allowed engineers to interact. This interaction was especially important for SKF's aspiration to increasingly become a solution provider for its customers: "It will take a long time for our competitors to catch up on this value proposition."

In 2013, GTCI focused on testing and providing advanced technical knowledge for design, process, product, and supplier validation. In addition, the center conducted bearing analysis and had a fully equipped laboratory for metallurgy and chemistry, mechatronics, and product investigation in order to: (1) develop significant critical mass in engineering knowledge; (2) develop the ability to quickly analyze and support the Indian market's needs; and (3) design components for the local market.

In 2013, due to the fact that GTCI still was in a phase of building its resources and competencies, engaging in learning through collaboration with other R&D units in SKF, 60 percent of GTCI's work was linked to SKF's global R&D activities, whereas the remaining 40 percent were related to work with Indian and Southeast Asian customers. The decision regarding what global projects GTCI should focus on was done based on priorities and workload leveling by various committees comprising of senior executives at the global level. The head of GTCI described it thus: "If there are ten projects, six global and four local, we have to prioritize. It's an allocation process. The committees have the bigger picture on how these ten projects will affect SKF globally, including India, and decide how we should prioritize."

In 2012 (that is, two years after the GTCI opened), the number of employed engineers at the center had increased from the initial 13 to 200. The CEO's vision – already communicated in 2012 – was to grow and reach 400 engineers in 2016 and as the senior vice president of R&D emphasized: "The CEO's vision was as usual ahead of all others. I think, but for his vision, we would have had a larger number of sites of smaller mass . . . He has been quite a visionary and quite inspirational."

In 2013, SKF inaugurated a Product Investigation Centre. The GTCI head stated: "The Product Investigation Centre and Mechatronics Lab are important and natural extensions of our offerings from GTCI. This initiative will provide a platform for delivering complete engineering solutions to our customers and leverage in-house capabilities." This center was offering customers services such as advanced and fast analysis of bearing field usage, post-endurance testing evaluation, and a mechatronics laboratory equipped with technically advanced equipment to help customers with services including new product development, firmware testing, and investigation of sensor-bearing field returns at the GTCI.

The future holds more change, as SKF has declared the intention to make the GTCI in India a Center of Excellence in the future. The senior vice president of R&D articulated:

> Our vision is to have everything covered in India – from basic research to applications engineering for specific business units' need in different markets. It will be coordinated with research that will not be done in other parts of the world. GTCI in India will become our center of excellence in some particular fields.

The head of GTCI speculated on future developments as well:

> For example, small city cars are in great demand in India. Therefore, globally, the decision might be taken that the wheel hub bearings for these cars will be conceptualized and developed in India. If a wheel bearing is for Brazil, the design rules would be formulated in India. That kind of granularity will come if we put this center in place and if we develop the necessary capability along the way and of course if the market globally moves in that direction.

In order to be able to reach this stage, however, the GTCI is required to develop its capabilities in new directions.

16.5 DISCUSSION

One of the key questions in international business research is how MNCs manage their international operations and assign different roles and responsibilities to their geographically dispersed units (Enright and

Subramanian, 2007). Most extant research on the configuration of the R&D activities in MNCs still indicates that, despite a continuous trend toward an increasing geographical dispersal of R&D operations and an increase in innovations from outside the home country (von Zedwith and Gassmann, 2002), most MNCs continue to have a strong preference for keeping core and high-level R&D activities in the home market (Benito et al., 2003).

However, in order to mitigate the underutilization of units in, for example, emerging markets and thereby take full advantage of capabilities and local market circumstances as access to qualified engineers to low costs, some MNCs are beginning to take actions to calibrate and focus the firm's R&D activities to other units (TAFTIE; 2011, Growth Analysis, 2013). One means to do so is to assign R&D responsibility for a certain set of activities. Thus, the evolution of mandate or charter (Birkinshaw and Hood, 1996; Birkinshaw and Morrison, 1995) is often gradual, and there is an extension of the responsibility over time, when it comes to the degree of difficulty in both task and geographical extension of responsibility. This evolution is also shown in the above case study of an R&D center in India. The case study demonstrates how the three mechanisms outlined in the literature review on subsidiary evolution affect development. Those three mechanisms – (1) headquarters assignment; (2) subsidiary choice; and (3) local environmental adaptation – were of great importance for the change in the subsidiary role over time.

The decision to invest in an extension of the role and responsibilities of an already existing technical development unit located at a manufacturing site in India was a direct result of a decision made by headquarters in Sweden. The decision was based on three realities; first, India was a fast-growing emerging market and, from a strategic standpoint, second it already possessed some basic capabilities of global requirements in the units since it was engaged in product development for the local market, and third the local market demands were becoming more advanced; thus, the unit was already slowly developing its capabilities in relation to local market actors.

Once the decision was made, headquarters took conscious actions towards supporting the further development of local capabilities by giving the responsibility to local managers to run the operations. Within a few years, the unit had acquired process, product, and management capabilities to be given the charter for all R&D activities of a small but – for the region – important product. What this demonstrates is that headquarters made a conscious strategic choice to utilize existing capabilities and knowledge of the local market demands when deciding to establish the R&D unit in close proximity to the manufacturing site, an act known to be used by MNCs in emerging markets. Brasant and Mani (2012), for example, use the concept of "knowledge-exploiting strategy" to describe the phenomenon. Besides, the European headquarters made a conscious decision to invest both financial and human resources in the transformation of the R&D center, from being a sole supplier of minor adapted products to the local market to, as time passed, supplying via a global technical center with the complete mandate for one product to a fully fledged R&D center with global responsibility of equal importance to other R&D centers in developed markets.

During the process, headquarters also demonstrated a high level of support for the unit and pressed strongly for its recognition by other units. The need to develop capabilities was also recognized by the R&D unit itself in order for it to deliver work in line with the global requirements in the MNC, to reach a level of credibility, and develop relationships to manufacture sites both inside and outside of India as well as to R&D centers globally. Thus, the Indian unit, with time, started to take more of its own initiatives (compare with Birkinshaw, 1997) to further extend its mandate in the MNC. Important means used to support the development was participation in training and education and collaborating with a well-known and approved consultancy firm that already had relations inside the MNC. But perhaps most important was the diligent work put into satisfactory deliveries to other units meeting the global requirements in the MNC.

An important reflection is that the mechanisms of interaction – headquarters assignment and subsidiary role – would either separately or together not have been sufficient for achieving the move for the Indian R&D center from providing low-value-added activities for the local market and being of low strategic importance towards providing high-value-added activities for the global market and being of high strategic importance. External pressures also matter. The demand in, as well as the importance of, the local market were crucial drivers behind the decision of the European headquarters to expand the mandate of the R&D center in India. But, of course, the demand from the local customers and the access to a highly educated, competent, and rather low-cost workforce also made it possible for the center to develop the capabilities necessary for handling the extended mandate. Moreover, the increase in capabilities then made it possible for the R&D center to push for a further increase in mandate.

An additional reflection is that the relative importance of the local market demands for the continued increase in charter played a decreasing importance, once the unit had developed its capabilities and to a higher extent started to do R&D work for the global market. It was only initially that the special characteristics of the local customers had a direct influence on the evolution of the R&D center, even in light of the fact that these special characteristics made headquarters realize the importance of having an R&D center in India.

16.6 CONCLUSIONS AND IMPLICATIONS

This chapter has presented a case study about the evolution in mandate for an R&D center located in an emerging market, and thereby an emerging innovation space in India. Our findings are consistent with other studies that have pointed to the importance of three mechanisms – headquarters assignment, subsidiary choice, and local environmental adaptation – for the change in charter for subsidiary units. This literature proposes that the local market demands – driving the subsidiaries' development of distinct capabilities – and relationships with actors in the local market are the main forces for the extension of subsidiary mandate with respect for the demands in the local market (Birkinshaw, 1996; Birkinshaw and Hood, 1998). This chapter suggests that the decision regarding the achievement of a global R&D mandate is to be seen as interplay between the exogenous and the endogenous factors.

In the case study, a shift in importance of the three mechanisms can be identified, in the different stages of the R&D center's evolution, even if the diligent maintenance of support from headquarters was important throughout the process. Their support was most likely to be explained as the result of the recognizable pattern of a global technology strategy among many MNCs to move R&D to emerging-market economies on the basis of cost and not just capabilities of a highly educated workforce. From this perspective, our case provides an important contemporary picture on MNCs' current R&D strategies as a product of an evolutionary process in which an increase in responsibility and charter is given to R&D units in emerging markets. This is a picture that is supported by both a European study on India as a hot spot for investments (TAFTIE, 2011) and a recent published report by Growth Analysis (2013). The case studied also illustrates that understanding MNCs' innovation activities in countries like India requires a dual foci, on business strategy and the emerging innovation space in India.

The two most important implications for the management of MNCs are the importance of diligent support throughout the evolution process of the R&D unit and the courage to delegate the management responsibility to local managers to run the process in the unit. It is evident from the case that a combination of market, technological, and managerial knowledge is crucial for the subsidiaries to achieve an extension of the mandate over time.

Further research is thus welcome in this area in order to fully understand the process behind the evolution of a subsidiary role in emerging markets. Thus, we propose a study that follows the evolution process in real time in order to avoid the limitations associated with retrospective case studies such as post-rationalization and forgetfulness of actions, thoughts, and decisions, which to some extent were encountered in this study. Of interest is also a study of the same process in different emerging-market economies in order to capture the possible differences relating to capabilities and local market demand. These more detailed empirical studies will also help nuance and develop theoretical and conceptual understanding.

NOTES

1. Authors are listed in alphabetical order. The authors acknowledge and are grateful for the financial support provided by the Foundation for Economic Research in West Sweden.
2. "Frugal engineering" is a term coined by Renault Chief Executive Carlos Ghosn to describe the competency of Indian engineers in developing products like Tata Motors' Nano.

REFERENCES

Basant, R. and S. Mani (2012), 'Foreign R&D centers in India: an analysis of their size, structure and implications', working paper, Indian Institute of Management Ahmedabad. Available at: http://www.econbiz.de/Record/foreign-r-d-centres-in-india-an-analysis-of-their-size-structure-and-implications-basant-rakesh/10009501719 (accessed 10 November 2013).

Benito, R.G., B. Grøgaard and R. Narula (2003), 'Environmental influences on MNE subsidiary roles: economic integration and the Nordic countries', *Journal of International Business Studies*, **34**, 443–56.

Birkinshaw, J. (1997), 'Entrepreneurship in multinational corporations: the characteristics of subsidiary initiatives', *Strategic Management Journal*, **18** (3), 207–29.

Birkinshaw, J. and N. Hood (1998), 'Multinational subsidiary evolution: capability and charter change in foreign-owned subsidiary companies', *Academy of Management Review*, **23** (4), 773–95.

Birkinshaw, J. and A.J. Morrison (1995), 'Configurations of strategy and structure in subsidiaries of multinational corporations', *Journal of International Business Studies*, **26** (4), 729–54.

Birkinshaw, J. and J. Ridderstråle (1999), 'Fighting the corporate immune system: a process study of subsidiary initiatives in multinational corporations', *International Business Review*, **8**, 149–180.

Birkinshaw, J. and T. Pedersen (2009), 'Strategy and management in MNE subsidiaries', in A. Rugman (ed.), *The Oxford Handbook of International Business*, 2nd edn, Oxford: Oxford University Press.

Birkinshaw, J., S. Young, and N. Hood (2005), 'Subsidiary entrepreneurship, internal and external competitive forces and subsidiary performance', *International Business Review*, **14** (2), 227–48.

Cantwell, J. and R. Mudambi (2005), 'Competence-creating subsidiary mandates', *Strategic Management Journal*, **26** (12), 1109–28.

Chen, H., C.-W. Hsu, and D. Caskey (2012), 'Internationalization of Taiwanese manufacturing firms: the evolution of subsidiary mandates and capabilities', *Asian Business & Management*, **12** (1), 37–60.

Choudhary, V., A. Kshirsagar and A. Narayanan (2012), 'How multinationals can win in India', *McKinsey Quarterly Report*, March, 1–7.

Dörrenbächer, C. and J. Gammelgaard (2010), 'Multinational corporations, interorganizational networks and subsidiary charter removals', *Journal of World Business*, **45** (3), 206–16.

Dossani, T. and M. Kenney (2007), 'The next wave of globalization: relocating service provision to India', *World Development*, **35** (5), 772–91.

Easton, G. (2010), 'One case is enough', working paper, Lancaster University Management School. Available at: http://www.research.lancs.ac.uk/portal/en/publications/one-case-study-is-enough(4465158f-e3b8-48ec-9da8-9c395d9d9a91).html (accessed 10 January 2014).

Eckert, S. and F. Rossmeissl (2007), 'Local heroes, regional champions or global mandates: empirical evidence on the dynamics of German MNC subsidiary roles in Central Europe', *Journal of East-West Business*, **13** (2–3), 191–218.

Eisenhardt, K.M. and M.E. Graebner (2007). 'Theory building from cases: opportunities and challenges', *Academy of Management Journal*, **50** (1), 25–32.

Enright, M.J. and V. Subramanian (2007), 'An organizing framework for MNC subsidiary typologies', *Management International Review*, **47** (6), 895–924.

The European Network of Innovation Agencies (TAFTIE) (2011), 'Innovation hot spots in India'. Available at: http://www.taftie.org/content/innovation-hot-spots-india-internationalisation-strategies-task-force-report-2011 (accessed 10 November 2013).

Galunic, D.C. and K.M. Eisenhardt (2001), 'Architectural innovation and modular corporate forms', *Academy of Management Journal*, **6**, 1229–49.

Growth Analysis (2013), 'Outsourcing and off shoring of R&D in global value chains', working paper 2013:10. Available at: http://www.tillvaxtanalys.se/sv/publikationer/working-paper-pm/working-paper-pm/2013-08-30-outsourcing-och-offshoring-av-fou-i-globala-vardekedjor.html (accessed 10 January 2014).

He, W. (2007), 'Examining the determinants of R&D investment in developing economies – an empirical study of US international R&D', paper presented at the Doctoral Consortium IIMA, 28–9 October, in: Basant, R. and Mani, S., *Op. cit*.

Kumar, N. and P. Puranam (2011), *India Inside: The Emerging Innovation Challenge to the West*, Boston, MA: Harvard Business Review Press.

Lampel, J. and A. Bhalla (2008), 'Embracing realism and recognizing choice in IT offshoring initiatives', *Business Horizons*, **51**, 429–40.

Langley, A. (1999), 'Strategies for theorizing from process data', *Academy of Management Review*, **24** (4), 691–710.

Lewin, A.Y., S. Massini and C. Peeters (2009), 'Why are companies offshoring

innovation? The emerging global race for talent', *Journal of International Business Studies,* **40** (6), 901–25.
Merriam, S.B. (1998), *Qualitative Research and Case Study Applications in Education,* San Francisco, CA: Jossey-Bass Publishers.
Mitra, R.M. (2007), 'India is emerging as a global R&D center', working paper no. R 2007:012, Swedish Institute of Growth Policy Studies, Östersund, Sweden. Available at: http://growthanalysis.se/download/18.56ef093c139bf3ef8902561/1349863309570/indias-mergence-as-a-global-r-and-d-center-07.pdf (accessed 10 November 2013).
Mrinalini, N. and S. Wakdikar (2008), 'Foreign R&D centers in India: is there a positive impact?', *Current Science*, **94** (4), 452–8.
Prahalad, C.K. and L. Dy (1981), 'An approach to strategic control in multinational companies', *Sloan Management Review*, **22** (4), 5–13.
Prahalad, C.K. (2004), *The Fortune at the Bottom of the Pyramid: Eradicating Poverty Through Profit*, Upper Saddle River, NJ: Wharton School Publishing.
Reddy, P. (2011), *Global Innovation in Emerging Economies*. New York: Routledge.
Schmiele, A. and A. Mangelsdorf (2009), 'Drivers of international R&D to Asian economies: a perspective from German companies', paper prepared for the 6th Asiatic International Conference Linkages in Innovation Systems: Global and Local Perspectives, 6–7 July, Hong Kong University of Science and Technology.
TIFAC (2006), 'FDI in the R&D sector: study for the pattern in 1998–2003', TIFAC and Academy of Business School Report, New Delhi: TIFAC.
Siggelkow, N. (2007), 'Persuasion with case studies', *Academy of Management Journal*, **50** (1), 20–4.
von Zedwith, M. and O. Gassmann (2002), 'Market versus technology drive in R&D internationalization: four different patterns of managing research and development', *Research Policy*, **31**, 569–88.
Yin, R. (1994), *Case Study Research: Design and Methods*, Thousand Oaks, CA: Sage.

17. Ramifications for Western firms navigating through innovation spaces in Asia

Maureen McKelvey

17.1 INTRODUCTION

This chapter proposes a conceptualization and discusses some ramifications of how qualitative and quantitative research strategies could be further developed, to understand what happens, as Western firms navigate through innovation spaces in Asia.[1]

This chapter proposes a conceptualization of innovation spaces, which provides a broader framing, of processes involving structural transformation of the economy, driven by entrepreneurs, multinational enterprises and public policy. Sections 17.2 and 17.3 provide a conceptualization and a theoretical background.

The underlying theoretical tradition focuses upon how and why actors develop knowledge and opportunities in processes influencing structural transformation. My previous work has stressed that the firms' innovative searches occur in an environment of co-evolution with market, technologies and institutions (McKelvey 1996; McKelvey 2004; McKelvey and Holmèn 2006; McKelvey 2014a), developing an evolutionary approach pioneered by Nelson and Winter (1982) and following a Schumpeterian paradigm (Schumpeter 1934, 1942).

Asia is proposed as embodying multiple levels and types of changes. Collectively, these economies have the potential to create radically new opportunities and thereby affect future pathways of global developments in innovation (see also Chapter 1). This leads to structural transformation, which is influenced by the development of new scientific and technological knowledge, new markets, and new organizational forms inside and outside the firm. In turn, for potential opportunities into realized innovations, the firms must access different innovation spaces, and also create flows of ideas, resources and so forth between different regional and national innovation systems. The impact, in the long run, is upon economic growth, and societal well-being. This is the broad framing, while the question addressed in this chapter is: How to make sense of these dynamic processes from the perspective of the firm, specifically of Western firms navigating through innovation spaces in Asia?

Section 17.4 addresses qualitative data, especially case studies for answering this question. However, as this book already offers multiple case studies in other chapters, this section focuses only on one important issue for Western firms navigating in innovation spaces in Asia, namely perceptions of risk and uncertainty. Specifically, this section uses a series of quotes from a series of interviews about managers' perceptions of technological and market opportunities. These quotes are then analysed within this chapter, as involving different kinds and degrees of uncertainty, which, if better understood, leads to predictions and advice for business managers of how to act while navigating.

Section 17.5 discusses the use of quantitative data to answer this question, and also provides some data to suggest the differing levels of opportunities available to Western firms in different Asian countries. Later work should rely upon larger datasets of macro trends, combined with micro-data of firms, and also use econometric and other techniques to test hypotheses. Section 17.5 discusses conclusions.

17.2 CONCEPTUALIZATION OF INNOVATION SPACES

This chapter focuses upon the consequence of this conceptualization, for better understanding what happens when Western firms are navigating through innovation spaces in Asia. The conceptualization draws from parallel work, where McKelvey (2014a: 4) proposes the following conceptualization of innovation spaces, including an extensive literature review to develop constituent elements theoretically.

Firms must navigate through an innovation space, in an evolutionary process, as they search and develop different types of business innovations. By definition, an innovation entails the unknown, as the concept focuses upon novelty of economic value.

An innovation space constitutes a geographical and virtual context, which affects innovation through a process of interaction of firms spanning industry as well as the regional/national institutional setting. Because innovation space can be analyzed as this interplay, the innovation space is not bounded by geography *per se*. Geography matters in that certain regions and countries tend to concentrate opportunities. However, firms can span across several spaces through networks and through activities

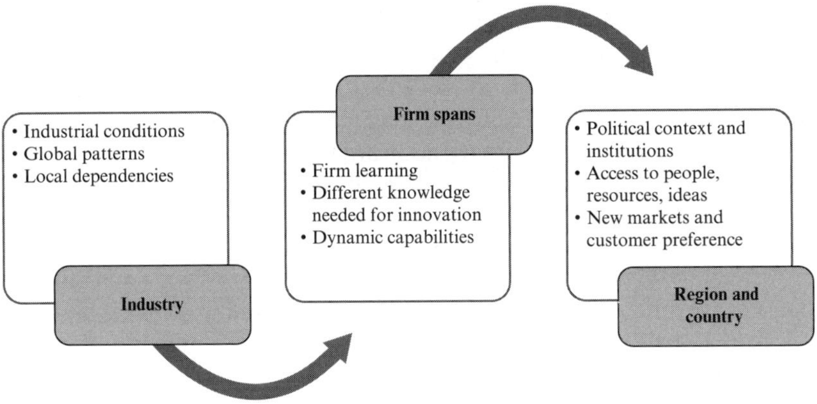

Source: McKelvey (2014a: 10)

Figure 17.1 Interactions creating the innovation space, as the firm spans between industry and region/country

in multiple countries and thereby the firm can configure its capabilities to link different types of innovation systems through the operations and capabilities of the firm.

The main dimensions to analyze within an innovation space reflect the innovative opportunities, which are created in this process of interaction of firm spanning industry as well as regional/national institutional setting. The innovation space can be thought of – and analysed – through three axes. The first axis is of creation of technological opportunities through scientific and technological knowledge. The second axis is the creation of market opportunities related to new market and customer knowledge. The third axis is the creation of productive opportunities, dependent upon using business knowledge of how to operate innovation in the firm locally and globally and leading to a reconfiguration of capabilities.

Figure 17.1 helps visualize what it means that firms are navigating through innovation spaces

Analysis at the firm level includes attention to firm learning, different types of knowledge needed for innovation and dynamic capabilities. The firm is shown as exploring – and spanning – the industry and the region/country level, in the sense that they must search this external environment in order to find relevant new knowledge and create new innovative opportunities.

The metaphor of 'navigating' comes from the idea that the firm is balancing routines but that new situations require that they are searching, or exploring unknown territory, and must discover new information along their pathway of development (Nelson and Winter 1982; Laursen 2012). New knowledge is required, whereby they must explore new areas, create maps, move forward and so forth. Theoretically, exploring new types of innovation spaces requires that the firm engage in search activities and solve new sets of problems, both of which generate new knowledge accumulated in the firm's capabilities.

The visualization in Figure 17.1 across firm-industry-regional/country also suggests that there will later be interdependencies, or co-evolution, between the firm, the industry and the different country contexts in which the firm is active. This metaphor of firms navigating innovation spaces also suggests the importance of directionality, spatial, and context, as they are exploring new geographical and virtual context.

As explored below, Western firms that are active in Asia may drive more major internal changes of the firm, under situations of more radical uncertainty, so that these search and learning processes may ultimately lead them to reconfigure their resources and capabilities for innovation.

17.3 THEORETICAL FRAMING

The above conceptualization draws upon different theoretical perspectives, briefly introduced here, including innovation management as specifically built upon dynamic capabilities; innovation systems; as well as different types of evolutionary and Schumpeterian economics theories in economics and economic geography.

First, innovation management tackles processes whereby 'an innovative outcome involves the successful application of new ideas, which results from organizational processes that combine various resources to that end' (Dodgson et al. 2014: 5). In the literature on theory of the firm as related to capabilities, and in contrast to assumptions in economics, firms are defined as heterogeneous in a way that impacts the outcomes – that is, there is no representative firm. Time is also important for explaining whether or not a firm is successful in exploring a new innovation space. The firm's internal combination of routines, resources, capabilities, and learning will also change over time, in response to internal and external conditions (Nelson 1996). This is related to the Penrose theory of the firm (Penrose 1959) and later developments, especially dynamic capabilities (Teece et al 1997; Teece 2007). A firm's search for novelty and exploration is also an attempt to balance routines and exploitation (Nelson and Winter 1982; Laursen 2012). Therefore, managers in the firm must act, and combine public and private knowledge, as well as other assets, in order to turn knowledge into business innovations (McKelvey, 2014b). The modern research on innovation management explicates how the internal capabilities of the firm rely upon the role of relationships and networks for the firm to search, identify and seize relevant business opportunities (Fagerberg et al. 2006; Dodgson et al. 2014).

Hence, this chapter relies upon these theoretical insights about the firm, in understanding how firms search for innovations, and navigate innovation spaces. Within an innovation space, the three types of opportunities can be analysed, and they relate to information to assess the emergence of knowledge and opportunities related to technological, market and productive ones (Holmèn et al 2007; McKelvey 2014a). This means that for analysing innovation spaces the firm must create these three types of opportunities, by reacting to changes in the environment (technological and market opportunities) and to reconfigure internally (productive opportunities).

Second, the literature on innovation systems was developed from different theoretical perspectives – primarily from institutional and evolutionary economics or from a systems perspective – with the aim to explain the role of knowledge, networks, learning and institutions in economic growth (McKelvey 1991; Nelson 1993; Cooke et al., 1997; McKelvey 1997; Edquist and McKelvey 2000; Lundvall et al. 2002; Malerba 2002; Edquist 2006; Malerba 2009). Innovation systems help define a set of interactions, agents and processes within regions, sectors and nations, and the approach has recently been applied to understanding emerging markets in terms of catch-up processes (Malerba and Nelson 2012). From the perspective of this chapter, the national and sector innovation system define arenas where the firm can access various types of resources, including financial, skilled labour, knowledge and so forth as illustrated in Figure 17.1.

Finally, in terms of evolutionary and Schumpeterian economics, capitalism is conceptualized as restless, with emergent properties of change endogenously created in the economy (Nelson 1996; Metcalfe 2002; Horst and Pyka 2004; Metcalfe 2008). Industrial dynamics is primarily driven through innovation in large companies and through entrepreneurship through new companies (Schumpeter 1934, 1942). Evolutionary economics focus upon dynamics and emergent systems as well as the peculiar role of knowledge and the appropriation of value of innovation.

Opportunities are generated endogenously through the dynamics of the system. Recent economic literature has examined to what extent Schumpeter was evolutionary and dynamic, as well as strived to define principles for evolutionary economics (McKelvey and Holmén 2006; Witt 2010; Andersen 2011). Metcalfe (2002; 2008) argues that the restless nature of capitalism has to do with knowledge, as the pre-eminent source of variation, and that markets are the primary arena of selection. In relation to this literature, this chapter focuses upon the micro-level of evolutionary economics. It addresses a firm's search for innovations, which are a key source of knowledge and of change in the economy, and it uses an evolutionary theoretical approach to define the conditions of firms navigating through innovation spaces.

Finally, another relevant application of evolutionary thinking is used to tackle issues related to proximity and space, even though the above definition includes both geographical and virtual context. Certainly, the conceptualization here could be further developed in relation to evolutionary economic geography (Boschma and Frenken 2006; Boschma and Martin 2010; Boschma 2012). Important ideas in economic geography stress the particular role of knowledge, networks and individuals to link different geographical spaces (Bathelt et al 2004) and the accumulation of particular sorts of knowledge and industries in regions (Cooke et al 1997; Asheim et al. 2006). Recent work by Crescenzi et al. (2012) uses panel data to analyse the geography of innovation in China and India, finding differences between the countries.

Boschma and Martin (2010: 6–7) stake out an ambitious range of scholarship for an evolutionary economic geography approach:

> The processes by which the economic landscape – the spatial organization of economic production, circulation, exchange, distribution and consumption – is transformed over time [. . .] spatialities of economic novelty (innovations, new firms, new industries, new networks), with how the spatial structure of the economy emerge from the micro-behaviours of economic agents (individuals, firms, organisations); with how, in the absence of central coordination or direction, the economic landscape exhibits self-organisation, and with how the processes of both path creation and path dependence interact to shape geography of economic development and transformation, and why and how such processes may themselves be place dependent.

The current chapter certainly shares an intellectual heritage and interest outlined in this manifesto for evolutionary economic geography, albeit with a limited topic to explore here.

Still, space is a complex concept, albeit the word here is used quite broadly, encompassing different levels as visualized in Figure 17.1. Within economic geography, different concepts related to the importance of local, regional, national, international levels have been put forward, which can be called geographical scales or level of analysis. Within this meaning, a recent special issue of *Regional Studies* tackled 'spaces of innovation' (Boschma 2012; Healy and Morgan 2012; Shearmur 2012). These papers help define how proximity affects innovation, as well as the use of spatial concepts to analyse innovation. More generally for future research, this focuses attention on why it matters that geographical contexts can differ in important ways at different scales.

This chapter and conceptualization of innovation spaces, however, does not concern all these issues of interest to economic geographers. This chapter is primarily concerned with drawing out the ramifications for understanding innovation spaces in Asia. In doing so, this chapter helps develop a theoretical view which can be applied to understand how and why micro-behaviour of firms as agents navigate through complex spatial and industrial dimensions, and in ways which impact their future development of innovations.

17.4 WHEN WESTERN MULTINATIONAL ENTERPRISES EXPLORE INNOVATION SPACES IN ASIA

Now let us turn to ramifications for understanding trends in Asia, including ideas for developing research strategies, here with a discussion of qualitative interpretive material based upon interviews. This book has many chapters with case studies of entrepreneurs, multinational enterprises and industries, providing detailed insights into one or more processes, as related to theoretical understanding (Chapters 5, 6, 7, 8, 9, 12, 14 and 15). Let us therefore leave these broad cases, and focus upon the specific issue of what risk and uncertainty means to Western firms, as they navigate innovation spaces in Asia.

Literature within innovation management suggests that firms need to distinguish between risk and uncertainty, as well as whether they are primarily in relation to market or technological opportunities (Dodgson et al. 2014). The underlying theoretical assumption is that these business managers are making decisions in a dynamic context with Knightian uncertainty, which implies that some future possibilities are truly uncertain, whereas others may be reduced to calculable risk (Knight 1921). Insights from managers – through interviews – can provide some valuable clues as to how they perceive different types, and different degrees, of uncertainty about market and technological opportunities.

So, what does risk, uncertainty and opportunity mean when we ask Western multinational enterprises with investments in Asia?

17.4.1 Developing Technology and Innovation in China

This section raises the question of what Western firms say they are searching and learning, when they try to engage in research and development in

Asia and develop new types of innovations relevant for Asia customers.[2] Asia has become an important venue for Western companies to develop technology and innovation, often through R&D, and this section focuses upon China. Based on this chapter and the conceptualization introduced above and further developed below, one needs to consider how the managers perceive technological opportunities as well as market opportunities. Western managers have to address two sides of the same coin – on the one hand, how they should act in China to develop technology and innovation and on the other hand, how the Chinese are innovating today.

Accessing innovation spaces in Asia through developing technology and innovation locally is happening across Asia, but the scale is particularly dramatic in China. The following quote from the Abrami et al. (2014:2) indicates the scale of the investment by foreign countries into developing technology and innovation in China:

> Whereas in 2004 there were some 600 foreign R&D centers in China, by 2010 that number had more than doubled, and their scale and strategic importance had increased. Pfizer moved its Asia headquarters to Shanghai that year. In 2011 Microsoft opened its Asia Pacific R&D center in Beijing, and General Motors opened an Advanced Technical Center comprising several engineering and design labs. Merck's Asia R&D headquarters in Beijing is scheduled to become operational in 2014. (Abrami et al. 2014:2)

According to Survey& (2013:11), the companies doing R&D in China for global markets are concentrated in the automobile sector, computing and telecommunication, consumer goods, health-care pharmaceuticals and industrial products. The trend of year-on-year increasing investment into China by Western multinational enterprises is expected to continue, at least for the next ten years.

Andersson and Thimell (2012) wrote their Master's thesis about how and why nine Swedish multinational companies, primarily in industrial sectors like automobile and engineering but also in food, could innovate in China. They conducted nine interviews, of 1-3 hours each, in China, and later transcribed. Table 1 uses quotes from their Master's thesis. The table reproduces their quotes, and it also provides indicative interpretation in terms of the type of opportunity (market and technological) and the degree of uncertainty (low or high) as implications for understanding what the manager meant that the firm needed to learn.

These quotes in Table 1 give indications that managers try to balance very conflicting interpretations of how to address the types of opportunities, and levels of uncertainty, associated with developing technology and innovation in China. They are trying to learn from the experience, but doing so is difficult, and often contradictory, both within and across companies.

Table 17.1 Examples of perceptions of opportunities and uncertainty when innovating in China

Quotes from interviews in China with Swedish managers	Interpretation about what the Western firm needs to learn
'When people say that Chinese people are not innovative, I believe they are wrong. They are innovative but in a different way. Of course they copy products but they do so in an efficient manner and adapt to the local context. Who says what we do is right?' (p. 46)	Having a focus upon speed during innovation is a positive attribute. The implication is that the customers demands speed in bringing new products to market.
'If you think about it, you can copy products really quick, that is easy, but do Chinese people come up with new innóvations. I don't think so. Copying products is not innovation. They do things faster since they (products) already existed.' (p. 23)	Having a focus upon speed means that technical development isn't so difficult. In contrast, radical changes and innovations are much harder. The implication is that technology suffers, when the focus is upon speed and market opportunities.
'The pace in the Chinese way of thinking can sometimes hamper the quality of product and processes, but the Western way of thinking is often too slow. There has to be a balance.' (p. 37)	Developing a balance between speed and quality is difficult but important. The implication is that Chinese and Western firms are at different ends of a continuum, and can learn from each other.
'The key to success in China is to have good relationship and communication with end customers and partners.' (p. 37)	Developing close relationships matters for business. The implication is that market opportunities can be developed, and uncertainty reduced, through close communication.

Source: All quotes from Anderson and Thimell (2012). Pages specified by quotes

Bruun and Bennett (2002:102) also analyse Scandinavian businesses moving into China, and they develop a nine-step path towards higher strategic importance for the MNC as a whole. From bottom to top, their proposed steps for an R&D unit of an MNC to gain additional responsibilities are: Assume responsibility for production; Maintain technical processes; Assume responsibility for procurement and local logistics; Make process improvement recommendations; Assume responsibility for process development; Make product improvement recommendations;

Assume responsibility for product development: Supply global markets; Become a global hub for product or process knowledge. Hence, the progression here moves from more simple production to being at the centre of relevant global knowledge networks, requiring learning locally as well as at the foreign headquarters.

The business press is also actively debating the pros and cons, and providing direct advice, for Western companies innovating in China. For example, McKinsey & Company (2013) addresses how the multinational enterprise can be successful in China, by recommending the following four steps. (1) Deeply understand Chinese customers. Multinationals need to use their capabilities to 'solve problems' for customers, but to do so, they need specific market knowledge about the demand and preferences of Chinese customers. (2) Retain local talent. In a highly competitive environment like China, talents change jobs often. Managers need global talent programmes, challenging innovation jobs, etc. to retain the best of the local talents. (3) Instill a culture of risk taking. Risk-taking is associated with innovation, but the organization should also examine how failure, accountability and 'saving face' affects individuals' willingness to take risks. (4) Promote collaboration. This is key for innovation. They suggest that one problem is when the project management processes focus upon building scale processes within manufacturing, rather than on innovating. Taken together, these steps will help develop human capital and appropriate structures which are needed, as exemplified by, 'Too often, we visit impressive R&D facilities, stocked with the latest equipment, that are almost empty because staffing them has proved difficult.'

The next section goes on to provide a conceptualization, of how to tackle the types of opportunities and degrees of uncertainty involved, in order to provide recommendations for firms' learning about new innovation spaces. Finding out more about how and why innovation occurs in China – and what Western companies can learn – remains of interest and the subject of many future debates.

17.4.2 Analysis to Match R&D Investments to Opportunities and Uncertainty in Asia

The previous section provides some understanding of what it means, how and why managers in multinational enterprises perceive technological and market opportunities when they move into a new innovation space in China. Moreover, this discussion, combined with theories from innovation management (Dodgson et al. 2014), is used here to make predications and see how these managers should act when navigating through different conditions in their geographical and virtual context.

Table 17.2 Four ideal types of how to match R&D investments to opportunities and uncertainty in Asia

	High level of uncertainty about technological opportunities in Asia	Low level of uncertainty about technological opportunities in Asia
High level of uncertainty about market opportunities in Asia	Invest in R&D in Asia (internal) Invest in open innovation in Asia (external)	Encourage participation by new customers into the firm – integrate Asian customers into innovation processes Let Asian customers test low-quality products
Low level of uncertainty about market opportunities in Asia	Due to trade-off, reduce quality in order to increase speed Retain quality and invest in R&D	Sell existing products imitate other products in Asian market

Table 17.2 provides an analysis involving predictions for how Western business managers moving into Asia should go about, as they engage in their search processes for innovation as well as some ideas given through the interviews about adaptive strategies.[3]

Table 17.2 defines four strategies – or ideal types – that firms can use to match their R&D investments when they face new conditions in the external environment in Asia. The figure divides the concept of uncertainty into the two dimensions related to technological opportunities and market opportunities.[4] It thus provides a matrix, consisting of two dimensions. The first dimension is whether the dynamics (and uncertainty) of developing the innovation primarily lie within the domain of 'technology' or are of 'market'. The second dimension is the degree of novelty, which can be divided into 'radical' versus 'incremental' change.

At the top-right box in Table 17.2, the firm faces a high level of market uncertainty but a low level of technological uncertainty. There is a slow pace of technical development, suggesting a focus upon incremental changes, but there is also a need to develop products and services for new customers. Examples of strategies are to encourage new groups of consumers to engage in innovation, a more specific one would be to allow Asian customers to test the merits of a prototype (lower-quality) product before market launch.

At the bottom-right box, the firm faces low levels of uncertainty in relation to both market and technology. As stated above, technology remains stable with incremental changes and the market demand is reasonably well defined. Examples of strategies are to focus upon existing product lines, and imitate competitors. They could slightly modify their existing global products to these markets, or alternatively, develop products and services in Asia and sell them globally.

At the bottom-left box of Table 17.2, the firm faces low levels of market uncertainty but high levels of technological uncertainty. While the market is reasonably well defined, they can expect rapid technological development, alternative standards, and rapid product life cycles (generations of products). They can either try to reduce quality in order to move products quicker into markets like their competitors in China or else retain quality but invest more into R&D at a faster pace.

At the top-left box, the firm faces high levels of market and technological opportunities. Both dimensions involve risks and uncertainty, often of a more radical nature. Examples of strategies are to invest in internal R&D in Asia and into external partnerships in Asia through open innovation. In the later stages of developing R&D units in Asia, as discussed in other chapters in this book, the literature about Western multinational enterprises suggests that they have to be more active in developing innovations for the market and also placing more advanced technical and engineering research and development there.

17.5 NEED FOR EXTENDED ANALYSIS THROUGH QUANTITATIVE DATA

Another research approach is through quantitative data and hypotheses testing, where more analysis is also needed. The main conceptual perspective proposed in this chapter follows the search activities of the firm, set within an institutional context influenced especially by national and sectoral innovation systems (Lall and Tuebal 1988; Fransman 1995; Mathews 2006, 2007; Malerba and Nelson 2012; Lee 2013). Industrial dynamics are driven by innovation and entrepreneurship, which are in turn dependent upon the emergence of different types of knowledge linked to opportunities, as explained above. This section therefore explores existing data and how it could be analysed and interpreted in relation to this framework.

Quantitative data is very important, and there is a need to develop frameworks and methods relevant here in future research. Chapter 1 proposed eight trends related to changes in the political economy in Asia that will impact innovation, while Figure 17.2 visualizes how these trends lead to new opportunities.

Theoretically, these trends should be seen as potential indicators of

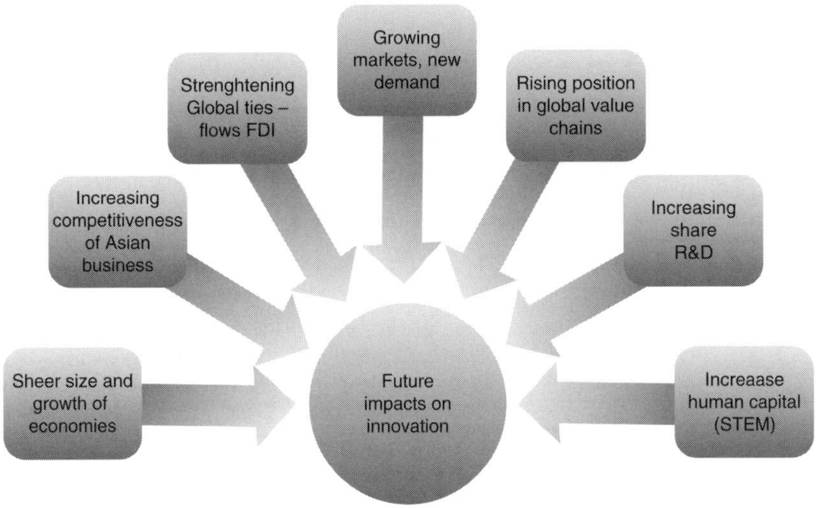

Figure 17.2 *Asia as innovation spaces rich in (potential) opportunities*

the vast increase in opportunities that Asia currently represents. To develop this thinking, quantitative data – where Chapter 1 provides some initial statistics – should be further developed and modelled over time, to examine impacts.

Thus, it would be important to find specific quantitative data to measure and study the impacts of different types of opportunities. From the perspective of the firm, they need to use their networks and organizational structure in order to explore the different opportunities possible in different geographical scales. The firm is looking into three-dimensional space, and they want to know where to navigate to (or from) as they search, explore and learn about technological opportunities and market opportunities.

Figure 17.3 provides a visualization of how four innovation spaces can be represented at the country level, using the three types of opportunities, respectively, of technological, market and productive. The values set for Asia, Europe, US and Africa, respectively, are fictive values, used to represent how the graph and understanding could be mapped into quantitative data in the future. More explicit work can be developed along these lines.

By suggesting that the firm must span different levels, one should also use quantitative data to better conceptualize these processes of ones involving dynamics and direction. While the Asian Development Bank includes 45 economies as Asian, Table 17.3 provides data on 11 countries to indicate how potential innovative opportunities in different national

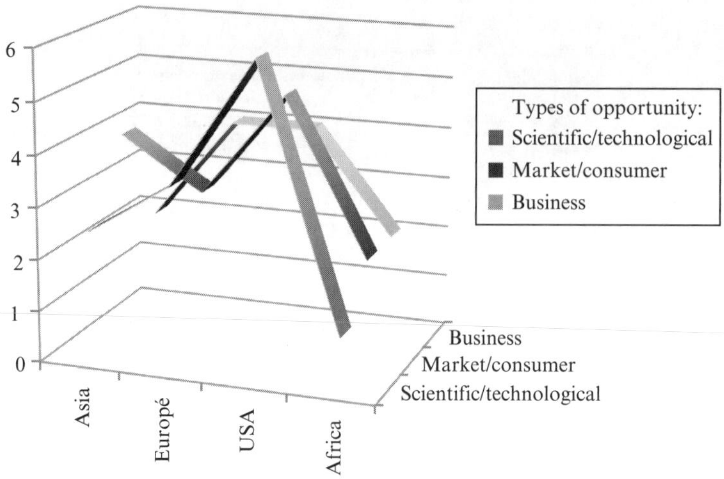

Source: (McKelvey 2014a: 12)

Figure 17.3 Representational visualization of how the firm may perceive innovation spaces

innovation systems could be mapped. This could add real values to the visualization of Figure 17.3 of exploration of three-dimensional space.

This table illustrates the diversity of countries within Asia – and also suggests different potentials. In the left column, the population is indicative of market potential, although these may be expanding mass and/or niche markets as explained in Chapter 1. The population ranges from 5 million in Singapore to 1.4 billion in China. Anyone interested in the historical – and projected future – relationships between population and growth should examine the visualization of statistics found at www.gapminder.org. In the middle column, similarly, annual economic growth is indicative of the possibility of market growth and access to capital, ranging from 0.5 per cent in Malaysia to 7.7 per cent in China. Finally, in the right column, gross domestic expenditures on R&D is indicative of technological opportunities as a measure of scientific and technological capabilities, also with a wide range from 0.1 in Indonesia to 3.7 per cent in South Korea.

Many additional statistics can be interpreted as well. Different organizations in society can finance and perform R&D, leading to somewhat different foci for development. According to The Organisation for Economic Co-operation and Development (OECD, 2013: 68), business is the dominate performer of R&D for all OECD countries, except New Zealand,

Table 17.3 Selected Asian country overview, by population, economic growth rates and GERD

	Population (rounded)	Growth rate of GDP 2013 (% per year)	GERD (Gross domestic expenditures on R&D as % of GD) by latest estimates (year)
China	1.4 billion	7.7	1.7% (2009)
Korea, Republic of	50 million	2.8	3.7% (2010)
India	1.3 billion	4.9	0.8% (2007)
Indonesia	249 million	5.8	0.1% (2009)
Japan	127 million	0.7	3.4% (200)
Malaysia	30 million	0.5	0.6% (2006)
Pakistan	184 million	3.6	0.5% (2009)
Singapore	5 million	4.1	2.4% (2009)
Sri Lanka	20 million	2.8	0.1% (2008)
Thailand	67 million	2.9	0.2% (2007)
Vietnam	89 million	5.4	–

Sources: Wikipedia (2014 for population); ADB (2014: Table A1) for statistics on GDP growth for all countries except Japan; BBC (2014) for Japan's growth; UN (2013: Table C.4.1) for GERD

and the importance of business R&D also holds for Malaysia, China, Singapore and the Philippines. For Indonesia and Vietnam, the government is the main financer. There is also the diversity of organizations to take into account. There are entrepreneurs and multinational enterprises (domestic and foreign) as well as state-owned enterprises (SOEs) or government-linked companies (GLCs) in Singapore and Malaysia. Moreover, foreign multinationals tend to do more R&D and more advanced R&D but domestic enterprises and SOEs seem to be increasing their share, especially in China.[5] Chapters 10, 11, 12, 13, 14 and 15 all address aspects of located and developing domestic (to Asia) and foreign multinational enterprises in terms of both innovative activities and of product innovations.

In order to further stimulate entrepreneurship, framework conditions in the innovation system matter at the country level. Typical problems in many Asian countries are: limited venture capital money and skills; shortage of technical personnel; inefficient government bureaucracy and high regulatory burden of starting a company (OECD 2009, 2013). Each specific country differs in terms of framework conditions, however, and some like China – also as demonstrated in Chapters 5, 6, 8 and 9 of this book – are hotspots of entrepreneurship in advanced sectors.

In terms of framework conditions, Malerba et al. (2015) analyses the rise of, and conditions for, knowledge intensive entrepreneurship in China, India and Russia. Patents are used to analyse the entrance, persistence and specialization of firms' patenting, as well as a discussion of framework conditions in each of these three countries. Similar to the conclusions of the current book (Chapter 18), Malerba et al. (2015) find that common factors include the role of economic growth, the increase in skills and education, the growth of R&D, the increase in the extent and variety of policy support, the rise of venture capital, the spread of information technology, and the role of returnees from advanced countries.

A final aspect where quantitative data is vital is to understand and quantify internationalization, as well as relative specialization in the global world. Following upon the discussions of how innovation spaces may be to Asia, inside Asia or from Asia, an important aspect is how statistics can be used to measure and indicate stocks and flows between countries, as well as how to interpret the results. One way is to analyse different types of cross-border investments into development (DDT), R&D and manufacturing, as found in Table 17.4.

Table 17.4 indicates significant differences in international activities

Table 17.4 Cross-border investments in DDT, R&D and manufacturing, 2003–2012

Sector	Design, development and testing	Research and development presented as number of projects (% share)	Manufacturing presented as number of projects (% share)
ICT / electronics	2443 (61%)	1301 (41%)	3017 (10%)
Transportation	382 (10%)	273 (9%)	5967 (19%)
Life sciences and chemicals	329 (8%)	802 (25%)	3857 (13%)
Services	228 (6%)	104 (3%)	188 (0.6%)
Industrial machinery, equipment and tools	203 (5%)	144 (5%)	2662 (9%)
Other sectors	3980 (10%)	523 (17%)	15133 (50%)

Sources: Chaminade et al. (2004: Table 2.3), based upon Castelli and Castellani (2013)

across different types of sectors, as also suggested in (Malerba and Nelson 2012). Information and communication technology (ICT) and electronics are particularly important for design, development and testing (DDT) and R&D, whereas other sectors constitute the primary group for internationalization of manufacturing. Sectoral differences provide rich sources of internationally comparative data, to analyse stocks and flows.

A related area is research on how statistics relate to global value chains (GVCs). According to (OECD 2013: 41): 'Global value chains are created by a series of intermediate steps across different countries which contribute to the product's value added.' The debate on GVCs has led to attempts to provide a more nuanced understanding of the position of each country, relative to statistics on imports, production and exports. Global value chains are thought to be particularly important for Asia, and also for tracking whether they move up the value chain over time. In interpretation of trade data, GVCs lead to a blurring of the concept of country of origin.

Future research should use econometric, social network analysis and other approach to analyse quantitative in order to test hypotheses over larger data sets.

17.6 CONCLUSIONS

This chapter has, firstly, proposed a conceptualization and theoretical framework for analysing what happens as Western firms navigate through innovation spaces in Asia. This conceptualization is then used to interpret some ramifications of what that means in terms of qualitative and quantitative research strategies, as well as some ideas about future research.

In this chapter, the focus has been on novelty through the creation of innovative opportunities by firms spanning the industry and regional/national context. The firms are the key organizations defined as retaining capabilities and attributes, although the whole notion of search and learning implies that the firm will learn over time, in response to changing external conditions. Finally, the selection processes described here are generally pre-market, in the sense that the search is for innovations, before market introduction. Future research needs to develop indicators and empirical insights, as well as further develop theoretically which mechanisms and processes tend to generate novelty, retain capabilities and attributes, and select among alternatives.

NOTES

1. I wish to thank the University of Gothenburg, and also please see the detailed acknowledgements for the book as a whole. I would like to thank participants at the 15th International Joseph A. Schumpeter Conference, 27–30 July 2014, Jena, Germany, for comments on the conceptual paper, and particularly comments by Tim Kastelle, Jorge Niosi, Bastian Rake, Morris Tuebal and Pier-Pablo Saviotti.
2. While this section does not directly draw upon their work, with the exception of reproducing some quotes, I would like to acknowledge that the University of Gothenburg has had Master's students in Innovation and Industrial Management MSc who have conducted a series of interviews during the last years with Western MNCs active in research and development in Asia. I would like to acknowledge the Master's students, Erik Andersson, Fredrik Thimell och Tatiana Kirilova, who performed these interviews. Moroever, a paper using the interview material prepared by the MSc students, was written as Johansson and McKelvey (2012). This joint paper between Johansson and McKelvey was accepted and presented at Globelics 2012. Globelics stands for the Global Network for the Economics of Learning, Innovation and Competence. The Tenth Globelics conference was held at Hangzhou, Zheijang Universtiy, 9-11 November 2012. The Globelics paper Johansson and McKelvey (2012) is entitled 'Developing new capabilities for innovation and development: A study of Swedish firms responses to tensions in China'. The main points were: (1) China, as an emerging market, is indeed characterized as a Schumpeterian environment, in the sense of having high levels of rapidly changing and uncertain environment; (2) Multiply tensions arise as an effect of setting up and operating a MNC unit in China. And (3) Tensions arise when the Swedish multinationals have R&D activities in China, and these can be aggregated into: Speed, cost, lower market segment and customer value vs quality; and Chinese vs Swedish ways of making decisions and managing the organization. This joint paper was later shortened and a revised version of the same paper (Johansson and McKelvey 2013) was presented at the workshop 'Business in Asia' in February 2013, held at University of Gothenburg, Sweden.
3. Table 17.2 provides a way to discuss the four ideal types, but it does not display four mutually exclusive stages, however, The firm can of course be engaged in development processes that both involve a high level of technical uncertainty as well as a low level of market uncertainty. They must combine the predictions in that case.
4. Productive opportunities are not included as they are assumed to be internal to the organization of the firm. Rather than talk about opportunities per se, the following discussion moves on to focus on the firm and the dimensions of their search while trying to develop an innovation.
5. See OECD (2013) for Malaysia and Singapore, especially Chapters 7 and 8; for China, see OECD (2009) and for Korea see OECD (2008).

REFERENCES

Abrami, R.A., W.C. Kirby and W. McFarlan (2014). 'China Can't Innovate', *Harvard Business Review*, March.

Andersson, E. and F. Thimell (2012). 'Need for Speed – Exploring Swedish High-Tech Subsidiaries' Innovation Process in China'. Master's thesis in MSc in Innovation and Industrial Management, Graduate School and Department of Economy and Society, University of Gothenburg.

Andersen, E.S. (2011). *Joseph A. Schumpeter: A Theory of Social and Economic Evolution*. London: Palgrave Macmillan.

Asheim, B., Cooke, P. and Martin, R. (2006). 'The rise of the cluster concept in regional analysis and policy: a critical assessment', in Asheim, B., Cooke, P. and Martin, R. (eds), *Clusters and Regional Development: Critical Reflections and Explorations*. Abingdon: Routledge, pp. 1–29.

Asian Development Bank (ADB) (2014). *Asian Development Outlook 2014: Fiscal Policy for Inclusive Growth*. Mandaluyong City, Philippines: Asian Development Bank.

Bathelt, H., Malmberg, A. and Maskell, P. (2004). 'Clusters and knowledge: local buzz, global pipelines and the process of knowledge creation', *Progress in Human Geography*, **28** (1), 31–56.

BBC (2014). 'Japan's deficit hits record as economic growth slows', available at: www.bbc.com/news/business-26510495 (accessed 20 October 2014).

Boschma, R. (2012). 'Proximity and innovation: a critical assessment', *Regional Studies*, **34** (3), 213–29.

Boschma, R. and Frenken, K (2006). 'Why is economic geography not an evolutionary science? Towards an evolutionary economics geography', *Journal of Economic Geography*, **6** (3), 273–302.

Boschma, R. and Martin, R. (2010). *The Handbook of Evolutionary Economic Geography*. Northampton, MA, USA and Cheltenham, UK: Edward Elgar.

Castelli, C. and Castellani, D. (2013). 'The internationalisation of R&D: sectoral and geographic patterns of crossborder investment', *Economia e Politica Industriale*, **1**, 127–43.

Brunn, P. and D. Bennett (2002). 'Transfer of technology to China: a Scandinavian and European perspective', *European Management Journal*, **20** (1), 98–106.

Chaminade, C., Castellani, D, and Plechero, M. (2014). *The Emergence of China and India as New Innovation Power Houses – Threat or Opportunity?* Globaliseringsforum rapport #6. Stockholm: Entreprenörskapsforum.

Cooke, P., Gomez Uranga M. and Etxebaria, G. (1997). 'Regional innovation systems: institutional and organizational dimensions', *Research Policy*, **26** (4–5), 475–91.

Crescenzi, R., Rodriquez-Pose, A. and Storper, M. (2012). 'The territorial dynamics of innovation in China and India', *Journal of Economic Geography*, **12** (5), 1055–85.

Dodgson, M., Gann D. and Phillips, N. (2014). *The Handbook of Innovation Management*. Oxford: Oxford University Press.

Edquist, C. (2006). 'Systems of Innovation. Perspectives and Challenges' in J. Fagerberg, D.C. Mowery and R.R. Nelson. (eds), *The Oxford Handbook of Innovation*. Oxford: Oxford University Press.

Edquist, C. and McKelvey, M. (eds) (2000). *Systems of Innovation: A Reference Collection*. Northampton, MA, USA and Cheltenham, UK: Edward Elgar.

Fagerberg, J., Mowery, D.C. and Nelson, R.R. (2006). *The Oxford Handbook of Innovation*. Oxford: Oxford University Press.

Fransman, M. (1995). *Japan's Computer and Communication Industry: The Evolution of Industrial Giants and Global Competitiveness*. Oxford: Oxford University Press.

Healy, A. and Morgan, K. (2012). 'Spaces of innovation: learning, proximity and the ecological turn', *Regional Studies*, **46** (8), 1041–53.

Holmèn, M., Magnusson, M. and McKelvey, M. (2007). 'What are innovative opportunities?', *Industry and Innovation*, **14**, 27–45.

Horst, H. and Pyka, A. (eds) (2004). *Companion on Neo-Schumpeterian Economics*. Northampton, MA, USA and Cheltenham, UK: Edward Elgar.
Johansson, M. and McKelvey, M. (2012). 'Developing new capabilities for innovation and development: a study of Swedish firms responses to tensions in China', paper presented at Tenth Globelics conference Hangzhou, Zheijang University, 9–11 November.
Johansson M. and McKelvey, M. (2013). 'Exploring China as a source of learning for Swedish multinationals in China', presented at the Business Strategy and Innovation Spaces in China and India workshop, February, at the Institute of Innovation and Entrepreneurship, University of Gothenburg, Sweden.
Knight, F. (1921). *Risk, Uncertainty and Profit*. Boston, MA: Houghton Mifflin.
Laursen, K., 2012. 'Keep searching and you'll find: what do we know about variety creation through firms' search activities for innovation?', *Industrial and Corporate Change*, **21** (5), 1181–220.
Lall, S and Tuebal, M. (1988). 'Market-stimulating technology policies in developing countries: a framework with examples from Southeast Asia', *World Development*, **26** (8) 1369–85.
Lee, K. (2013). *Schumpeterian Analysis of Economic Catch-Up: Knowledge, Path-Creation and the Middle Income Trap*. Cambridge: Cambridge University Press.
Lundvall, B.Å., Johnson, B., Andersen, E.S. and Dalum, B. (2002). 'National systems of production, innovation and competence building', *Research Policy*, **31** (2), 213–31.
Mathews, J. (2006). 'Dragon multinationals: new players in 21st century globalization', *Asia Pacific Journal of Management*, **23** (1), 5–27.
Mathews J.A. (2007). 'Catch-up strategies and the latecomer effect in industrial development', *New Political Economy*, **11** (3), 313–35.
Malerba, F. (2002). 'Sectoral systems of innovation and production', *Research Policy*, **31**, 247–64.
Malerba, F. (2009). *Sectoral Systems of Innovation: Concepts, Issues and Analyses of Six Major Sectors*. Cambridge: Cambridge University Press.
Malerba, F. and Nelson, R.R. (2012). *Economic Development as a Learning Process Across Sectoral Systems*. Northampton, MA, USA and Cheltenham, UK: Edward Elgar.
Malerba, M., Sunil, M., Valero, S., Wu, X. and Yudanov, A. (2015). 'The growth of knowledge intensive entrepreneurship in China, India and Russia: quantitative analysis and general overview', in Malerba, F., Caloghirou, Y., McKelvey, M. and Radosevic, S. (eds), *Dynamics of Knowledge Intensive Entrepreneurship: Business Strategy and Public Policy*. Abingdon: Routledge, forthcoming.
McKelvey, M. (1991), 'How do national systems of innovation differ? A critical analysis of Porter, Freeman, Lundvall and Nelson', in Hodgson, G.M. and Screpanti, E. (eds), *Rethinking Economics: Markets, Technology and Economic Evolution*. Aldershot, UK and Brookfield, VT, USA: Edward Elgar, pp. 117–37.
McKelvey, M. (1996). *Evolutionary Innovations: The Business of Biotechnology*. Oxford: Oxford University Press.
McKelvey, M (1997). 'Coevolution in commercial genetic engineering', *Industrial and Corporate Change*, **6** (3), 503–32.
McKelvey, M. (2004). 'How and why dynamic selection regimes affect the firm's innovative search activities', *Innovation: Management, Policy and Practice*, **6** (1), 3–24.
McKelvey, M. (2014a). 'Firms navigating through innovation spaces.' Chapter

presented at the International Joseph Alois Schumpeter Society conference, the 15th ISS Conference, 27–30 July, Jena, Germany.
McKelvey, M. (2014b). 'Science, technology and business innovation', in Dodgson, M., Gann D., and Phillips, N. (2014), *The Handbook of Innovation Management*. Oxford: Oxford University Press, pp. 69–82.
McKelvey, M. and M. Holmèn (2006). *Flexibility and Stability in the Innovating Economy*. Oxford: Oxford University Press.
Metcalfe, S. (2002). 'Knowledge of growth and growth of knowledge', *Journal of Evolutionary Economics*, **12** (1), 3–13.
Metcalfe, S. (2008). *Restless Capitalism – The Evolutionary Nature of Capitalism*. Princeton, NJ: Princeton University Press.
Nelson, R. and Winter, S. (1982). *An Evolutionary Theory of Economic Change*. Cambridge, MA: Belknap Press of Harvard University Press.
Nelson, R.R. (1993). *National Systems of Innovation: A Comparative Analysis*. Oxford: Oxford University Press.
Nelson, R.R. (1996). *The Sources of Economic Growth*. Cambridge, MA: Harvard University Press.
OECD (2008). *OECD Reviews of Innovation Policy: China*. Paris: OECD.
OECD (2009). *OECD Reviews of Innovation Policy: South Korea*. Paris: OECD.
OECD (2013). *OECD Reviews of Innovation Policy: Innovation in Southeast Asia*. Paris: OECD.
Penrose, E.T. (1959). *The Theory of the Growth of the Firm*. New York: John Wiley
Schumpeter, J., (1934). *The Theory of Economic Development*. Cambridge, MA: Harvard University Press.
Schumpeter, J.A. (1942). *Capitalism, Socialism and Democracy*. New York: Harper & Brothers.
Shearmur, R. (2011). 'Innovation, regions and proximity: from neo-regionalism to spatial analysis', *Regional Studies*, **45** (9), 1225–43.
Survey& (2013). 'An emerging innovation power: 2013 Chinese innovation survey', http://www.strategyand.pwc.com/global/home/what-we-think/reports-white-papers/article-display/emerging-innovation-power.
Teece, D. J., Pisano, G. and Shuen, A. (1997). 'Dynamic capabilities and strategic management', *Strategic Management Journal*, **18** (7), 509–33.
Teece, D. (2007). 'Explicating dynamic capabilities: the nature of (sustainable) enterprise performance', *Strategic Management Journal*, **28** (13), 1319–50.
UN (2013). *Statistical Yearbook for Asia and the Pacific*. Geneva, Switzerland: United Nations.
Wikipedia (2014). 'Population'. Available at: http://en.wikipedia.org/wiki/Population (accessed 1 August 2014).
Witt, U. (2010). 'How evolutionary is Schumpeter's theory of economic development?', *Industry and Innovation*, **9** (2), 7–22.

18. Conclusions

Maureen McKelvey and Sharmistha Bagchi-Sen

18.1 INTRODUCTION

This book, *Innovation Spaces in Asia: Entrepreneurs, Multinational Enterprises and Policy*, provides insight into a timely topic.[1] Asia is posed to have a major impact upon global innovation processes and outcomes. The underlying proposition of this book can be summarized as follows: Asia is currently and will continue to impact innovation in the future. The ongoing trends will become a force for global change that will affect the direction and rate of future business innovation to deliver relevant services and products, globally. The rise of new business opportunities in Asia is also impacting social and business innovation, and thereby also impacting firm strategy and public policy in many other countries.

This book contributes to the literature by analyzing: (1) various aspects of "how" and "where" of innovation in Asia; (2) how and why doing business in Asia involves new challenges, such as entrepreneurship, the rise of competition from homegrown multinationals and an increase in customer demands; and (3) how firms are navigating this new geographical context that is not only dynamic and complex but also has characteristics and emerging trends which may be different from the national contexts within which Western multinationals emerged and operate.

Innovation spaces are created dynamically and over time, in interactive processes among entrepreneurs, multinational enterprises and policy.

18.2 CREATING INNOVATION SPACES

Networks and knowledge flows that underlie geographical space and the operational space of firms are included in the conceptualization of innovation space. This and the subsequent section draw upon chapters in this book to address how entrepreneurs, multinational enterprises and policy create innovation spaces. These are emerging spaces, linking individuals with the external environment, to access knowledge, resources

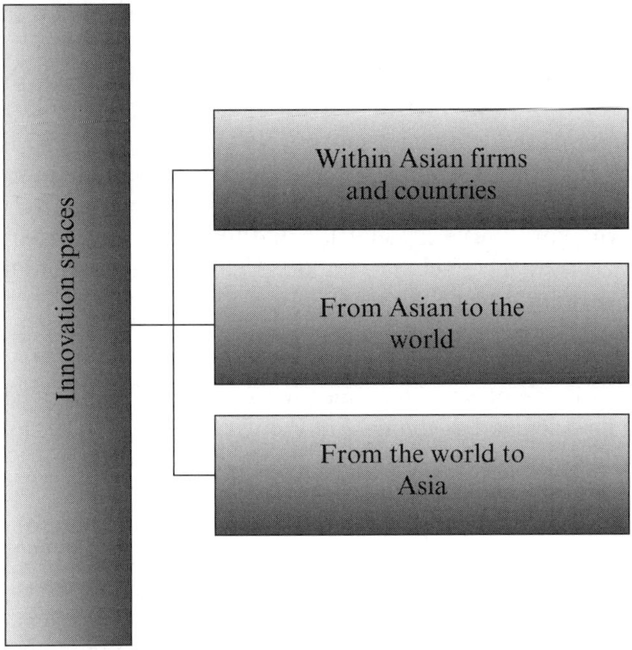

Figure 18.1 *Creation of innovation spaces within Asian firms and countries, from Asia to the world, and from the world to Asia*

and markets. We identify three aspects of how innovation spaces in Asia function, which link national contexts with global ones in different ways.[2]

Figure 18.1 proposes three ways of addressing innovation spaces as linking individuals, companies and the flow of resources, institutions and markets, in particular geographical locations and scales.

The first captures the ongoing processes within specific Asian countries and Asian firms. The development of innovation spaces within Asian firms and countries have been addressed within Chapters 2 to 9. All chapters include aspects of internationalization of knowledge and technology and global linkages, so they are not only about internal national innovation systems. These chapters address the dynamic development of innovation spaces from the perspective of the ongoing processes within specific Asian countries and Asian firms, specifically China, India and South Korea. These chapters address ongoing processes involving entrepreneurs, public policy and the role of institutions. Empirical insights are provided about the pathway of development of the national institutions such as policies related to innovation, intellectual property rights as well as about the dynamics of entrepreneurship in advanced sectors, including emphasis on the role of technology, customers, financing, public policy and global linkages.

The second is from Asia to the world, which captures the movement as Asian firms are active in Western countries to access resources and markets. How companies move from Asia to the world is a topic addressed more explicitly in Chapters 10 to 14. These chapters address the flows of creating innovation spaces from the perspective of Asian firms moving into Western countries, and focus upon the large multinational enterprises. These chapters address the stereotype that large firms from Asia tend not to be innovative, only focusing on low cost and not on investing in technology and innovation. The theoretical framing is primarily around theories of multinational enterprises, and how they apply to such companies from emerging markets as well as the need to adjust to new demand at the base of the pyramid. Empirical insights are given through, for example, Indian pharmaceutical companies' international acquisitions, which demonstrate that emerging market MNEs can exploit certain advantages at the industry and national levels; for example, they quickly adjust to regulations in the European Union (EU) and the US to access knowledge and markets. Other insights are given through Indonesian multinationals as they move into nearby markets, and the ability of Chinese multinationals to tap into Sweden's innovative country in automobiles and telecommunication. Finally, one chapter addresses the conceptualization of how multinational enterprises should respond to new demands, when customers make up a huge potential market but with very limited financial resources.

The third is the world moving into Asia to access innovative ideas or resources and often to tap into domestic resources and markets. How companies from the rest of the world move into Asia to access and create new innovation spaces is addressed in Chapters 13, 15, 16 and 17. They thus address innovation spaces from the perspective of Western firms moving into Asian countries, and demonstrate some of the advantages and challenges of doing so. These chapters are focused upon India and Sri Lanka. Clearly, moving and shifting the firm into Asia is not easily done, and so these chapters also address the stereotype that Western firms can easily move to outsource customer development, technological development and research and development (R&D) to Asia. These chapters address what happens over time when companies outsource certain business activities (for example, software) and the impact of a new major market like India on the conceptualization of business innovation within a Western company.

In terms of methodology, many different sources of data and analysis can be undertaken. Most chapters in this book are qualitative, and provide case studies of the dynamics of very specific processes over time between individuals, firms and innovation spaces. One advantage of case studies is that they enable us to study processes where the boundaries between these concepts of innovation, opportunities and entrepreneurship are blurred.

The most obvious place where definitions become blurred is when a process view is taken. Entrepreneurs are identified as spanning different spatial scales as well as different types of organizations, as in the case of returnees to China and nanotechnology in Korea and China. For multinational enterprises engaging in innovation, as for SKF in India, the case study indicates the feedback loops between internal firm decisions, institutions and processes that may in turn require organizational and marketing changes to develop a new R&D unit with appropriate capabilities and mandate. Sometimes, and especially in developing countries, the chapters about Indonesia and India also suggest that the borderline from business innovation to social innovation and entrepreneurship is also blurred, for example, when company shifts direction to meet an important societal need through a low-cost and accessible product. In Asia in particularly, the innovations and R&D labs may be provide either one component in an international value chain or global innovation network, or be seen as a way to source lower price R&D with similar standards by accessing national innovation systems, such as shown in the software cases in Bangalore, India and Sri Lanka.

A key characteristic of the book is the rich empirical understanding of the dynamic processes, involving the strategic decisions of firms and entrepreneurs with the broader socio-economic environment in terms of institutions, markets, knowledge and innovation systems. Some chapters demonstrate how firms are developing maps of these changing contexts and trying to move towards goals, with new organizational practices, new partners and new features of products and services. Other chapters focus upon specific topics of how innovation spaces in specific countries in Asia are developing, such as new institutional framework for patents and new ways of providing global access to that knowledge. Innovation managers have many choices to make, such as how to outsource certain business activities, where to incorporate the new demands of customers in expanding markets as well as whether and how to affect the pathway of development of the national institutions and markets. Innovation managers may need to decide whether to source and access scientific and technological knowledge as well as market and business knowledge by locating or relocating in Asia or whether to access knowledge remotely through licensing of their patents or hiring skilled personnel or through networks. In addition to researchers, innovation managers and students in courses about innovation management as well as entrepreneurship are interested in these topics. Theoretically, chapters in this book provide a new perspective in line with recent streams of literature in innovation and entrepreneurship, international business and strategy, evolutionary economics and economic geography.

18.3 REVISITING THE KEY CONCEPTS

This book provides an overall perspective, while each chapter makes its own more specific contribution. Collectively, the chapters in this book contribute to an understanding of various themes that help define and develop how innovation spaces in Asia are created over time. In doing, the theoretical perspective helps with an overall ambition of this book, which is to conceptually analyze the links the micro-behavior and processes of firms with the macrostet processes related to the socio-economic-political context, in order to understand how structural transformation occurs.

18.3.1 Entrepreneurs and Innovation

Innovation and entrepreneurship have become buzzwords, used in many different contexts, often with different meaning. This section provides some basic definitions, while readers interested in more detailed analysis are referred to books such as Dodgson et al. (2014) and Fagerberg et al. (2006). Many studies within innovation and entrepreneurship perhaps started with tangible manufacturing products and high-tech industries like biotechnology or nanotechnology, but the empirical studies and conceptualizations have moved far beyond this, to include a range of empirical settings and issues, such as services, low-tech industries and user-driven innovations.

One concept is entrepreneurship, where Carlsson (2013) provides an analysis of "The Evolving Domain of Entrepreneurship," which explains how the definitions and field of entrepreneurship research have changed over time including micro and macro phenomenon. For example, one definition is: "The entrepreneur is someone who specializes in taking responsibility for and making judgmental decisions that affect the location, the form, and the use of goods, resources, or institutions" (Hébert and Link, 1989: 47). Much more broadly, Venkataraman (1997) argues that:

> Our field is fundamentally concerned with understanding how, in the absence of current markets for future goods and services, these goods and services manage to come into existence. Thus, entrepreneurship as a scholarly field

> seeks to understand how opportunities to bring into existence "future" goods and services are discovered, created, and exploited, by whom, and with what consequences.

This broad definition of the field is usually studied empirically at a much narrower range of phenomena. The main levels of empirical analysis of entrepreneurship are between the individual person (entrepreneur) and the formation of the venture (company). Even though the person and venture involve different levels of analysis, they are deeply intertwined in that modern literature also stresses that the traits and characteristics of the entrepreneur will affect the future development, and performance of the company, as for the specific case of knowledge-intensive entrepreneurship (McKelvey and Lassen 2013). Other literature examines aspects of how change impacts existing organizations, such as the concept of entrepreneurial orientation (Lumpkin and Dess 1996) or that of corporate entrepreneurship and intrapreneurship.

For innovations, one widely diffused and operationalized definition of innovation is The Organisation for Economic Co-operation and Development (OECD) definition (Table 18.1) (Oslo manual for measuring innovation), which is also used to gather official statistics and also for cross-national work such as the Community Innovation Surveys (CIS).

There are different aspects of the definition, with a common one being novelty of economic value. The dimension of "economic value" is also what distinguishes an innovation from the earlier stage of an idea, patent, technology or invention, and this has implications for measurement and

Table 18.1 OECD definition of various types of innovations

Type of innovation	Definition
Product innovation	A good or service that is new or significantly improved. This includes significant improvements in technical specifications, components and materials, software in the product, user friendliness or other functional characteristics
Process innovation	A new or significantly improved production or delivery method. This includes significant changes in techniques, equipment and/or software.
Marketing innovation	A new marketing method involving significant changes in product design or packaging, product placement, product promotion or pricing.
Organizational innovation	A new organizational method in business practices, workplace organization or external relations.

analysis in that a product innovation must be introduced to a market, whereas a patent is an indicator of the technological frontier.

The field of innovation management deals with developing processes, involving coordinating resources and capabilities into organizational structures and networks, as well as motivating individuals and teams to deliver the expected positive benefits, under conditions of uncertainty. Dodgson et al. (2014: 5) state that:

> We are content with the widely accepted definition as the successful application of new ideas but believe for analytical and practical purposes, the definition of innovation management has to be more nuanced [. . .] Innovation is both an outcome and a process, a fact and an act. An innovative outcome involves the successful application of new ideas, which results from organizational processes that combine various resources to that end.

Studying entrepreneurs and innovation can be done in different ways. Business innovation refers to arenas where both entrepreneurs and multinational enterprises are the main actors and where profitability is an organizational goal. Moreover, innovation spaces matter because the firm is closely tied to the environment, through, for example, knowledge networks and contact with customers to aid in the interpretation of market opportunities. The boundaries of the firm are fluid and changing, in response to internal and external conditions in the environment (Teece et al. 1997; Teece 2007). From the perspective of entrepreneurial as well as a resource and capabilities perspective, firms must search for new opportunities, and thus their search for solutions to their technical and market problems (Shane 2000; Laursen 2012) and in relation to the creation of new opportunities through the innovation system (Holmèn et al. 2007; McKelvey 2014). Bell and Pavitt (1993) stressed the role of multinational enterprises in defining the technological capabilities of countries.

In many ways, this modern literature draws upon classical ideas of the generation of knowledge and opportunity at the level of the economic system. There are dynamics which link entrepreneurs and innovation at the economic system level, because innovation disrupts equilibrium in the economy. Schumpeter (1934: 66) defined innovation as the "introduction of new goods . . . new methods of production . . . the opening of new markets . . . the conquest of new sources of supply . . . and the carrying out of a new organization of any industry." These five types of innovations range from internal to the firm to industrial dynamics, because they define five types of opportunities, which can be created and exploited by firms. Hence, the links between a firm and the surrounding environment give rise to new opportunities that enable the growth of firms, industries and regions (Schumpeter 1934).

Relevant phenomenon about innovation spaces in Asia discussed in this book include:

- Entrepreneurs in these advanced sectors have global network linkages to access international resources and knowledge, and many have international experience, but the development of their entrepreneurial ventures are also influenced by their national innovation system. Knowledge-intensive entrepreneurship within high tech industries is analyzed in several chapters in China and one in Korea (Chapters 6–9).
- Acquiring talents related to entrepreneurship and innovation may also mean acquiring strategic capabilities related to innovation in industries dependent upon advanced technology. The analysis includes the automobile industry and telecommunication, as Chinese firms move into Europe through acquisitions and hiring from competitors (Chapter 12).
- Outsourcing of software development by both large and small Western companies can lead to the development of more advanced innovative skills in the emerging market (Chapters 14 and 15). For an entrepreneurial firm that becomes a multinational enterprise with significant innovation abroad, access to skilled labor, preferably at a lower cost, may be crucial for decisions to out-source (Chapter 15).
- Entrepreneurs may need to shape products appropriate for niche markets in the emerging market, which are also influenced by a range of socio-political factors, as demonstrated in the case of the electrical bike in China (Chapter 5).

18.3.2 Multinational Enterprises and Foreign Direct Investment

Another important concept is multinational enterprises (MNEs) and their relationship to foreign direct investment (FDI). The international business perspective proposes that the specific role of the multinational enterprise and its decisions in the global world is worthy of study, especially that of FDI (Amann and Cantwell 2012; Tolentino 2010). Dunning's eclectic paradigm provides a theoretical framework addressing why companies decide to invest in foreign markets or enter them and which kind of involvement they choose. Within this framework, a company will decide to engage in foreign markets via foreign direct investment under three conditions: (1) if that company possesses a specific competitive advantage (for example, technology, skills), which provides a company with an edge over its competitors in the host country, (2) if the internalization of intermediary markets provides benefits and (3) if the location in the host country provides locational advantages (for example, lower labor cost). The potential advantage has to outweigh overcoming barriers of entry, unfamiliarity with the host market as well as potential social and economic costs resulting in adaption costs in the foreign environment (liability of foreignness) (Dunning 1980; Zaheer 1995; Oetzel and Doh 2009; Hennart 2010). Typical motivations for FDIs are related to asset exploitation (based on competitive advantage), market orientation, as well as asset augmentation (including resource procurement) (Dunning 2003).

Similar theories have also been applied to dragon multinationals (Dunning 2006; Mathews 2006). Interestingly enough, MNEs from emerging markets are now changing their position in the international context. Multinationals from developing countries are developing their capabilities to innovate and produce (Goldstein 2007; Sauvant et al. 2010; Nayak 2011), and they try to compete through new globalization strategies (Hennart 2012). Moreover, with the rise of emerging-market MNEs (EMMNEs), we can see their purchase of companies globally (Boateng et al. 2008; Yeung and Liu 2008).

The changes identified in existing literature can lead us to conceptualize the changes underway as follows. On the one hand, foreign MNEs are entering the Asian markets with not only high-end products but also sometimes trying to make major adjustments to capture the base-of-the-pyramid (BOP) consumers. On the other hand, Indian and Chinese companies are exploiting their home-based advantage of low-cost production to expand, but also accessing significant scientific and technology knowledge and becoming more innovative in technology. As multinationals from Asia emerged as global competitors, Western MNEs took note of the competitive advantage that gave EMMNEs an edge – this edge transcended cheap labor availability and included resources such as the ability of these EMMNEs to utilize various local alliances and intelligence to adapt for the market. The wealth accumulation attracted Western MNEs and at the same time, this growth encouraged EMMNEs to undertake FDI in Europe and the US for the first time.

Relevant phenomenon about innovation spaces in Asia discussed in this book include:

- BOP innovations for the poorest segment seem vital but difficult. A variety of innovation strategies and modes of interaction are possible. One option for Western MNE is to enter into alliances with non-governmental organizations (NGOs) to develop attributes that are attractive for the Asian market (Chapter 13).

- Potentially, MNEs from emerging markets can innovate across the value chain in (1) R&D as well as in (2) manufacturing, (3) marketing and service after sales and the (4) strategic alliances with new mediators such as NGOs. This may in turn lead to different pathways of internationalization, exemplified in Indian pharmaceuticals and Indonesian multinationals (Chapters 10–11; Chapter 13).
- Building up capabilities in R&D in an emerging market may take time, and needs to have the full commitment of headquarters, even if subsidiaries are also active in these processes. The case of SKF in R&D in India is discussed (Chapter 16).
- EMMNEs can develop and adapt products due to specific knowledge of domestic consumer preference, which helps them sell in other emerging markets (Chapter 11).
- EMMNEs can enter the US and Europe to acquire assets that are valuable for innovation, such as acquire companies, patents, and human capital (Chapter 10; Chapter 12).

18.3.3 Policy and National Institutions

Policy and national institutions are together an important aspect of developing an innovation space. Previous literature developing the concept of innovation systems has stressed the importance of institutions, networks and knowledge bases at the regional, sectoral and national levels, known as innovation systems (McKelvey 1991; Nelson 1993; Cooke et al. 1997; Edquist and McKelvey 2000; Lundvall et al. 2002; Edquist 2006; Malerba 2009). In this book, policy and national institutions set the framing conditions for entrepreneurs and multinational enterprises, and more than that.

Policy and national institutions tend to come into the analysis of most chapters, one way or another, particularly through the impacts upon how they actively encourage (and occasionally work against) the development of scientific and technological capabilities and of innovation. These approaches to discussing institutions and policy enable therefore a more systemic, and dynamic view, of how different elements in the system relate to each other.

Innovation policies have a significant influence on the development of institutions with the potential ability to reward innovative behavior. Legal frameworks inhibiting imitation for example can enable innovative activities by securing profits for innovators, and political consistency and stability is an important factor in reducing uncertainty (Lundvall et al.

2002). Policies can also be crucial for the provision of input factors within an innovation system (for example, higher education system) affecting overall innovative performance (Nelson 1993). Moreover, institutions that are shaped by public policy have significant regulatory influence on innovative activities by imposing barriers or easing access to markets for innovative products. The modes for testing and approval procedures for pharmaceuticals, and policy support for health-care innovations, for example, can have influence the process of innovation as well as the decision to innovate (Martin et al. 2006).

National policies have a multidimensional and complex influence on innovation by not only affecting direct inputs but also by influencing the socio-economic system comprised by cultural, economic, social and economic aspects in which the innovation system is embedded (Edquist 2006).

Relevant phenomenon about innovation spaces in Asia discussed in this book include:

- Overall public policy in China entails catch-up with leading countries, and public investments and policy can affect the overall model of development, but there are pros and cons with the chosen model of opulent innovation (Chapter 2).
- Several case studies indicate positive effects, when Chinese policy encourages entrepreneurship through favorable conditions for highly skilled returnees from abroad as well as specific policies to develop an attractive local environment for advanced sectors (Chapters 6–7; Chapter 9).
- The development of institutional framework of intellectual property rights is useful for understanding dynamic competitive conditions and specializations in technologies and sectors (Chapters 3–4).
- Sometimes, regulation for public safety may retard the development of a new sector while the lack of enforcement of regulation or overturning of previous regulation may help promote diffusion. This was seen in the case of China of the eco-innovation industry in e-bikes (Chapter 5).
- Public policy – especially liberalization but also providing more favorable macro conditions – seems to help stimulate innovation and internationalization. This is suggested in the cases of pharmaceutical industry expanding from India and in cases of Indonesian multinationals moving abroad (Chapters 10–11).

18.4 FUTURE RESEARCH

An underlying proposition of this book is that Asia is and will impact innovation in the future. The ongoing structural transformation will push innovation into new directions for social and economic change. This force for global change will affect the type, direction and rate of future business innovation to deliver relevant services and products, globally.

This book opens up many lines of future research, to develop a range of existing literature along somewhat new trajectories, in line with streams of modern literature discussed above. Emerging markets require that entrepreneurs and multinational enterprises from, or moving into Asia, must develop new resources and capabilities. Four themes for future research follow.

A first theme is *firms based in innovation spaces in Asia as they function globally and locally*. This theme should further address the development of sectors and firms – both entrepreneurs and multinational enterprises – working within specific national contexts. Asia includes many countries, each with specific political economy situations in terms of national innovation systems and pathways of development. Hence, further exploring the dynamic interactions between firms and national context, as they drive industrial dynamics is needed in order to understand how Asia as an innovation space.

A second theme is *negotiation over emerging markets as arenas for innovation*. The shift underway in innovation spaces in Asia goes beyond firms and customers. We propose that the economies of emerging markets represent new arenas involving NGOs, governments, foundations and other actors which are active in developing and diffusing base of pyramid innovations. Analyzing their network and arenas for stakeholder engagement in these arenas would help better understand how new spaces of innovation are evolving in emerging markets.

A third theme is *entrepreneurs working within innovation spaces within specific countries in Asia*. This theme focuses upon entrepreneurs who have projects or are working with their own company, with more understanding of the national context and global linkages. These companies are usually started within one country, with an analysis here on the dynamics of starting a venture and managing public policy or customer demand. The companies chosen do differ, in that some focus upon their local market whereas others focus upon the global arena, and a question is how the external environment influences development.

A fourth theme is called *co-evolution of opportunities and conflicts, when firms expand into innovation spaces in Asia*. This theme focuses upon what happens when different types of firms, for different reasons, move into an explicit attempt to draw upon the innovation spaces in Asia. The co-evolution and conflicts of values, perceptions and market niches may have to do with the nature of re-conceptualizing innovative activities in services, or the difficulties of changing the direction of the company with a move to Asia.

NOTES

1. We wish to thank the financiers, as found in the detailed acknowledgments for the book as a whole.
2. Note that these three aspects are not absolutely separate categories, and most focus on dynamics and flows of resources, people, investments, and so on, into and out of Asia.

REFERENCES

Amann, E. and Cantwell, J. (2012). *Innovative Firms in Emerging Market Countries*. Oxford: Oxford University Press.
Bell, M., and Pavitt, K. (1993). 'Technological accumulation and industrial growth: contrasts between developed and developing countries', *Industrial and Corporate Change*, **2** (2), 157–210.
Boateng, A., Qian, W. and Y. Tianle. (2008). 'Cross-border M&A by Chinese firms: an analysis of strategic motivations and performance', *Thunderbird International Business Review*, **50**, 259–70.
Carlsson, B. Braunerhjelm, P. McKelvey, M. Olofsson, C. Persson, L. and Ylinenpää, H. (2013). 'The evolving domain of entrepreneurship research', *Small Business Economics*, **41**, 913–30.
Cooke, P., Gomez Uranga, M. and Etxebaria, G. (1997). 'Regional innovation systems: institutional and organizational dimensions', *Research Policy*, **26** (4–5), 475–91.
Dodgson, M., Gann D. and Phillips, N. (2014). *The Handbook of Innovation Management*. Oxford: Oxford University Press.
Dunning, J.H. (1980). 'Toward an eclectic theory of international production: some empirical tests', *Journal of International Business Studies*, **11**, 9–31.
Dunning, J.H. (2003). 'Some antecedents of internalization theory', *Journal of International Business Studies*, **34**, 103–15.
Dunning, J.H. (2006). 'Comment on dragon multinationals: new players in 21st century globalization', *Asia Pacific Journal of Management*, **23**, 139–41.
Edquist, C. (2006). 'Systems of innovation: perspectives and challenges', in Fagerberg, J., Mowery, D.C. and Nelson, R.R. (eds) (2006), *The Oxford Handbook of Innovation*. Oxford: Oxford University Press, pp. 181–208.
Edquist, C. and McKelvey, M. (eds) (2000). *Systems of Innovation: A Reference Collection*. Northampton, MA, USA and Cheltenham, UK: Edward Elgar.
Fagerberg, J., Mowery, D.C. and Nelson, R.R. (2006). *The Oxford Handbook of Innovation*. Oxford: Oxford University Press.

Goldstein, A. (2007). *Multinational Companies from Emerging Economies*. New York: Palgrave Macmillan.
Hennart, J. (2012). 'Emerging market multinationals and the theory of the multinational enterprise', *Global Strategy Journal*, **2**, 168–87.
Hébert, R.F. and Link, A.N. (1989). 'In search of the meaning of entrepreneurship', *Small Business Economics*, **1** (1), 39–49.
Holmèn, M., Magnusson, M. and McKelvey, M. (2007). 'What are innovative opportunities?', *Industry and Innovation*, **14** (1), 27–45.
Laursen, K. (2012). 'Keep searching and you'll find: what do we know about variety creation through firms' search activities for innovation?', *Industrial and Corporate Change*, **21** (5) 1181–220.
Lundvall, B.Å., Johnson, B., Andersen, E.S. and Dalum, B. (2002). 'National systems of production, innovation and competence building', *Research Policy*, **31** (2), 213–31.
Lumpkin, G.T. and Dess, G.G. (1996). 'Clarifying the entrepreneurial orientation construct and linking it to performance', *Academy of Management Review*, **12** (1), 135–72.
Malerba, F. (2009). *Sectoral Systems of Innovation: Concepts, Issues and Analyses of Six Major Sectors*. Cambridge: Cambridge University Press.
McKelvey, M. (1991). 'How do national systems of innovation differ? A critical analysis of Porter, Freeman, Lundvall and Nelson', in G.M. Hodgson and E. Screpanti (eds), *Rethinking Economics: Markets, Technology and Economic Evolution*. Aldershot, UK and Brookfield, VT, USA: Edward Elgar, pp. 117–37.
McKelvey, M. and Lassen, A.H. (2013). *Managing Knowledge Intensive Entrepreneurship*. Northampton, MA, USA and Cheltenham, UK: Edward Elgar.
McKelvey, M. (2014). 'Firms navigating through innovation spaces', chapter presented at the International Joseph Alois Schumpeter Society conference, the 15th ISS Conference, 27–30 July, Jena, Germany.
Martin, P., Abraham, J., Davis, C. and Kraft, A. (2006). 'Understanding the "productivity crisis" in the pharmaceutical industry: over-regulation or lack of innovation?', in A. Webster (ed.), *New Technologies in Health Care: Challenge, Change and Innovation*. Basingstoke: Palgrave Macmillan, pp. 177–93.
Mathews, J. (2006). 'Dragon multinationals: new players in 21st century globalization', *Asia Pacific Journal of Management*, **23** (1), 5–27.
Nayak, A. (2011). *Indian Multinationals: The Dynamic Growth in a Developing Country Context*. New York: Palgrave Macmillan.
Nelson, R.R. (1993). *National Systems of Innovation: A Comparative Analysis*. Oxford: Oxford University Press.
Oetzel, J. and Doh, J.P. (2009). 'MNEs and development: a review and reconceptualization', *Journal of World Business*, **44** (2), 108–20.
Sauvant, K., Pradhan, J., Chatterjee, A. and Harley, B. (2010). *The Rise of Indian Multinationals*. New York: Palgrave Macmillan.
Schumpeter, J. (1934). *The Theory of Economic Development*. Cambridge, MA: Harvard University Press.
Shane, S., (2000). 'Prior knowledge and the discovery of entrepreneurial opportunities', *Organization Science*, **11** (4), 448–69.
Teece, D.J., Pisano, G. and Shuen, A. (1997). 'Dynamic capabilities and strategic management', *Strategic Management Journal*, **18** (7), 509–33.
Teece, D.J. (2007). 'Explicating dynamic capabilities: the nature of (sustainable) enterprise performance', *Strategic Management Journal*, **28** (13), 1319–50.

Tolentino, P. (2010). 'Home country macroeconomic factors and outward FDI of China and India', *Journal of International Management*, **16**, 102–20.
Venkataraman, S, (1997). 'The distinctive domain of entrepreneurship research', in J. Katz and R. Brockhaus (eds), *Advances in Entrepreneurship, Firm Emergence and Growth*, vol 3. Greenwich, CT: JAI Press, pp. 119–38.
Yeung, H.W.C. and Liu, W. (2008). 'Globalizing China: the rise of mainland firms in the global economy', *Eurasian Geography and Economics*, **49** (1), 57–86.
Zaheer, S. (1995). 'Overcoming the liability of foreignness', *Academy of Management Journal*, **38** (2), 341–63.

Index